PRINTED TEST AND INSTRUCTOR'S RESOURCE GUIDE

THOMAS J. SCHICKER

INTERMEDIATE ALGEBRA CONCEPTS AND APPLICATIONS

SIXTH EDITION

Marvin L. Bittinger
Indiana University—Purdue University at Indianapolis

David J. Ellenbogen
Community College of Vermont

Addison Wesley

Boston San Francisco New York
London Toronto Sydney Tokyo Singapore Madrid
Mexico City Munich Paris Cape Town Hong Kong Montreal

Reproduced by Addison-Wesley from camera-ready copy supplied by the author.

Copyright © 2002 Pearson Education, Inc.

All rights reserved. No part of this publication may be reproduced, stored in a retrieval system, or transmitted, in any form or by any means, electronic, mechanical, photocopying, recording, or otherwise, without the prior written permission of the publisher. Printed in the United States of America.

ISBN   0-201-73490-7

1 2 3 4 5 6 7 8 9 10 PHCC 05 04 03 02 01

TABLE OF CONTENTS

I. **ALTERNATE TESTS, FORMS A, B, C, D, E, F, G, and H** 1

There are eight <u>NEW</u> alternate test forms for each chapter. They are not repeats of the Fifth Edition. Alternate Test Forms A, B, C, D, E, and F are organized with questions in the same topic order as the Chapter Tests in the text. Alternate Test Forms G and H are multiple choice tests.

There are eight NEW alternate test forms for the final examination. Alternate Test Forms A, B, and C of the final examinations are organized by chapters. On Forms D, E, and F, the questions are organized by type. For example, all the problem solving is together. Test Forms G and H are multiple choice tests. If a pretest for the book is desired, instructors may use Form A, B, or C of the final examination for this purpose.

Synthesis questions occur at the end of each chapter test. These questions are separated from the rest of the test by a solid line and are meant to be more challenging like those problems found in the last part of each exercise set. The synthesis questions have been placed at the end to make it easy to omit them if the instructor wishes to do so.

Chapter 1	1
Chapter 2	21
Chapter 3	53
Chapter 4	73
Chapter 5	105
Chapter 6	125
Chapter 7	145
Chapter 8	165
Chapter 9	185
Chapter 10	205
Chapter 11	225
Final Examination	245

II. **ANSWER KEYS FOR ALTERNATE TESTS, FORMS A, B, C, D, E, F, G, and H** 297

III. **EXTRA PRACTICE SHEETS** 385

These sheets provide extra drill on topics in the text. The instructor can use them for lecture examples or additional homework assignments. Students will find them to be an excellent review for tests. They provide an excellent source of practice and reteaching for students who have done poorly on a test and who are going to retest.

Section(s)		
1.2	Addition and Subtraction of Real Numbers	385
1.4	Solving Problems	387
1.6	Properties of Exponents	389
2.1	Graphing Linear Equations	391
3.2	Solving Systems of Linear Equations	395
3.3	Solving Applications: Systems of Two Equations	397
3.4 and 3.5	Solving Systems of Equations in Three Variables	399
4.1	Solving Inequalities with Both Principles	401
4.3	Solving Equations and Inequalities with Absolute Value	403
4.4	Inequalities in Two Variables	405
5.3 - 5.7	Factoring Polynomials	409
5.8	Applications of Polynomial Equations	411
6.1	Multiplying and Dividing Rational Expressions	413
6.2	Addition and Subtraction of Rational Expressions	415
6.3	Simplifying Complex Rational Expressions	417
6.4 and 6.5	Solving Rational Equations Including Problem Solving	419
6.6	Division of Polynomials	421
6.7	Synthetic Division	423
7.1 and 7.2	Radical Expressions and Rational Numbers as Exponents	425
7.3 and 7.4	Multiplying, Dividing, and Simplifying Radical Expressions	427
7.6	Solving Radical Equations	429
7.7	Geometric Applications	431
8.2	Solving Quadratic Equations Using the Quadratic Formula	433
8.1 and 8.3	Solving Problems Using Quadratic Equations	435
8.5	Solving Equations Reducible to Quadratic	437
8.6	Graphing Quadratic Functions	439

	8.9	Polynomial and Rational Inequalities	443
	9.6	Solving Exponential and Logarithmic Equations	445
	10.4	Solving Nonlinear Systems of Equations	447
	11.4	The Binomial Theorem	449
IV.	ANSWER KEYS FOR EXTRA PRACTICE SHEETS		451
V.	CORRELATION GUIDE		469
VI.	VIDEO INDEX		474

Special thanks are extended to Patty Slipher for her job of checking the manuscript.

CHAPTER 1

TEST FORM A

NAME_____

CLASS_____ SCORE_____ GRADE_____

ANSWERS

1. Translate to an algebraic expression: Five times the product of two numbers.

2. Evaluate $a^3 + 2b^2 + 4c \div ba$ for $a = -2$, $b = 10$, and $c = 5$.

3. Find the area of a triangle when the base is 4 cm and the height is 5.3 cm.

Perform the indicated operation.

4. $-10 + (-35)$

5. $-18.1 + 2.9$

6. $5.87 + (-0.15)$

7. $15.8 - 58.2$

8. $-15.4 - 31.1$

9. $-2.5(4.7)$

10. $-\dfrac{1}{3} - \left(-\dfrac{3}{5}\right)$

11. $-\dfrac{3}{5}\left(-\dfrac{10}{9}\right)$

12. $\dfrac{-15.3}{-5.1}$

13. $\dfrac{1}{2} \div \left(-\dfrac{3}{8}\right)$

14. Simplify: $6 + (3-5)^2 - 8 \div 2^3 \cdot 7$.

15. Use a commutative law to write an expression equivalent to $8x + y$.

Combine like terms.

16. $2y - 8y + 13y$

17. $a^2b - 3ab^2 + 12ab^2 + 8a^2b + 1$

18. Simplify: $11x - 2(3x - 9) + 14$

Solve. If the solution set is $\mathbb{R}$ or $\varnothing$, classify the equation as an identity or a contradiction.

19. $12x - 37 = 31x + 20$

20. $3t - (16 - 5t) = 8(t - 2)$

21. Solve for m: $a = \dfrac{mg - T}{m}$

22. Nisha's scores on five tests are 68, 72, 75, 85, and 91. What must Nisha score on the last test so that her average will be 76?

1._____
2._____
3._____
4._____
5._____
6._____
7._____
8._____
9._____
10._____
11._____
12._____
13._____
14._____
15._____
16._____
17._____
18._____
19._____
20._____
21._____
22._____

CHAPTER 1

TEST FORM A

NAME_____

ANSWERS

23._____

24._____

25._____

26._____

27._____

28._____

29._____

30._____

31._____

32._____

33._____

34._____

35._____

36._____

37._____

23. Find three consecutive even integers such that the sum of three times the first, two times the second, and the largest is 140.

Simplify. Do not use negative exponents in the answer.

24. $-9(x-2)-8(x+3)$

25. $2b-[9-2(7b-11)]$

26. $(2x^{-5}y^{-2})(-7x^{-3}y^4)$

27. -5^{-2}

28. $(-9xy^{-4})^{-2}$

29. $\left(\dfrac{3x^3 y^{-2}}{6y^{-3}}\right)^2$

30. $(2xy^9)^0$

Simplify and write the answer in scientific notation. Use the correct number of significant digits.

31. $(8.34 \times 10^{-2})(2.05 \times 10^{-5})$

32. $\dfrac{1.8 \times 10^8}{3.1 \times 10^{-2}}$

33. $\dfrac{6.5 \times 10^{-8}}{2.3 \times 10^{-5}}$

34. The average distance between the sun and the planet Mercury is 5.8×10^7 km. About how far does Mercury travel in one orbit around the sun? (Assume a circular orbit.)

Simplify.

35. $(5x^{2a} y^{b+1})^{2c}$

36. $\dfrac{-102a^{x+9}}{51a^{x-7}}$

37. $\dfrac{(-34x^{x-3} y^{y-1})(5x^{x+1} y^{y-3})}{(-3x^{x-1} y^{y-3})(17x^{x+3} y^{y-2})}$

CHAPTER 1

TEST FORM B

NAME_____

CLASS_____ SCORE_____ GRADE_____

ANSWERS

1. Translate to an algebraic expression: Twice a number more than five times another number.

2. Evaluate $b^2 + 4a - 6c \div ba$ for $a = -2$, $b = 10$, and $c = 5$.

3. Find the area of a triangle when the base is 5 cm and the height is 6.3 cm.

Perform the indicated operation.

4. $-11 + (-32)$

5. $-17.2 + 3.8$

6. $6.98 + (-1.26)$

7. $24.9 - 47.3$

8. $-26.3 - 22.2$

9. $-3.7(5.6)$

10. $-\dfrac{2}{3} - \left(-\dfrac{3}{7}\right)$

11. $-\dfrac{5}{7}\left(-\dfrac{9}{10}\right)$

12. $\dfrac{-24.8}{-6.2}$

13. $\dfrac{2}{3} \div \left(-\dfrac{4}{9}\right)$

14. Simplify: $7 - (4-7)^2 + 9 \div 3^2 \cdot 6$.

15. Use a commutative law to write an expression equivalent to $x + 5y$.

Combine like terms.

16. $3y - 10y + 16y$

17. $9a^2b - 5ab^2 + 10ab^2 - 7a^2b + 2$

18. Simplify: $12x - 3(2x - 8) + 3$

Solve. If the solution set is ℝ or ∅, classify the equation as an identity or a contradiction.

19. $13x + 56 = 30x + 22$

20. $6t - (14 - 6t) = 12(t - 1)$

21. Solve for G: $g = G\dfrac{M}{R^2}$

22. Naoki's scores on five tests are 78, 80, 81, 85, and 86. What must Naoki score on the last test so that her average will be 80?

1._____
2._____
3._____
4._____
5._____
6._____
7._____
8._____
9._____
10._____
11._____
12._____
13._____
14._____
15._____
16._____
17._____
18._____
19._____
20._____
21._____
22._____

CHAPTER 1

TEST FORM B

NAME_____

ANSWERS

23. _____

24. _____

25. _____

26. _____

27. _____

28. _____

29. _____

30. _____

31. _____

32. _____

33. _____

34. _____

35. _____

36. _____

37. _____

23. Find three consecutive odd integers such that the sum of three times the first, two times the second, and the largest is 62.

Simplify. Do not use negative exponents in the answer.

24. $-3(x-3)-7(x+4)$

25. $3b - [\,11 - 3(7b-7)\,]$

26. $(3x^{-6}y^{-3})(8x^{-2}y^5)$

27. -10^{-3}

28. $(-3x^2 y^{-3})^{-3}$

29. $\left(\dfrac{6x^4 y^{-8}}{-18y^{-6}}\right)^2$

30. $(4x^2 y^8)^0$

Simplify and write the answer in scientific notation. Use the correct number of significant digits.

31. $(7.45 \times 10^{-3})(3.14 \times 10^{-4})$

32. $\dfrac{4.2 \times 10^7}{2.9 \times 10^{-3}}$

33. $\dfrac{3.2 \times 10^{-6}}{5.6 \times 10^{-9}}$

34. The average distance between the sun and the planet Venus is 1.1×10^8 km. About how far does Venus travel in one orbit around the sun? (Assume a circular orbit.)

Simplify.

35. $(4x^{3a} y^{b-1})^{3c}$

36. $\dfrac{-96a^{x+8}}{32a^{x-6}}$

37. $\dfrac{(-36x^{x-2} y^{y-2})(4x^{x+2} y^{y-2})}{(-2x^{x-2} y^{y-2})(9x^{x+2} y^{y-3})}$

4

CHAPTER 1   NAME_____

TEST FORM C   CLASS_____ SCORE_____ GRADE_____

1. Translate to an algebraic expression: Five times the sum of two different numbers.

2. Evaluate $c^3 - b^2 - b \div ac$ for $a = -2$, $b = 10$, and $c = 5$.

3. Find the area of a triangle when the base is 6 cm and the height is 3.5 cm.

Perform the indicated operation.

4. $-12 + (-29)$

5. $-16.3 + 4.7$

6. $7.09 + (-2.37)$

7. $33.1 - 36.4$

8. $-37.2 - 13.3$

9. $-4.9(6.5)$

10. $-\dfrac{3}{4} - \left(-\dfrac{4}{5}\right)$

11. $-\dfrac{7}{9}\left(-\dfrac{3}{14}\right)$

12. $\dfrac{-36.5}{-7.3}$

13. $\dfrac{3}{4} \div \left(-\dfrac{5}{16}\right)$

14. Simplify: $8 + (5-9)^2 - 2 \div 2^2 \cdot 5$.

15. Use a commutative law to write an expression equivalent to $7x + y$.

Combine like terms.

16. $4y - 12y + 19y$

17. $2a^2b + 7ab^2 - 8ab^2 + 6a^2b + 3$

18. Simplify: $13x - 4(x - 7) - 12$

Solve. If the solution set is $\mathbb{R}$ or $\varnothing$, classify the equation as an identity or a contradiction.

19. $14x + 5 = 29x + 35$

20. $9t - (12 - 7t) = 8\left(2t - \dfrac{3}{2}\right)$

21. Solve for P: $W = 2d^2k\left(1 - \dfrac{P}{2}\right)$

22. Tony's scores on five tests are 80, 85, 85, 90, and 95. What must Tony score on the last test so that her average will be 85?

ANSWERS

1._____
2._____
3._____
4._____
5._____
6._____
7._____
8._____
9._____
10._____
11._____
12._____
13._____
14._____
15._____
16._____
17._____
18._____
19._____
20._____
21._____
22._____

CHAPTER 1

TEST FORM C

NAME_____

ANSWERS

23. Find three consecutive even integers such that the sum of three times the first and five times the second is equal to seven times the largest.

Simplify. Do not use negative exponents in the answer.

24. $-7(x-4)-6(x+5)$

25. $4b-[13-4(7b-6)]$

26. $(4x^{-7}y^{-4})(-9x^3y^6)$

27. -7^{-2}

28. $(-8x^3y^{-2})^{-2}$

29. $\left(\dfrac{12x^5y^{-7}}{3x^4y^{-6}}\right)^2$

30. $(6x^3y^7)^0$

Simplify and write the answer in scientific notation. Use the correct number of significant digits.

31. $(6.56 \times 10^{-4})(4.23 \times 10^{-3})$

32. $\dfrac{3.0 \times 10^6}{5.3 \times 10^{-4}}$

33. $\dfrac{4.7 \times 10^{-10}}{4.1 \times 10^{-7}}$

34. The average distance between the sun and the planet Earth is 1.5×10^8 km. About how far does Earth travel in one orbit around the sun? (Assume a circular orbit.)

Simplify.

35. $(3x^{4a}y^{b+2})^{4c}$

36. $\dfrac{-88a^{x+7}}{22a^{x-5}}$

37. $\dfrac{(-38x^{x-1}y^{y-3})(3x^{x+3}y^{y-1})}{(-6x^{x-3}y^{y-1})(19x^{x+1}y^{y-2})}$

23._____

24._____

25._____

26._____

27._____

28._____

29._____

30._____

31._____

32._____

33._____

34._____

35._____

36._____

37._____

CHAPTER 1

TEST FORM D

NAME_____

CLASS_____ SCORE_____ GRADE_____

ANSWERS

1. Translate to an algebraic expression: Five less than twice the product of two numbers.

2. Evaluate $a^2 - 3c^2 - b \div ca$ for $a = -2$, $b = 10$, and $c = 5$.

3. Find the area of a triangle when the base is 7 cm and the height is 2.5 cm.

Perform the indicated operation.

4. $-13 + (-26)$

5. $-15.4 + 5.6$

6. $8.10 + (-3.48)$

7. $42.2 - 25.5$

8. $-16.2 - 11.4$

9. $-5.1(7.4)$

10. $-\dfrac{3}{2} - \left(-\dfrac{2}{5}\right)$

11. $-\dfrac{9}{4}\left(-\dfrac{5}{18}\right)$

12. $\dfrac{-50.4}{-8.4}$

13. $\dfrac{4}{5} \div \left(-\dfrac{6}{25}\right)$

14. Simplify: $9 - (2-7)^2 + 3 \div 2^2 \cdot 4$.

15. Use a commutative law to write an expression equivalent to $x + 4y$.

Combine like terms.

16. $5y - 14y + 22y$

17. $8a^2b + 9ab^2 - 6ab^2 - 5a^2b + 1$

18. Simplify: $14x - 5(2x - 6) - 5$

Solve. If the solution set is $\mathbb{R}$ or $\varnothing$, classify the equation as an identity or a contradiction.

19. $15x + 17 = 28x - 22$

20. $10t - (10 - 6t) = 4(4t - 3)$

21. Solve for v: $t_B - t_A = \dfrac{r}{c - v}$

22. Jenna's scores on five tests are 83, 91, 93, 95, and 98. What must Jenna score on the last test so that her average will be 90?

1._____
2._____
3._____
4._____
5._____
6._____
7._____
8._____
9._____
10._____
11._____
12._____
13._____
14._____
15._____
16._____
17._____
18._____
19._____
20._____
21._____
22._____

CHAPTER 1

TEST FORM D

NAME_____

ANSWERS

23. Find three consecutive odd integers such that the sum of five times the first and seven times the second is equal to ten times the largest.

23._____

Simplify. Do not use negative exponents in the answer.

24._____

24. $-5(x+5)-5(x-6)$ 25. $5b-[15-3(8b-9)]$

25._____

26. $(5x^{-8}y^{-5})(7x^2y^{-4})$ 27. -5^{-3}

26._____

28. $(-4x^4y^{-1})^{-3}$ 29. $\left(\dfrac{4x^2y^{-6}}{-20y^{-3}}\right)^2$

27._____

30. $(8x^4y^6)^0$

28._____

Simplify and write the answer in scientific notation. Use the correct number of significant digits.

29._____

31. $(5.67 \times 10^{-5})(5.32 \times 10^{-2})$

30._____

32. $\dfrac{6.4 \times 10^5}{4.1 \times 10^{-5}}$

31._____

33. $\dfrac{5.0 \times 10^{-8}}{3.8 \times 10^{-11}}$

32._____

34. The average distance between the sun and the planet Mars is 2.3×10^8 km. About how far does Mars travel in one orbit around the sun? (Assume a circular orbit.)

33._____

34._____

Simplify.

35._____

35. $(2x^{5a}y^{b-2})^{5c}$

36._____

36. $\dfrac{-70a^{x+6}}{14a^{x-4}}$

37._____

37. $\dfrac{(-40x^{x+1}y^{y-2})(2x^{x+2}y^{y+1})}{(-4x^{x-2}y^{y+1})(8x^{x-1}y^{y-1})}$

CHAPTER 1

TEST FORM E

NAME_____

CLASS_____ SCORE_____ GRADE_____

ANSWERS

1. Translate to an algebraic expression: Two times the product of two numbers.

2. Evaluate $a^3 + 2b^2 + 4c \div ba$ for $a = -3$, $b = 12$, and $c = 4$.

3. Find the area of a triangle when the base is 8 cm and the height is 7.4 cm.

Perform the indicated operation.

4. $-14 + (-23)$

5. $-14.5 + 6.5$

6. $9.21 + (-4.59)$

7. $51.3 - 14.6$

8. $-27.3 - 32.5$

9. $-6.3(7.4)$

10. $-\dfrac{3}{5} - \left(-\dfrac{5}{2}\right)$

11. $-\dfrac{6}{7}\left(-\dfrac{5}{18}\right)$

12. $\dfrac{-28}{-5.6}$

13. $\dfrac{5}{4} \div \left(-\dfrac{3}{8}\right)$

14. Simplify: $2 + (1 - 7)^2 - 4 \div 3^2 \cdot 3$.

15. Use a commutative law to write an expression equivalent to $6x + y$.

Combine like terms.

16. $6y - 16y + 25y$

17. $3a^2b - 11ab^2 + 4ab^2 + 4a^2b - 2$

18. Simplify: $15x - 5(3x - 5) + 10$

Solve. If the solution set is $\mathbb{R}$ or $\varnothing$, classify the equation as an identity or a contradiction.

19. $16x - 18 = 27x - 7$

20. $8t - (8 - 5t) = 13(t - 1)$

21. Solve for v: $b = \dfrac{a}{B\left(1 - k\dfrac{v}{c}\right)}$

22. Anya's scores on five tests are 86, 91, 91, 90, and 67. What must Anya score on the last test so that her average will be 85?

1._____
2._____
3._____
4._____
5._____
6._____
7._____
8._____
9._____
10._____
11._____
12._____
13._____
14._____
15._____
16._____
17._____
18._____
19._____
20._____
21._____
22._____

CHAPTER 1

NAME_____

TEST FORM E

ANSWERS

23. Find three consecutive even integers such that ten times the first plus twenty times the second minus fifteen times the largest is 100.

23._____

Simplify. Do not use negative exponents in the answer.

24._____

24. $-6(x+6)-4(x-7)$ 25. $6b-[17-2(8b-10)]$

25._____

26. $(5x^{-7}y^{-6})(-8x^{-4}y^{-5})$ 27. -9^{-2}

26._____

28. $(-7x^5y)^{-2}$ 29. $\left(\dfrac{30xy^{-5}}{6y^{-3}}\right)^3$

30. $(10x^5y^5)^0$

27._____

Simplify and write the answer in scientific notation. Use the correct number of significant digits.

28._____

31. $(4.78\times 10^{-5})(6.41\times 10^{-2})$

29._____

32. $\dfrac{5.2\times 10^4}{7.5\times 10^{-6}}$

30._____

33. $\dfrac{2.9\times 10^{-10}}{6.9\times 10^{-7}}$

31._____

34. The average distance between the sun and the planet Jupiter is 7.8×10^8 km. About how far does Jupiter travel in one orbit around the sun? (Assume a circular orbit.)

32._____

33._____

Simplify.

34._____

35. $(2x^{6a}y^{b+3})^{6c}$

35._____

36. $\dfrac{-66a^{x+5}}{11a^{x-3}}$

36._____

37. $\dfrac{(-42x^{x+2}y^{y-1})(2x^{x+1}y^{y+2})}{(-3x^{x-1}y^{y+2})(7x^{x-2}y^{y+1})}$

37._____

CHAPTER 1   NAME_____

TEST FORM F   CLASS_____ SCORE_____ GRADE_____

1. Translate to an algebraic expression: Twice a number increased by five times another number.

2. Evaluate $b^2 + 4a - 6c \div ab$ for $a = -3$, $b = 12$, and $c = 4$.

3. Find the area of a triangle when the base is 9 cm and the height is 8.4 cm.

Perform the indicated operation.

4. $-15 + (-20)$

5. $-13.6 + 7.4$

6. $0.32 + (-5.60)$

7. $69.4 - 93.7$

8. $-35.4 - 23.6$

9. $-7.5(6.5)$

10. $-\dfrac{3}{7} - \left(-\dfrac{7}{2}\right)$

11. $-\dfrac{2}{3}\left(-\dfrac{3}{8}\right)$

12. $\dfrac{-18.8}{-4.7}$

13. $\dfrac{4}{3} \div \left(-\dfrac{5}{6}\right)$

14. Simplify: $3 - (2 - 7)^2 + 5 \div 3^2 \cdot 2$.

15. Use a commutative law to write an expression equivalent to $x + 3y$.

Combine like terms.

16. $7y - 18y + 28y$

17. $7a^2b - 13ab^2 + 2ab^2 - 3a^2b - 3$

18. Simplify: $16x - 4(2x - 4) + 7$

Solve. If the solution set is $\mathbb{R}$ or $\varnothing$, classify the equation as an identity or a contradiction.

19. $17x - 10 = 26x - 46$

20. $4t - (6 - 4t) = 24\left(\dfrac{t}{3} - \dfrac{1}{4}\right)$

21. Solve for k: $NP = \dfrac{RT}{6\pi k D}$

22. Igor's scores on five tests are 70, 84, 78, 83, and 75. What must Igor score on the last test so that his average will be 80?

ANSWERS

1._____
2._____
3._____
4._____
5._____
6._____
7._____
8._____
9._____
10._____
11._____
12._____
13._____
14._____
15._____
16._____
17._____
18._____
19._____
20._____
21._____
22._____

CHAPTER 1　　　　　　　　　　　　　　NAME_____

TEST FORM F

ANSWERS

23. Find three consecutive odd integers such that five times the first plus ten times the second minus seven times the largest is equal to 240.

23._____

Simplify. Do not use negative exponents in the answer.

24._____

24. $-4(x+7) - 3(x-8)$　　　25. $7b - [19 - 3(9b - 8)]$

25._____

26. $(4x^{-6}y^{-7})(9x^{-5}y^{-6})$　　27. -4^{-3}

26._____

28. $(-2x^{-3}y^2)^{-3}$　　29. $\left(\dfrac{4x^3 y^{-7}}{-16y^{-2}}\right)^3$

27._____

30. $(12x^6 y^4)^0$

28._____

Simplify and write the answer in scientific notation. Use the correct number of significant digits.

29._____

31. $(3.89 \times 10^{-4})(7.50 \times 10^{-3})$

30._____

32. $\dfrac{8.6 \times 10^3}{6.3 \times 10^{-7}}$

31._____

33. $\dfrac{7.8 \times 10^{-9}}{1.1 \times 10^{-6}}$

32._____

34. The average distance between the sun and the planet Saturn is 1.4×10^9 km. About how far does Saturn travel in one orbit around the sun? (Assume a circular orbit.)

33._____

34._____

Simplify.

35._____

35. $(3x^{7a} y^{b-3})^{4c}$

36._____

36. $\dfrac{-56a^{x+4}}{8a^{x-2}}$

37._____

37. $\dfrac{(-44x^{x+3} y^{y+1})(3x^{x-1} y^{y+3})}{(-2x^{x+1} y^{y+3})(11x^{x-3} y^{y+2})}$

CHAPTER 1　　　　　　　　　　　　　　　　　　NAME_____

TEST FORM G　　　　　　　　　　　　CLASS_____ SCORE_____ GRADE_____

ANSWERS

1. Translate to an algebraic expression: Twice the sum of two numbers.

   a) $4xy$　　b) $2xy$　　c) $2(x+y)$　　d) $2x+y$

   1._____

2. Evaluate $c^3 - b^2 - b \div ac$ for $a = -3$, $b = 12$, and $c = 4$.

   a) -56　　b) 128　　c) 72　　d) -64

   2._____

3. Find the area of a triangle when the height h is 10 ft and the base b is 4.8 ft.

   a) 12 ft^2　　b) 48 ft^2　　c) 40 ft^2　　d) 24 ft^2

   3._____

4. Add: $-16 + (-17)$.

   a) -33　　b) -23　　c) 33　　d) -1

   4._____

5. Add: $1.43 + (-6.71)$.

   a) -8.14　　b) -5.28　　c) 8.14　　d) 5.28

   5._____

6. Subtract: $78.5 - 82.8$.

   a) -4.3　　b) -161.3　　c) -3.4　　d) 4.3

   6._____

7. Subtract: $-\frac{4}{5} - \left(-\frac{5}{3}\right)$.

   a) $-\frac{37}{15}$　　b) $\frac{13}{15}$　　c) $\frac{12}{15}$　　d) $-\frac{13}{15}$

   7._____

8. Multiply: $-8.7(5.6)$.

   a) -43.24　　b) -48.72　　c) -41.3　　d) 43.24

   8._____

9. Multiply: $-\frac{3}{7}\left(-\frac{5}{12}\right)$.

   a) $-\frac{15}{84}$　　b) $-\frac{5}{28}$　　c) $\frac{5}{28}$　　d) $\frac{35}{36}$

   9._____

CHAPTER 1

TEST FORM G

NAME_____

ANSWERS

10._____

10. Divide: $\dfrac{-11.4}{-3.8}$.

    a) -15.2 b) -7.6 c) -3 d) 3

11._____

11. Divide: $\dfrac{3}{2} \div \left(-\dfrac{7}{18}\right)$.

    a) $\dfrac{7}{12}$ b) $-\dfrac{14}{54}$ c) $-\dfrac{27}{7}$ d) $-\dfrac{7}{12}$

12._____

12. Simplify: $4 + (1-9)^2 - 6 \div 2^2 \cdot 8$.

    a) $\dfrac{17}{8}$ b) -68 c) 56   d) 64

13._____

13. Use a commutative law to write an expression equivalent to $5x + y$.

    a) $y + 5x$ b) $5y + x$ c) $5(x+y)$ d) $y5 + x$

14._____

14. Combine like terms: $8y - 20y + 31y$.

    a) $-19y$ b) $-43y$ c) $-12y$ d) $19y$

15._____

15. Combine like terms: $4a^2b + 15ab^2 - 18ab^2 + 2a^2b - 1$.

    a) $6a^2b + 3ab^2 + 1$ b) $6a^2b - 3ab^2 + 1$
    c) $6a^2b + 3ab^2 - 1$ d) $6a^2b - 3ab^2 - 1$

16._____

16. Simplify: $17x - 3(x-3) - 8$.

    a) $14x + 17$ b) $20x + 1$ c) $14x + 1$ d) $20x - 17$

17._____

17. Solve: $18x + 5 = 25x + 5$.

    a) 5   b) 0   c) 10   d) -5

18._____

18. Solve: $9t - (4 + 3t) = 4\left(\dfrac{3}{2}t - 2\right)$.

    a) $\varnothing$ b) -3 c) $\mathbb{R}$ d) 3

CHAPTER 1
TEST FORM G

NAME_____

ANSWERS

19. Solve for N: $K = \dfrac{R}{N} Bv - P$.

    a) $N = \dfrac{RBv}{K+P}$ b) $N = \dfrac{K+P}{RBv}$

    c) $N = \dfrac{RB}{v(K+P)}$ d) $N = \dfrac{v(K+P)}{RB}$

    19._____

20. Jamie's scores on five tests are 68, 67, 73, 65, and 82. What must Jamie score on his sixth test so that his average will be 75?

    a) 95   b) 93   c) 79   d) 86

    20._____

21. Find three consecutive even integers such that the middle integer subtracted from the sum of four times the largest and two times the smallest is 284.

    a) 56, 58, 60   b) 38, 40, 42
    c) 54, 56, 58   d) 42, 44, 46

    21._____

22. Simplify: $8b - [21 - 4(8b - 5)]$.

    a) $-24b + 1$ b) $40b - 41$
    c) $-40b + 41$ d) $24b - 41$

    22._____

23. Simplify: $(3x^{-5}y^{-8})(-5x^4 y^{-2})$. Do not use negative exponents.

    a) $-15xy^{10}$ b) $-\dfrac{15}{xy^{10}}$ c) $-2x^{20}y^{16}$ d) $-\dfrac{15y^{16}}{x^{20}}$

    23._____

24. Simplify: $(-6x^{-2}y^3)^{-2}$. Do not use negative exponents.

    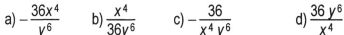

    a) $-\dfrac{36x^4}{y^6}$ b) $\dfrac{x^4}{36y^6}$ c) $-\dfrac{36}{x^4 y^6}$ d) $\dfrac{36y^6}{x^4}$

    24._____

CHAPTER 1

TEST FORM G

NAME_____

ANSWERS

25. Simplify: $\left(\dfrac{12x^2 y^{-2}}{4y^{-1}}\right)^3$. Do not use negative exponents.

a) $\dfrac{3x^5}{y^6}$ b) $27x^6 y^3$ c) $\dfrac{3x^6}{y^3}$ d) $\dfrac{27x^6}{y^3}$

25._____

26. Simplify using scientific notation and correct number of significant figures:

$$(2.91 \times 10^{-3})(8.69 \times 10^{-4}).$$

a) 2.53×10^{-6} b) 2.53×10^{-8}
c) 6.53×10^{-6} d) 4.53×10^{-8}

26._____

27. Simplify using scientific notation and correct number of significant figures:

$$\dfrac{7.4 \times 10^2}{9.7 \times 10^{-8}}.$$

a) 1.4×10^9 b) 7.6×10^9 c) 6.7×10^9 d) 1.6×10^{-6}

27._____

28. The average distance between the sun and the planet Uranus is 2.9×10^9 km. About how far does Uranus travel in one orbit around the sun? (Assume a circular orbit.)

a) 1.8×10^{12} km b) 8.1×10^{10} km
c) 1.8×10^{10} km d) 1.8×10^{-10} km

28._____

29. Simplify: $(4x^{8a} y^{b+4})^{3c}$.

a) $16^c x^{14ac} y^{3bc+12c}$ b) $64^{3c} x^{24ac} y^{3bc+3c}$
c) $16^c x^{24ac} y^{3bc+4c}$ d) $64^c x^{24ac} y^{3bc+12c}$

29._____

30. Simplify: $\dfrac{-48a^{x+3}}{6a^{x-1}}$.

a) $-8a^4$ b) $8a^{-4}$ c) $-8a^2$ d) $-8a^{-4}$

30._____

CHAPTER 1   NAME_____

TEST FORM H   CLASS_____ SCORE_____ GRADE_____

1. Translate to an algebraic expression: Five times the product of two numbers minus two.

    a) $2 - 5xy$ b) $5xy - 2$ c) $5(x + y)$ d) $5x + y - 2$

2. Evaluate $a^2 - 3c^2 - b \div ca$ for $a = -3$, $b = 12$, and $c = 4$.

    a) -30 b) 128 c) 72 d) -64

3. Find the area of a triangle when the height h is 11 ft and the base b is 5.8 ft.

    a) 31.9 ft^2 b) 48 ft^2 c) 63.8 ft^2 d) 15.9 ft^2

4. Add: $-17 + (-14)$.

    a) -3 b) -23 c) 33 d) -31

5. Add: $2.54 + (-7.82)$.

    a) -9.16 b) -6.28 c) -5.28 d) 5.28

6. Subtract: $87.6 - 71.9$.

    a) 14.7 b) 16.3 c) 6.3 d) 15.7

7. Subtract: $-\dfrac{2}{5} - \left(-\dfrac{5}{2}\right)$.

    a) $-\dfrac{4}{25}$ b) $\dfrac{29}{10}$ c) $\dfrac{21}{10}$ d) $-\dfrac{21}{10}$

8. Multiply: $-1.9(4.7)$.

    a) 6.3 b) -6.3 c) -8.93 d) 3.6

9. Multiply: $-\dfrac{3}{4}\left(-\dfrac{2}{9}\right)$.

    a) $-\dfrac{1}{6}$ b) $\dfrac{1}{6}$ c) $\dfrac{5}{6}$ d) $\dfrac{11}{6}$

ANSWERS

1._____

2._____

3._____

4._____

5._____

6._____

7._____

8._____

9._____

CHAPTER 1

TEST FORM H

NAME_____

ANSWERS

10._____

10. Divide: $\dfrac{-5.8}{-2.9}$.

   a) 2   b) –7.6   c) –2   d) 3

11._____

11. Divide: $\dfrac{2}{5} \div \left(-\dfrac{9}{10}\right)$.

   a) $\dfrac{4}{3}$ b) $-\dfrac{4}{9}$ c) $-\dfrac{9}{4}$ d) $-\dfrac{20}{9}$

12._____

12. Simplify: $5 - (6-8)^2 + 7 \div 3^3 \cdot 9$.

   a) $\dfrac{3}{8}$ b) $\dfrac{10}{3}$ c) $\dfrac{8}{3}$ d) $\dfrac{3}{10}$

13._____

13. Use a commutative law to write an expression equivalent to $x + 2y$.

   a) $y - 2x$ b) $2y - x$ c) $2y + x$ d) $y + 2x$

14._____

14. Combine like terms: $9y - 22y + 34y$.

   a) $-19y$ b) $21y$ c) $-12y$ d) $19y$

15._____

15. Combine like terms: $6a^2b + 17ab^2 - 20ab^2 - a^2b - 2$.

   a) $5a^2b + 3ab^2 + 2$ b) $5a^2b - 3ab^2 - 2$
   c) $7a^2b + 3ab^2 - 2$ d) $7a^2b - 3ab^2 - 2$

16._____

16. Simplify: $18x - 2(2x - 2) - 9$.

   a) $14x - 5$ b) $20x + 1$ c) $14x + 1$ d) $20x - 17$

17._____

17. Solve: $19x + 6 = 24x - 19$.

   a) -5 b) 0   c) 10   d) 5

18._____

18. Solve: $14t - (2 - 2t) = 2(8t - 1)$.

   a) $\varnothing$ b) $\mathbb{R}$ c) 7   d) -8

CHAPTER 1 NAME_____

TEST FORM H

	ANSWERS

19. Solve for v: $K = \dfrac{R}{N} Bv - P$.

   a) $v = \dfrac{RBN}{K+P}$ b) $v = \dfrac{K+P}{RBN}$

   c) $v = \dfrac{N(K+P)}{RB}$ d) $v = \dfrac{RB(K+P)}{N}$

19._____

20. Phelan's scores on five tests are 88, 85, 92, 87, and 888. What must Phelan score on his sixth test so that his average will be 90?

   a) 95   b) 100   c) 79   d) 86

20._____

21. Find three consecutive odd integers such that the middle integer subtracted from the sum of four times the smallest and two times the largest is 321.

   a) 63, 65, 67   b) 38, 40, 42
   c) 54, 56, 58   d) 67, 69, 71

21._____

22. Simplify: $9b - [23 - 2(9b - 12)]$.

   a) $-27b - 1$ b) $18b - 1$
   c) $-18b + 47$ d) $27b - 47$

22._____

23. Simplify: $(2x^{-4}y^{-9})(6x^5 y^3)$. Do not use negative exponents.

   a) $12xy^{12}$ b) $-\dfrac{12}{x^{20}y^{10}}$ c) $\dfrac{12x}{y^6}$ d) $-\dfrac{12y}{x^6}$

23._____

24. Simplify: $(-3x^{-1}y^4)^{-3}$. Do not use negative exponents.

   a) $-\dfrac{3x^3}{y^6}$ b) $\dfrac{x^3}{9y^6}$ c) $-\dfrac{27}{x^3 y^{12}}$ d) $-\dfrac{x^3}{27y^{12}}$

24._____

CHAPTER 1   NAME_____

TEST FORM H

ANSWERS

25._____

25. Simplify: $\left(\dfrac{10x^8 y^{-4}}{-5y^{-3}}\right)^2$. Do not use negative exponents.

a) $\dfrac{2x^{16}}{y^2}$ b) $4x^{16}y^6$ c) $\dfrac{4x^{16}}{y^2}$ d) $\dfrac{2x^6}{y^2}$

26._____

26. Simplify using scientific notation and correct number of significant figures:

$$(1.12 \times 10^{-2})(9.78 \times 10^{-5}).$$

a) 9.2×10^{-3} b) 1.1×10^{-8}
c) 1.1×10^{-6} d) 9.2×10^3

27._____

27. Simplify using scientific notation and correct number of significant figures:
$$\dfrac{1.8 \times 10^3}{8.5 \times 10^{-8}}.$$

a) 2.1×10^{10} b) 7.6×10^{-5} c) 2.1×10^9 d) 1.6×10^{-9}

28._____

28. The average distance between the sun and the planet Neptune is 4.5×10^9 km. About how far does Uranus travel in one orbit around the sun? (Assume a circular orbit.)

a) 2.8×10^{10} km   b) 8.2×10^{10} km
c) 1.8×10^{10} km   d) 2.8×10^{-10} km

29._____

29. Simplify: $(5x^{9a}y^{b-4})^{2c}$.

a) $25^c x^{14ac} y^{2bc+6c}$ b) $25^c x^{18ac} y^{2bc-8c}$
c) $5^c x^{11ac} y^{4bc-4c}$ d) $10^c x^{18ac} y^{2bc+6c}$

30._____

30. Simplify: $\dfrac{-36a^{x+2}}{4a^x}$.

a) $-9a^3$ b) $9a^{-3}$ c) $9a^2$ d) $-9a^2$

20

CHAPTER 2
TEST FORM A

NAME_____

CLASS_____ SCORE_____ GRADE_____

Determine whether the ordered pair is a solution of the given equation.

1. $(0, 1)$, $5x + 3y = 3$
2. $(2, -3)$, $-3p + 2q = -10$

Graph.

3. $y = -2x + 1$

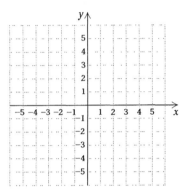

4. $y = -x^2 + 4$

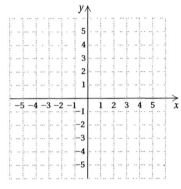

5. $f(x) = 3$

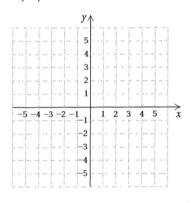

6. $9 - x = 15$

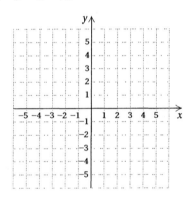

7. For the following graph of f, determine **(a)** $f(3)$; **(b)** the domain of f; **(c)** any x value for which $f(x) = 2$; and **(d)** the range of f.

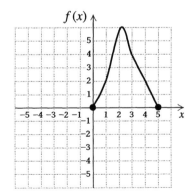

ANSWERS

1._____

2._____

See graph.
3._____

See graph.
4._____

See graph.
5._____

See graph.
6._____

7.a._____

b._____

c._____

d._____

21

CHAPTER 2

TEST FORM A

NAME_____

ANSWERS

8. _____

9. _____

10. _____

11. _____

12. _____

13. _____

14. _____

15. _____

8. The function $S(t) = 0.8t + 15.2$ can be used to estimate the total Canadian sales of sandals, in millions of dollars, t years after 1995. Predict the total Canadian sales of sandals in 2010.

9. In 1996, 38 fist-fights broke out in Shakeytown over the public use of cellular phones, and in 2000 there were 122 such fights. Estimate the number of fights over cell phones in Shakeytown in 1999.

Find the slope and the y-intercept.

10. $f(x) = \frac{2}{5}x - 13$

11. $-8y - 3x = 18$

Find the slope of the line containing the following points. If the slope is undefined, state so.

12. $(-2, 0)$ and $(2, 6)$

13. $(-2.9, 1.7)$ and $(-1.8, 3.9)$

14. Find the rate of change for the graph below. Use appropriate units.

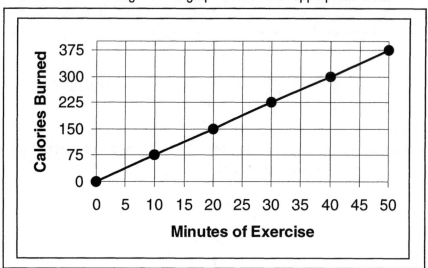

15. Find a linear function whose graph has slope -3 and y-intercept $(0, -4)$.

22

CHAPTER 2

TEST FORM A

NAME_____

ANSWERS

16. Graph using intercepts:

    $-8x + 3y = 6$.

17. Solve $x + 1 = \frac{1}{2}x$ graphically.

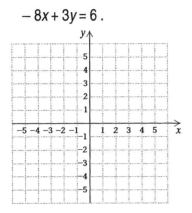
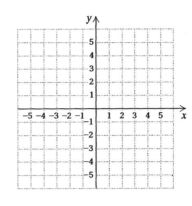

16. See graph.

17. See graph.

18. _____

18. Which of these are linear equations?

   (a) $2x - 3 = 0$
   (b) $3x - 4y = 10$
   (c) $5x - 6y^2 = 1$

19. _____

19. Find an equation of the line, in point–slope form, with slope 15 and containing the point $(3, 4)$.

20. _____

20. Using function notation, write a slope-intercept equation for the line containing $(1, 3)$ and $(2, 5)$.

21. _____

Without graphing, determine whether each pair of lines is parallel, perpendicular, or neither.

22. _____

21. $y + 9 = 5x$
    $-5x + y = -2$

22. $y = -5x + 2$
    $5y - x = 3$

23. _____

Find an equation of the line...

24. _____

23. containing $(3, 4)$ and parallel to the line $9x - 2y = 18$.

24. containing $(3, 4)$ and perpendicular to the line $9x - 2y = 18$.

CHAPTER 2

TEST FORM A

NAME_____

ANSWERS

25.a._____

b._____

c._____

26.a._____

b._____

27._____

28._____

25. Find the following, given that $g(x) = -9x - 5$ and $h(x) = x^2 + 8$.

    (a) $h(-2)$
    (b) $(g \cdot h)(3)$
    (c) The domain of h/g

26. If you rent a car for one day and drive it 350 mi, the cost is $75. If you drive it 500 mi, the cost is $90. Let $C(m)$ represent the cost, in dollars, of driving m miles.

    (a) Find a linear function that fits the data.
    (b) Use the function to find how much it will cost you to rent the car for one day and drive it 250 mi.

27. The graph of the function $f(x) = mx + b$ contains the points $(r, 2)$ and $(5, s)$. Express s in terms of r if the graph is parallel to the line $8x - y = 10$.

28. Given that $f(x) = 9x^2 + 8$ and $g(x) = 2x - 6$, find an expression for $h(x)$ so that the domain of $f/g/h$ is $\{x \mid x \in \mathbb{R} \text{ and } x \neq 3 \text{ and } x \neq \frac{5}{8}\}$

CHAPTER 2

TEST FORM B

NAME_____

CLASS_____ SCORE_____ GRADE_____

Determine whether the ordered pair is a solution of the given equation.

1. $(0, 2)$, $7x + 8y = 16$
2. $(2, 3)$, $-4p + 3q = 2$

Graph.

3. $y = 2x - 1$
4. $y = x^2 - 5$

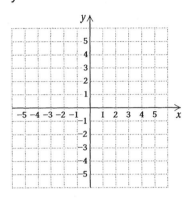
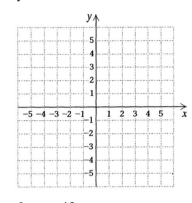

5. $f(x) = -2$
6. $8 - x = 12$

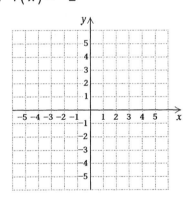
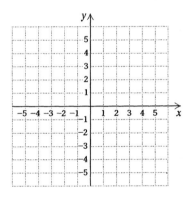

7. For the following graph of f, determine **(a)** $f(2)$; **(b)** the domain of f; **(c)** any x value for which $f(x) = 1$; and **(d)** the range of f.

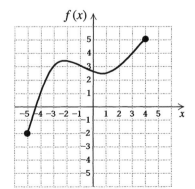

ANSWERS

1._____

2._____

3. See graph._____

4. See graph._____

5. See graph._____

6. See graph._____

7.a._____

b._____

c._____

d._____

CHAPTER 2   NAME_____

TEST FORM B

ANSWERS

8. The function $S(t) = 2.3t + 2.5$ can be used to estimate the total Estonian sales of chai, in millions of dollars, t years after 1997. Predict the total Estonian sales of chai in 2005.

8._____

9. In 1995, each household in Lawntown spent an average of 10 hours per week cutting grass. That time increased steadily to 16 hours per week in 2000. Estimate the average number of hours each Lawntown household spent on grass cutting in 1997.

9._____

Find the slope and the y-intercept.

10. $f(x) = -\dfrac{3}{7}x + 12$ 11. $-6y + 5x = -3$

10._____

Find the slope of the line containing the following points. If the slope is undefined, state so.

11._____

12. $(5, -1)$ and $(8, 5)$ 13. $(3.1, 2.5)$ and $(3.1, -1.8)$

14. Find the rate of change for the graph below. Use appropriate units.

12._____

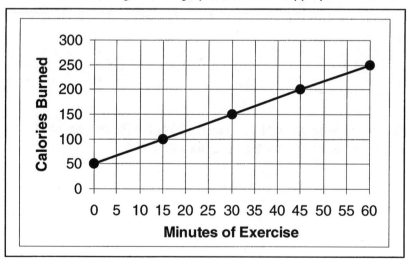

13._____

14._____

15. Find a linear function whose graph has slope 8 and y-intercept $(0, -5)$.

15._____

CHAPTER 2

NAME_____

TEST FORM B

16. Graph using intercepts:

$8x + 5y = -20$.

17. Solve $x + 2 = -\frac{1}{2}x$ graphically.

ANSWERS

16. See graph. _____

17. See graph. _____

18. _____

18. Which of these are linear equations?

(a) $3x - 4 = 0$
(b) $6x - 7y^2 = 2$
(c) $4x - 5y = 11$

19. _____

19. Find an equation of the line, in point–slope form, with slope -13 and containing the point $(5, 7)$.

20. _____

20. Using function notation, write a slope-intercept equation for the line containing $(2, -2)$ and $(3, -4)$.

21. _____

Without graphing, determine whether each pair of lines is parallel, perpendicular, or neither.

21. $2y + 8 = 4x$
    $-8x + 4y = -3$

22. $3y = -6x + 4$
    $6y + 3x = 5$

22. _____

23. _____

Find an equation of the line...

24. _____

23. containing $(3, -4)$ and parallel to the line $8x - 3y = 24$.

24. containing $(3, -4)$ and perpendicular to the line $8x - 3y = 24$.

CHAPTER 2

TEST FORM B

NAME _____

ANSWERS

25. Find the following, given that $g(x) = -8x - 6$ and $h(x) = x^2 + 7$.

    (a) $h(-2)$
    (b) $(g \cdot h)(3)$
    (c) The domain of h/g

25.a. _____

b. _____

c. _____

26. If you rent a car for one day and drive it 200 mi, the cost is $75. If you drive it 300 mi, the cost is $95. Let $C(m)$ represent the cost, in dollars, of driving m miles.

    (a) Find a linear function that fits the data.
    (b) Use the function to find how much it will cost you to rent the car for one day and drive it 400 mi.

26.a. _____

b. _____

27. The graph of the function $f(x) = mx + b$ contains the points $(r, 3)$ and $(6, s)$. Express s in terms of r if the graph is parallel to the line $7x - 2y = 12$.

27. _____

28. Given that $f(x) = 8x^2 + 7$ and $g(x) = 3x - 7$, find an expression for $h(x)$ so that the domain of $f/g/h$ is $\{x \mid x \in \mathbb{R}$ and $x \neq \frac{7}{3}$ and $x \neq \frac{6}{7}\}$

28. _____

CHAPTER 2 NAME_____

TEST FORM C CLASS_____ SCORE_____ GRADE_____

Determine whether the ordered pair is a solution of the given equation.

1. $(-1, 3)$, $3x + 4y = 9$
2. $(1, -4)$, $-5p + 4q = 21$

Graph.

3. $y = -x + 3$

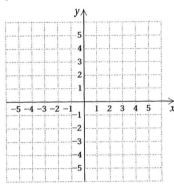

4. $y = \frac{1}{4}x^2 - 3$

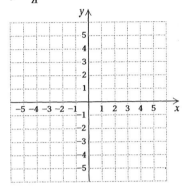

5. $f(x) = 5$

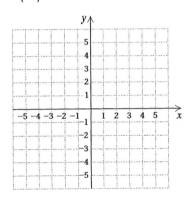

6. $7 - x = 10$

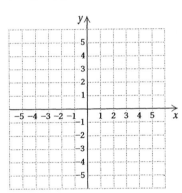

7. For the following graph of f, determine **(a)** $f(2)$; **(b)** the domain of f; **(c)** any x value for which $f(x) = 3$; and **(d)** the range of f.

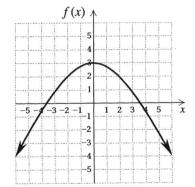

ANSWERS

1._____

2._____

See graph.
3._____

See graph.
4._____

See graph.
5._____

See graph.
6._____

7.a._____

b._____

c._____

d._____

29

CHAPTER 2

TEST FORM C

NAME _____

ANSWERS

8. The function $S(t) = 308.7 - 4.1t$ can be used to estimate the total Bahamian sales of snow-tires, in thousands of dollars, t years after 1990. Predict the total Bahamian sales of snow-tires in 2003.

8. _____

9. Zeke spent $50 on coffee in May and increased his spending on coffee to $74 in September. Estimate his coffee spending in July.

9. _____

Find the slope and the y-intercept.

10. $f(x) = -\dfrac{4}{5}x + 11$ 11. $-4y - 7x = -5$

10. _____

Find the slope of the line containing the following points. If the slope is undefined, state so.

12. $(0, 4)$ and $(-2, -2)$ 13. $(3.2, 1.2)$ and $(5.8, 1.2)$

11. _____

14. Find the rate of change for the graph below. Use appropriate units.

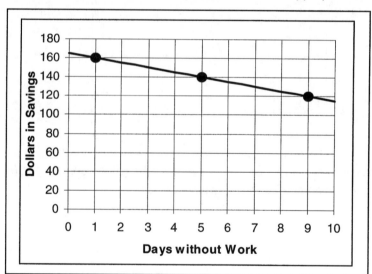

12. _____

13. _____

14. _____

15. Find a linear function whose graph has slope -1 and y-intercept $(0, 6)$.

15. _____

CHAPTER 2

TEST FORM C

16. Graph using intercepts:
    $-6x + 5y = -30$.

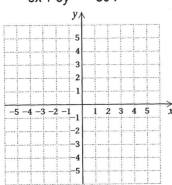

17. Solve $x - 2 = 2x$ graphically.

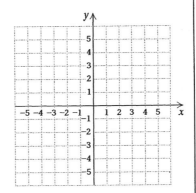

18. Which of these are linear equations?

    (a) $4x - 5 = 0$
    (b) $5x - 6y = 12$
    (c) $7x - 8y^2 = 3$

19. Find an equation of the line, in point–slope form, with slope 11 and containing the point $(4, -5)$.

20. Using function notation, write a slope-intercept equation for the line containing $(-3, 4)$ and $(2, -6)$.

Without graphing, determine whether each pair of lines is parallel, perpendicular, or neither.

21. $3y - 12 = 3x$
    $-3x + 4y = 8$

22. $5y = -7x + 6$
    $7y - 5x = 7$

Find an equation of the line...

23. containing $(-3, 4)$ and parallel to the line $7x - 4y = 28$.

24. containing $(-3, 4)$ and perpendicular to the line $7x - 4y = 28$.

ANSWERS

16. See graph.

17. See graph.

18. _____

19. _____

20. _____

21. _____

22. _____

23. _____

24. _____

CHAPTER 2

TEST FORM C

NAME_____

ANSWERS

25. Find the following, given that $g(x) = -7x - 7$ and $h(x) = x^2 + 6$.

    (a) $h(-2)$
    (b) $(g \cdot h)(3)$
    (c) The domain of h/g

25.a._____

b._____

c._____

26. If you rent a car for one day and drive it 50 mi, the cost is $45. If you drive it 300 mi, the cost is $120. Let $C(m)$ represent the cost, in dollars, of driving m miles.

    (a) Find a linear function that fits the data.
    (b) Use the function to find how much it will cost you to rent the car for one day and drive it 210 mi.

26.a._____

b._____

27. The graph of the function $f(x) = mx + b$ contains the points $(r, 4)$ and $(7, s)$. Express s in terms of r if the graph is parallel to the line $6x - 3y = 14$.

27._____

28. Given that $f(x) = 7x^2 + 6$ and $g(x) = 4x - 7$, find an expression for $h(x)$ so that the domain of $f/g/h$ is $\{x \mid x \in \mathbb{R} \text{ and } x \neq \frac{7}{4} \text{ and } x \neq \frac{7}{6}\}$.

28._____

CHAPTER 2　　　　　　　　　　　　　　　　　　NAME_____

TEST FORM D　　　　　　　　　　　CLASS_____ SCORE_____ GRADE_____

Determine whether the ordered pair is a solution of the given equation.

1. $(-1, 4)$, $4x + 5y = 16$　　　2. $(1, 4)$, $-6p + 4q = -10$

Graph.

3. $y = x - 3$　　　4. $y = -\dfrac{1}{4}x^2 + 4$

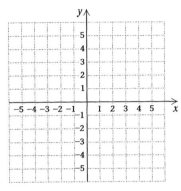
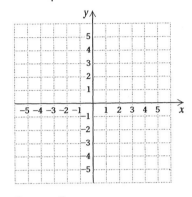

5. $f(x) = -4$　　　6. $6 - x = 7$

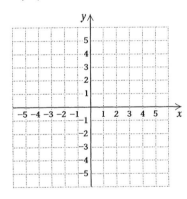
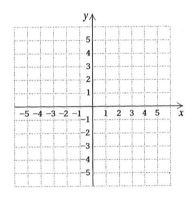

7. For the following graph of f, determine (a) $f(2)$; (b) the domain of f; (c) any x value for which $f(x) = 2$; and (d) the range of f.

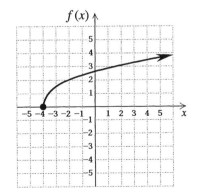

ANSWERS

1._____

2._____

See graph.
3._____

See graph.
4._____

See graph.
5._____

See graph.
6._____

7.a._____
　b._____
　c._____
　d._____

33

CHAPTER 2
TEST FORM D

NAME_____

ANSWERS

8. The function $S(t) = 8.9 - 0.6t$ can be used to estimate the total Costa Rican sales of bow-ties, in thousands of dollars, t years after 1995. Predict the total Costa Rican sales of bow-ties in 2009.

8._____

9. Cassandra spent $85 on gasoline in July and decreased her monthly spending on gas to $65 in January. Estimate her spending on gasoline in October.

9._____

Find the slope and the y-intercept.

10. $f(x) = -\frac{5}{7}x - 10$

11. $-2y + 9x = 36$

10._____

Find the slope of the line containing the following points. If the slope is undefined, state so.

12. $(6, -5)$ and $(7, -12)$

13. $(0.5, 10.6)$ and $(-0.6, 7.3)$

11._____

14. Find the rate of change for the graph below. Use appropriate units.

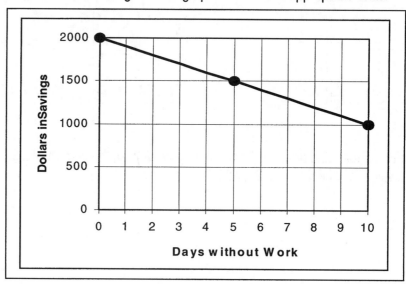

12._____

13._____

14._____

15. Find a linear function whose graph has slope 4 and y-intercept $(0, 7)$.

15._____

CHAPTER 2

TEST FORM D

NAME_____

16. Graph using intercepts:
    $6x + 7y = 24$.

17. Solve $x - 1 = -2x$ graphically.

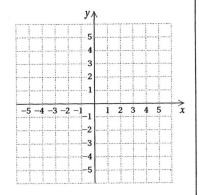

ANSWERS

16. See graph._____

17. See graph._____

18._____

18. Which of these are linear equations?

   (a) $5x - 6 = 0$
   (b) $8x - 9y^2 = 4$
   (c) $6x - 7y = 13$

19._____

19. Find an equation of the line, in point–slope form, with slope -9 and containing the point $(6, -8)$.

20._____

20. Using function notation, write a slope-intercept equation for the line containing $(-2, -5)$ and $(1, 1)$.

21._____

Without graphing, determine whether each pair of lines is parallel, perpendicular, or neither.

22._____

21. $4y - 5 = 2x$
    $-2x + 4y = -2$

22. $7y = -8x + 8$
    $8y - 7x = 9$

23._____

Find an equation of the line...

23. containing $(-3, -4)$ and parallel to the line $6x - 5y = 30$.

24._____

24. containing $(-3, -4)$ and perpendicular to the line $6x - 5y = 30$.

CHAPTER 2

TEST FORM D

NAME_____

ANSWERS

25.a._____

 b._____

 c._____

26.a._____

 b._____

27._____

28._____

25. Find the following, given that $g(x) = -6x - 8$ and $h(x) = x^2 + 5$.

    (a) $h(-2)$
    (b) $(g \cdot h)(3)$
    (c) The domain of h/g

26. If you rent a car for one day and drive it 250 mi, the cost is $102. If you drive it 400 mi, the cost is $147. Let $C(m)$ represent the cost, in dollars, of driving m miles.

    (a) Find a linear function that fits the data.
    (b) Use the function to find how much it will cost you to rent the car for one day and drive it 100 mi.

27. The graph of the function $f(x) = mx + b$ contains the points $(r, 5)$ and $(8, s)$. Express s in terms of r if the graph is parallel to the line $5x - 4y = 16$.

28. Given that $f(x) = 6x^2 + 5$ and $g(x) = 5x - 8$, find an expression for $h(x)$ so that the domain of $f/g/h$ is $\{x \mid x \in \mathbb{R} \text{ and } x \neq \frac{8}{5} \text{ and } x \neq \frac{9}{4}\}$.

CHAPTER 2   NAME_____

TEST FORM E   CLASS_____ SCORE_____ GRADE_____

Determine whether the ordered pair is a solution of the given equation.

ANSWERS

1. $(1, 5)$, $6x + 2y = 16$

2. $(0, -5)$, $-6p + 3q = -14$

Graph.

1._____

3. $y = -\dfrac{1}{2}x + 2$

4. $y = -x^2 - 1$

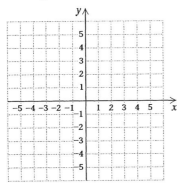
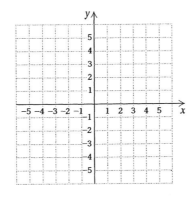

2._____

5. $f(x) = 1$

6. $5 - x = 4$

3. See graph._____

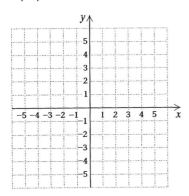
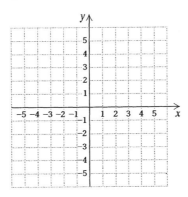

4. See graph._____

5. See graph._____

7. For the following graph of f, determine **(a)** $f(2)$; **(b)** the domain of f; **(c)** any x value for which $f(x) = 1$; and **(d)** the range of f.

6. See graph._____

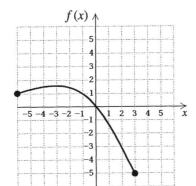

7.a._____

b._____

c._____

d._____

37

CHAPTER 2
TEST FORM E

NAME_____

ANSWERS

8. The function $S(t) = 1.7t + 9.2$ can be used to estimate the total Romanian sales of footstools, in millions of dollars, t years after 1998. Predict the total Romanian sales of footstools in 2005.

8._____

9. In 1990, 105,000 people ate at The Loony Spoon, and 215,000 customers ate there in 1998. Estimate the number of people served at The Loony Spoon in 1993.

9._____

Find the slope and the y-intercept.

10. $f(x) = -\frac{5}{2}x + 9$

11. $4y - 11x = 22$

10._____

Find the slope of the line containing the following points. If the slope is undefined, state so.

12. $(1, 1)$ and $(1, 3)$

13. $(8.7, -2.1)$ and $(7.6, 2.3)$

11._____

14. Find the rate of change for the graph below. Use appropriate units.

12._____

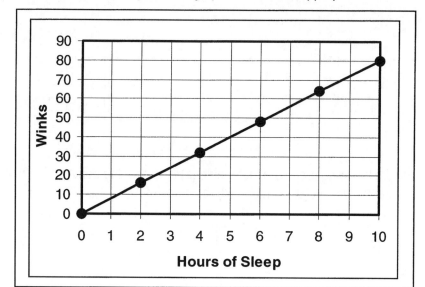

13._____

14._____

15. Find a linear function whose graph has slope -10 and y-intercept $(0, -8)$.

15._____

CHAPTER 2

TEST FORM E

16. Graph using intercepts:
$-4x + 7y = 20$.

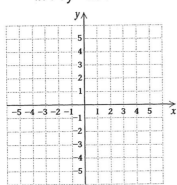

17. Solve $x + 4 = 3x$ graphically.

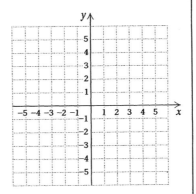

ANSWERS

16. See graph.

17. See graph.

18. _____

18. Which of these are linear equations?

 (a) $6x - 7 = 0$
 (b) $7x - 8y = 14$
 (c) $9x - 10y^2 = 5$

19. _____

19. Find an equation of the line, in point–slope form, with slope 7 and containing the point $(-5, 6)$.

20. _____

20. Using function notation, write a slope-intercept equation for the line containing $(-1, 2)$ and $(-2, 1)$.

21. _____

Without graphing, determine whether each pair of lines is parallel, perpendicular, or neither.

22. _____

21. $5y + 10 = x$
 $-5x + 25y = -12$

22. $9y = -9x + 10$
 $10y + 10x = 9$

23. _____

Find an equation of the line...

23. containing $(5, 2)$ and parallel to the line $5x - 6y = 30$.

24. _____

24. containing $(5, 2)$ and perpendicular to the line $5x - 6y = 30$.

CHAPTER 2

TEST FORM E

NAME_____

ANSWERS

25.a._____

b._____

c._____

26.a._____

b._____

27._____

28._____

25. Find the following, given that $g(x) = -5x - 9$ and $h(x) = x^2 + 4$.

   (a) $h(-2)$
   (b) $(g \cdot h)(3)$
   (c) The domain of h/g

26. If you rent a car for one day and drive it 100 mi, the cost is $50. If you drive it 450 mi, the cost is $120. Let $C(m)$ represent the cost, in dollars, of driving m miles.

   (a) Find a linear function that fits the data.
   (b) Use the function to find how much it will cost you to rent the car for one day and drive it 250 mi.

27. The graph of the function $f(x) = mx + b$ contains the points $(r, 6)$ and $(9, s)$. Express s in terms of r if the graph is parallel to the line $4x - 5y = 18$.

28. Given that $f(x) = 5x^2 + 4$ and $g(x) = 6x - 9$, find an expression for $h(x)$ so that the domain of $f/g/h$ is $\{x \mid x \in \mathbb{R} \text{ and } x \neq \frac{3}{2} \text{ and } x \neq \frac{1}{3}\}$

40

CHAPTER 2

TEST FORM F

NAME_____

CLASS_____ SCORE_____ GRADE_____

ANSWERS

Determine whether the ordered pair is a solution of the given equation.

1. $(1, 6)$, $7x + 3y = 25$
2. $(0, 5)$, $-5p + 2q = 8$

Graph.

3. $y = \dfrac{1}{2}x - 4$
4. $y = x^2 + 2$

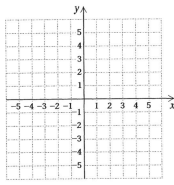
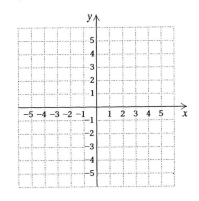

5. $f(x) = 0$
6. $4 - x = 1$

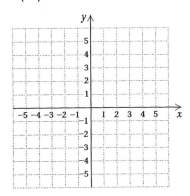
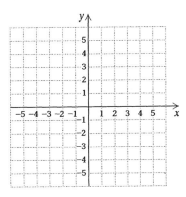

7. For the following graph of f, determine **(a)** $f(2)$; **(b)** the domain of f; **(c)** any x value for which $f(x) = 1$; and **(d)** the range of f.

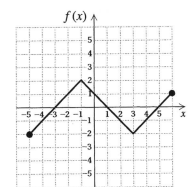

1._____

2._____

See graph.
3._____

See graph.
4._____

See graph.
5._____

See graph.
6._____

7.a._____

b._____

c._____

d._____

41

CHAPTER 2
TEST FORM F

NAME_____

ANSWERS

8._____

9._____

10._____

11._____

12._____

13._____

14._____

15._____

8. The function $S(t) = 1.2t + 11.5$ can be used to estimate the total Sudanese sales of comic books, in thousands of dollars, t years after 1999. Predict the total Sudanese sales of comic books in 2004.

9. Tom broke $270 worth of dishes in his first month as a waiter, and by his eighth month he broke only $60 worth of dishes. Estimate the worth of the dishes Tom broke in his third month as a waiter.

Find the slope and the y – intercept.

10. $f(x) = \dfrac{7}{3}x + 8$

11. $6y + 13x = -78$

Find the slope of the line containing the following points. If the slope is undefined, state so.

12. $(-3, 5)$ and $(-4, 5)$

13. $(-0.6, 12)$ and $(0.5, 10.9)$

14. Find the rate of change for the graph below. Use appropriate units.

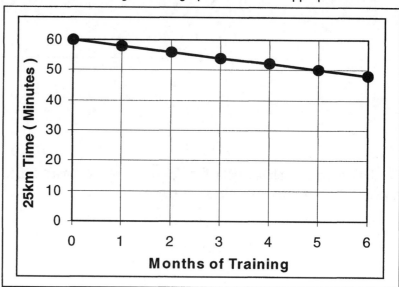

15. Find a linear function whose graph has slope 5 and y-intercept $(0, -9)$.

CHAPTER 2

TEST FORM F

16. Graph using intercepts: $4x + 9y = -12$.

17. Solve $x - 4 = -3x$ graphically.

ANSWERS

16. See graph._____

17. See graph._____

18._____

18. Which of these are linear equations?

   (a) $7x - 8 = 0$
   (b) $10x - 11y^2 = 6$
   (c) $8x - 9y = 15$

19._____

19. Find an equation of the line, in point–slope form, with slope -5 and containing the point $(-7, 8)$.

20._____

20. Using function notation, write a slope-intercept equation for the line containing $(-2, 6)$ and $(-3, 7)$.

21._____

Without graphing, determine whether each pair of lines is parallel, perpendicular, or neither.

22._____

21. $6y - 4 = -x$
    $2x + 10y = 5$

22. $11y = -10x - 3$
    $10y - 11x = 7$

23._____

Find an equation of the line...

23. containing $(5, -2)$ and parallel to the line $4x - 7y = 28$.

24._____

24. containing $(5, -2)$ and perpendicular to the line $4x - 7y = 28$.

CHAPTER 2

TEST FORM F

ANSWERS

25.a._____

b._____

c._____

26.a._____

b._____

27._____

28._____

25. Find the following, given that $g(x) = -4x - 1$ and $h(x) = x^2 + 3$.

   (a) $h(-2)$
   (b) $(g \cdot h)(3)$
   (c) The domain of h/g

26. If you rent a car for one day and drive it 250 mi, the cost is $60. If you drive it 300 mi, the cost is $65. Let $C(m)$ represent the cost, in dollars, of driving m miles.

   (a) Find a linear function that fits the data.
   (b) Use the function to find how much it will cost you to rent the car for one day and drive it 750 mi.

27. The graph of the function $f(x) = mx + b$ contains the points $(r, 7)$ and $(1, s)$. Express s in terms of r if the graph is parallel to the line $3x - 6y = 20$.

28. Given that $f(x) = 4x^2 + 3$ and $g(x) = 7x - 10$, find an expression for $h(x)$ so that the domain of $f/g/h$ is $\{x \mid x \in \mathbb{R} \text{ and } x \neq \frac{10}{7} \text{ and } x \neq 1\}$

CHAPTER 2   NAME_____

TEST FORM G   CLASS_____ SCORE_____ GRADE_____

ANSWERS

1. Which ordered pair is a solution of the equation $3x + y = -1$?

    a) $(2, 5)$ b) $(2, -5)$ c) $(-2, 5)$ d) $(5, 2)$

1._____

2. Which graph represents $-\frac{2}{3}x - 2$?

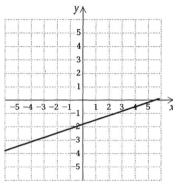
a)

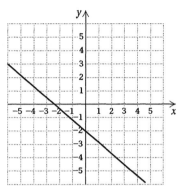
b)

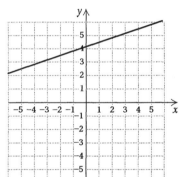
c)

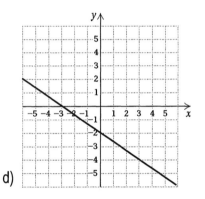
d)

2._____

3. For the following graph of f, determine the domain of f.

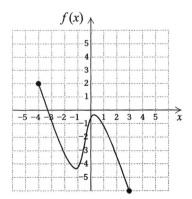

    a) $\{x \mid -4 \leq x \leq 3\}$ b) $\{x \mid -3 \leq x \leq 4\}$
    c) $\{x \mid -6 \leq x \leq 2\}$ d) $\{x \mid -4 \leq x \leq -3\}$

3._____

CHAPTER 2

TEST FORM G

NAME_____

ANSWERS

4._____

4. The function $S(t) = 139.2 - 11.5t$ can be used to estimate the total Cuban sales of windshield-wiper fluid, in thousands of dollars, t years after 1990. Predict the total Cuban sales of windshield-wiper fluid in 2010.

   a) $72,300   b) $37,200   c) $33,700   d) $3700

5._____

5. Find the slope and y-intercept of $f(x) = -\frac{5}{4}x - 7$.

   a) $-\frac{5}{4}; (0, 7)$ b) $-\frac{5}{4}; (0, -7)$

   c) $\frac{5}{4}; (0, -7)$ d) $-\frac{4}{5}; (0, -7)$

6._____

6. Find the slope and y-intercept of $8y - 15x = -40$.

   a) $\frac{15}{8}; (0, -5)$ b) $-\frac{15}{8}; (0, -8)$

   c) $-\frac{15}{4}; (0, -5)$ d) $-\frac{5}{4}; (0, 5)$

7._____

7. Find the slope of the line containing the points $(4, -3)$ and $(6, 5)$.

   a) 4   b) -4 c) 0.25   d) -0.25

8._____

8. Find a linear function whose graph has slope -7 and y-intercept $(0, 8)$.

   a) $f(x) = -7x - 8$ b) $f(x) = 7x + 8$
   c) $f(x) = -8x + 7$ d) $f(x) = -7x + 8$

9._____

9. Janis could bench-press 100 lb in February, eventually increasing her bench-press to 180 lb in August. Estimate the weight Janice was able to bench-press in May.

   a) 120 lb   b) 150 lb   c) 140 lb   d) 160 lb

CHAPTER 2

TEST FORM G

10. Which equation is NOT linear?

    a) $x = 3y - 6$ b) $4y + 2x - 5 = 3x - 9$

    c) $x = 3y^2 - 1$ d) $y = -1$

    10._____

11. Find an equation in point slope form of the line with slope 3 and containing $(-6, -7)$.

    a) $y - 7 = 3(x - 6)$ b) $y + 7 = 3(x + 6)$

    c) $y + 7 = -3(x + 6)$ d) $y + 6 = 3(x + 7)$

    11._____

12. Which line is parallel to $7y - 13 = -2x$

    a) $2x + 7y = 7$ b) $7x + 2y = 2$

    c) $2x - 7y = 7$ d) $2x + 7y = -7$

    12._____

13. Find an equation of the line containing $(-5, 2)$ and perpendicular to the line $3x - 8y = 24$.

    a) $y = \frac{8}{3}x + \frac{34}{3}$ b) $y = -\frac{8}{3}x - \frac{34}{3}$

    c) $y = -\frac{3}{8}x - \frac{34}{3}$ d) $y = \frac{8}{3}x - \frac{34}{3}$

    13._____

14. Given that $g(x) = -3x - 2$ and $h(x) = x^2 + 2$, find $(g \cdot h)(3)$.

    a) -55 b) -121 c) 121 d) 0

    14._____

ANSWERS

CHAPTER 2

TEST FORM G

NAME_____

ANSWERS

15. If you rent a car for one day and drive it 200 mi, the cost is $95. If you drive it 375 mi, the cost is $165. How much it will cost you to rent the car for one day and drive it 150 mi.

15._____

  a) $ 85   b) $65   c) $125   d) $75

16. Find the intercepts of $-2x + 9y = -8$.

  a) $(4, 0)$ and $\left(0, -\frac{8}{9}\right)$ b) $\left(\frac{8}{9}, 0\right)$ and $(0, -4)$

  c) $\left(-\frac{8}{9}, 0\right)$ and $(0, 4)$ d) $(-4, 0)$ and $\left(0, \frac{8}{9}\right)$

16._____

17. The graph of the function $f(x) = mx + b$ contains the points $(r, 8)$ and $(2, s)$. Express s in terms of r if the graph is parallel to the line $2x - 7y = 22$.

17._____

  a) $s = \frac{2}{7}r + \frac{52}{7}$ b) $s = \frac{7}{2}r + \frac{52}{7}$

  c) $s = \frac{7}{2}r + \frac{52}{2}$ d) $s = \frac{2}{7}r - \frac{52}{7}$

18. Given that $f(x) = 3x^2 + 2$ and $g(x) = 8x - 11$, find an expression for $h(x)$ so that the domain of $f/g/h$ is $\{x \mid x \in \mathbb{R}$ and $x \neq \frac{11}{8}$ and $x \neq 3\}$

18._____

  a) $h(x) = x + 3$ b) $h(x) = 3x - 3$

  c) $h(x) = x - 3$ d) $h(x) = 3x + 3$

CHAPTER 2   NAME_____

TEST FORM H   CLASS_____ SCORE_____ GRADE_____

1. Which ordered pair is a solution of the equation $2x + 3y = 16$?

   a) $(2, 5)$ b) $(2, 4)$ c) $(-2, 4)$ d) $(4, 2)$

2. Which graph represents $\frac{2}{3}x + 3$?

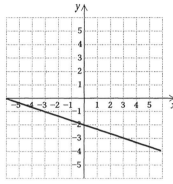
a)

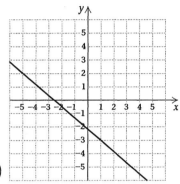
b)

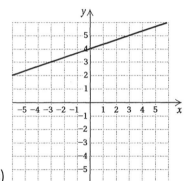
c)

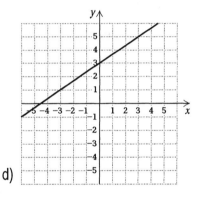
d)

3. For the following graph of f, determine the domain of f.

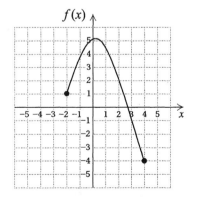

a) $\{x \mid -2 \leq x \leq 3\}$ b) $\{x \mid -3 \leq x \leq 5\}$
c) $\{x \mid -2 \leq x \leq 4\}$ d) $\{x \mid -4 \leq x \leq -3\}$

ANSWERS

1._____

2._____

3._____

CHAPTER 2   NAME_____

TEST FORM H

ANSWERS

4. The function $S(t) = 3.2t + 50.7$ can be used to estimate the total Mexican sales of shoehorns, in thousands of dollars, t years after 1996. Predict the total Mexican sales of shoehorns in 2007.

4._____

    a) $29,300   b) $39,200   c) $92,300   d) $9200

5. Find the slope and y–intercept of $f(x) = \frac{7}{5}x - 6$.

5._____

    a) $\frac{7}{5}; (0, -6)$ b) $-\frac{7}{5}; (0, -6)$

    c) $\frac{7}{5}; (-6, 0)$ d) $-\frac{5}{7}; (0, -6)$

6._____

6. Find the slope and y–intercept of $10y + 17x = 7$.

    a) $-\frac{17}{10}; \left(0, -\frac{7}{10}\right)$ b) $-\frac{17}{10}; \left(0, \frac{7}{10}\right)$

    c) $\frac{17}{10}; \left(0, \frac{7}{10}\right)$ d) $-\frac{7}{10}; \left(0, \frac{17}{10}\right)$

7._____

7. Find the slope of the line containing the points $(-2, 7)$ and $(1, -2)$.

    a) -3 b) -4 c) 3 d) 4

8._____

8. Find a linear function whose graph has slope 2 and y–intercept $(0, -11)$.

    a) $f(x) = -2x - 11$ b) $f(x) = 2x + 11$

    c) $f(x) = -11x + 2$ d) $f(x) = 2x - 11$

9._____

9. There were 30 absences from Fahid's electronics class during the first week. Absenteeism dropped steadily to 6 absences during the ninth week. Estimate the number of absences during the fourth week.

    a) 22   b) 10   c) 14   d) 18

CHAPTER 2

TEST FORM H

10. Which equation is NOT linear?

   a) $x = 3y^3 - 6$
   b) $4y + 2x - 5 = 3x - 9$
   c) $x = 3y - 1$
   d) $x = -12$

   10._____

11. Find an equation in point-slope form of the line with slope -1 and containing $(-8, -10)$.

   a) $y - 10 = x - 8$
   b) $y + 10 = -(x + 8)$
   c) $y + 8 = -(x + 10)$
   d) $y + 10 = 8 - x$

   11._____

12. Which line is parallel to $8y - 1 = -3x$

   a) $3x + 8y = 3$
   b) $3x + 2y = 8$
   c) $8x - 3y = 3$
   d) $2x + 8y = -3$

   12._____

13. Find an equation of the line containing $(-5, -2)$ and perpendicular to the line $2x - 9y = 18$.

   a) $y = \frac{9}{2}x + \frac{49}{2}$
   b) $y = -\frac{49}{2}x - \frac{9}{2}$
   c) $y = -\frac{9}{2}x - \frac{49}{2}$
   d) $y = \frac{9}{2}x - \frac{49}{2}$

   13._____

14. Given that $g(x) = -2x - 3$ and $h(x) = x^2 + 1$, find $(g \cdot h)(3)$.

   a) 72
   b) -72
   c) -90
   d) 90

   14._____

ANSWERS

CHAPTER 2

TEST FORM H

NAME_____

ANSWERS

15._____

15. If you rent a car for one day and drive it 250 mi, the cost is $111. If you drive it 600 mi, the cost is $216. How much it will cost you to rent the car for one day and drive it 450 mi.

   a) $ 185   b) $171   c) $228   d) $177

16. Find the intercepts of $2x + 11y = 4$.

   a) $(-2, 0)$ and $\left(0, \frac{4}{11}\right)$ b) $\left(\frac{4}{11}, 0\right)$ and $(0, -2)$

   c) $(2, 0)$ and $\left(0, \frac{4}{11}\right)$ d) $(-4, 0)$ and $\left(0, \frac{11}{2}\right)$

16._____

17. The graph of the function $f(x) = mx + b$ contains the points $(r, 9)$ and $(3, s)$. Express s in terms of r if the graph is parallel to the line $x - 8y = 24$.

17._____

   a) $s = -\frac{1}{8}r + \frac{75}{8}$ b) $s = \frac{1}{8}r + \frac{1}{8}$

   c) $s = \frac{75}{8}r + \frac{1}{8}$ d) $s = \frac{75}{8}r - \frac{1}{8}$

18. Given that $f(x) = 2x^2 + 1$ and $g(x) = 9x - 12$, find an expression for $h(x)$ so that the domain of $f / g / h$ is $\{x \mid x \in \mathbb{R} \text{ and } x \neq \frac{4}{3} \text{ and } x \neq \frac{4}{9}\}$

18._____

   a) $h(x) = 9x + 4$ b) $h(x) = 4x - 9$

   c) $h(x) = 4x - 3$ d) $h(x) = 9x - 4$

CHAPTER 3

TEST FORM A

Solve, if possible, using the substitution method.

1. $x + 3y = 10$,
   $3x - 3y = 5$

2. $7x - 5y = 8$
   $y = x - 2$

Solve, if possible, using the elimination method.

3. $x - 3y = 27$
   $2x + 3y = -9$

4. $2y + 3x = 2$
   $4x + 6y = -\dfrac{7}{3}$

5. The perimeter of a rectangle is 24. The length of the rectangle is eight less than three times the width. Find the dimensions of the rectangle.

6. Between her home mortgage, car loan, and credit card bill, Sepia is $115,000 in debt. Each month, Sepia's credit card accumulates 1.5% interest, her car loan 1% interest, and her mortgage 0.8% interest. After one month, her total accumulated interest is $975. The interest on Sepia's mortgage was $700 more than the interest on her car loan. Find the amount of each of these three debts.

Solve. If the system's equations are dependent or if there is no solution, then state this.

7. $-2x + 3y + z = 3$,
   $3x - y + 2z = 2$,
   $x + 2y - 3z = 1$

8. $x + 5y + 2z = -1$,
   $5x + 2y + z = 2$,
   $2x + y + 5z = 3$

9. $y + 2z = 17$,
   $-2y + z = -4$,
   $-x + 3y + 2z = 23$

10. $-y + 2z = 13$,
    $-2x + z = -2$,
    $-x - 2y = 7$

Solve using matrices.

11. $5x - 4y = -3$,
    $8x + 6y = 14$

12. $3x + 4y + 2z = 29$,
    $-3x + 4y + 2z = 11$,
    $3x - 4y + 2z = -3$

ANSWERS

1. _____

2. _____

3. _____

4. _____

5. _____

6. _____

7. _____

8. _____

9. _____

10. _____

11. _____

12. _____

CHAPTER 3

TEST FORM A

NAME_____

ANSWERS

13._____

14._____

15._____

16._____

17._____

18. a._____

    b._____

    c._____

    d._____

    e._____

19._____

20._____

Evaluate.

13. $\begin{vmatrix} 5 & 3 \\ -2 & 4 \end{vmatrix}$

14. $\begin{vmatrix} 1 & 2 & 3 \\ 3 & 1 & 2 \\ 2 & 3 & 1 \end{vmatrix}$

Solve using Cramer's rule.

15. $3x - 7y = -50$,
    $5x + 2y = -15$

16. An electrician, a carpenter, and a plumber are hired to work on a house. The electrician earns $20 per hour, the carpenter $25 per hour, and the plumber $30 per hour. On the first day on the job, the three worked a total of 19.5 hours and earned a total of $472.50. If the electrician worked 1.5 more hours than the carpenter did, then how many hours did the plumber work?

17. Find the equilibrium point for the demand and supply functions
    $D(p) = 76 - 7p$ and $S(p) = 54 + 4p$

18. Unethicon Inc. produces cloned hens. For the first year, the fixed set-up costs are $15,000. The variable costs for cloning each hen are $10. The revenue from each hen is $35. Find the following.
    a) The total cost $C(x)$ of producing x hens.
    b) The total revenue $R(x)$ from the sale of x hens.
    c) The total profit $P(x)$ from the production and sale of x hens.
    d) The profit or loss from production and sale of 300 hens; of 900 hens.
    e) The break-even point.

19. The graph of the function $f(x) = mx + b$ contains the points $(-15, -5)$ and $(3, 7)$. Find m and b.

20. At a county fair, an adult's ticket sold for $6.00, a senior citizen's ticket for $4.50, and a child's ticket for $2.00. On Tuesday, the number of adults' and senior citizens' tickets sold was 270 more than the number of children's tickets sold. The number of adults' tickets sold was 20 more than three times the number of senior citizens' tickets sold. Total receipts from the ticket sales were $2,670.00. How many of each type of ticket were sold?

CHAPTER 3

TEST FORM B

NAME _____

CLASS _____ SCORE _____ GRADE _____

Solve, if possible, using the substitution method.

1. $2x + 5y = 6,$
   $3x - 5y = 13$

2. $2x + 3y = -6$
   $y = 4x + 3$

Solve, if possible, using the elimination method.

3. $2x + 3y = -4$
   $x + 6y = -29$

4. $3y + 2x = 9$
   $4x - 5y = -\dfrac{34}{3}$

5. The perimeter of a rectangle is 64. The length of the rectangle is eight more than twice the width. Find the dimensions of the rectangle.

6. Between his home mortgage, car loan, and credit card bill, Sean is $168,000 in debt. Each month, Sean's credit card accumulates 1.5% interest, his car loan 1% interest, and his mortgage 0.8% interest. After one month, his total accumulated interest is $1410. The interest on Sean's mortgage was $1100 more than the interest on his credit card. Find the amount of each of these three debts.

Solve. If the system's equations are dependent or if there is no solution, then state this.

7. $-x + y + 3z = -2,$
   $3x - 3y + z = 26,$
   $-x + 3y - z = -16$

8. $6x + 5y + 4z = 43,$
   $3x + 2y + z = 16,$
   $x - 2y + 3z = 8$

9. $3y - z = 11,$
   $-2x + 3z = 0,$
   $3x + y - 2z = 15$

10. $2y - 3z = 12,$
    $3x - 2y = -9,$
    $-2x + 3z = -4$

Solve using matrices.

11. $6x + 5y = 17,$
    $9x - 7y = 11$

12. $-x + 3y - 5z = 17,$
    $x + 3y - 5z = 15,$
    $x - 3y - 5z = 33$

ANSWERS

1. _____

2. _____

3. _____

4. _____

5. _____

6. _____

7. _____

8. _____

9. _____

10. _____

11. _____

12. _____

CHAPTER 3   NAME_____

TEST FORM B

ANSWERS

Evaluate.

13. $\begin{vmatrix} 5 & 3 \\ 2 & -4 \end{vmatrix}$

14. $\begin{vmatrix} 2 & 1 & 3 \\ 3 & 2 & 1 \\ 1 & 3 & 2 \end{vmatrix}$

13._____

Solve using Cramer's rule.

14._____

15. $-3x + 7y = -40$,
    $5x + 2y = 12$

15._____

16. An electrician, a carpenter, and a plumber are hired to work on a house. The electrician earns $20 per hour, the carpenter $25 per hour, and the plumber $30 per hour. On the first day on the job, the three worked a total of 23.25 hours and earned a total of $571.25. If the electrician worked 2 more hours than the plumber did, then how many hours did the carpenter work?

16._____

17._____

17. Find the equilibrium point for the demand and supply functions
    $D(p) = 87 - 8p$ and $S(p) = 48 + 5p$

18.a._____

b._____

c._____

d._____

e._____

18. Unethicon Inc. produces cloned sheep. For the first year, the fixed set-up costs are $24,000. The variable costs for cloning each sheep are $15. The revenue from each sheep is $45. Find the following.
   a) The total cost $C(x)$ of producing x sheep.
   b) The total revenue $R(x)$ from the sale of x sheep.
   c) The total profit $P(x)$ from the production and sale of x sheep.
   d) The profit or loss from production and sale of 250 sheep; of 1000 sheep.
   e) The break-even point.

19. The graph of the function $f(x) = mx + b$ contains the points

19._____

   $(5, 7)$ and $(2, -2)$. Find m and b.

20. At a county fair, an adult's ticket sold for $6.00, a senior citizen's ticket for $4.50, and a child's ticket for $2.00. On Tuesday, the number of adults' and senior citizens' tickets sold was 330 more than the number of children's tickets sold. The number of adults' tickets sold was 30 more than three times the number of senior citizens' tickets sold. Total receipts from the ticket sales were $4,155.00. How many of each type of ticket were sold?

20._____

CHAPTER 3   NAME _____

TEST FORM C   CLASS _____ SCORE _____ GRADE _____

ANSWERS

Solve, if possible, using the substitution method.

1. $5x + 2y = 8,$
   $3x - 2y = -4$

2. $3x - 4y = 10$
   $y = -2x + 5$

1. _____

Solve, if possible, using the elimination method.

2. _____

3. $x + 6y = -22$
   $3x - 2y = 34$

4. $3y - 4x = 2$
   $3x - y = \dfrac{7}{12}$

3. _____

5. The perimeter of a rectangle is 84. The length of the rectangle is two less than three times the width. Find the dimensions of the rectangle.

4. _____

6. Between her home mortgage, car loan, and credit card bill, Yona is $40,000 in debt. Each month, Yona's credit card accumulates 1.5% interest, her car loan 1% interest, and her mortgage 0.8% interest. After one month, her total accumulated interest is $375. The interest on Yona's car loan was $10 more than the interest on her mortgage. Find the amount of each of these three debts.

5. _____

6. _____

7. _____

Solve. If the system's equations are dependent or if there is no solution, then state this.

8. _____

7. $-3x + 2y + z = 14,$
   $2x + y - 3z = -21,$
   $x - 3y + 2z = 7$

8. $-2x + 7y - 2z = 3,$
   $5x - 2y + 7z = 11,$
   $-2x + 5y - 2z = 1$

9. _____

9. $4x + 3z = -13,$
   $2y - z = 7,$
   $2x - 4y + z = -13$

10. $3x + 4z = 32,$
    $-3y + 3z = 33,$
    $4x - 4y = 40$

10. _____

Solve using matrices.

11. _____

11. $7x + 6y = 32,$
    $-x + 8y = 22$

12. $2x - 4y + 6z = 24,$
    $-2x + 4y + 6z = 48,$
    $2x + 4y - 6z = -16$

12. _____

57

CHAPTER 3

TEST FORM C

NAME_____

ANSWERS

Evaluate.

13. $\begin{vmatrix} 6 & 9 \\ -5 & 2 \end{vmatrix}$

14. $\begin{vmatrix} 4 & -3 & 5 \\ -4 & 3 & 5 \\ 4 & 3 & -5 \end{vmatrix}$

13._____

14._____

Solve using Cramer's rule.

15. $3x + 7y = 12$,
    $-5x + 2y = 21$

15._____

16. An electrician, a carpenter, and a plumber are hired to work on a house. The electrician earns $20 per hour, the carpenter $25 per hour, and the plumber $30 per hour. On the first day on the job, the three worked a total of 21.5 hours and earned a total of $532.50. If the carpenter worked 1.5 more hours than the electrician did, then how many hours did the plumber work?

16._____

17._____

17. Find the equilibrium point for the demand and supply functions
    $D(p) = 98 - 9p$ and $S(p) = 53 + 6p$

18.a._____

18. Unethicon Inc. produces cloned cows. For the first year, the fixed set-up costs are $33,250. The variable costs for cloning each cow are $22. The revenue from each cow is $57. Find the following.
    a) The total cost $C(x)$ of producing x cows.
    b) The total revenue $R(x)$ from the sale of x cows.
    c) The total profit $P(x)$ from the production and sale of x cows.
    d) The profit or loss from production and sale of 575 cows; of 1500 cows.
    e) The break-even point.

b._____

c._____

d._____

e._____

19._____

19. The graph of the function $f(x) = mx + b$ contains the points
    $(5, 1)$ and $(-10, -11)$. Find m and b.

20._____

20. At a county fair, an adult's ticket sold for $6.00, a senior citizen's ticket for $4.50, and a child's ticket for $2.00. On Tuesday, the number of adults' and senior citizens' tickets sold was 155 more than the number of children's tickets sold. The number of adults' tickets sold was 35 more than three times the number of senior citizens' tickets sold. Total receipts from the ticket sales were $1647.00. How many of each type of ticket were sold?

CHAPTER 3

TEST FORM D

NAME_____

CLASS_____ SCORE_____ GRADE_____

Solve, if possible, using the substitution method.

1. $3x + y = 16$,
   $2x - y = -5$

2. $x + 2y = -5$,
   $y = -3x - 2$

Solve, if possible, using the elimination method.

3. $3x - 2y = 35$,
   $2x - 4y = 34$

4. $4y - 3x = 4$,
   $6x + 5y = \dfrac{33}{4}$

5. The perimeter of a rectangle is 88. The length of the rectangle is two more than twice the width. Find the dimensions of the rectangle.

6. Between his home mortgage, car loan, and credit card bill, Joel is $115,000 in debt. Each month, Joel's credit card accumulates 1.8% interest, his car loan 1.2% interest, and his mortgage 1% interest. After one month, his total accumulated interest is $1210. The interest on Joel's car loan was $30 more than the interest on his credit card. Find the amount of each of these three debts.

Solve. If the system's equations are dependent or if there is no solution, then state this.

7. $2x - 2y + z = -6$,
   $x + 2y - z = -12$,
   $-x + 2y - 2z = 2$

8. $x + 3y - z = -10$,
   $3x + 2y + 3z = 0$,
   $-x + 3y + z = 10$

9. $3x + 5y = 30$,
   $2y + 3z = 13$,
   $2x + 3y + 5z = 26$

10. $5x - 4z = 2$,
    $4x - 5y = -28$,
    $4y + 5z = 1$

Solve using matrices.

11. $-8x + 7y = -10$,
    $2x + 9y = 24$

12. $2x + 3y + 5z = -3$,
    $2x - 3y + 5z = -3$,
    $-2x + 3y + 5z = -7$

ANSWERS

1._____

2._____

3._____

4._____

5._____

6._____

7._____

8._____

9._____

10._____

11._____

12._____

CHAPTER 3

TEST FORM D

NAME_____

ANSWERS

13._____

14._____

15._____

16._____

17._____

18. a._____

b._____

c._____

d._____

e._____

19._____

20._____

Evaluate.

13. $\begin{vmatrix} 6 & 9 \\ 5 & -2 \end{vmatrix}$

14. $\begin{vmatrix} 2 & -3 & 4 \\ 4 & -2 & 3 \\ 2 & -4 & 3 \end{vmatrix}$

Solve using Cramer's rule.

15. $3x + 7y = -8$,
    $5x - 2y = 14$

16. An electrician, a carpenter, and a plumber are hired to work on a house. The electrician earns $25 per hour, the carpenter $27 per hour, and the plumber $33 per hour. On the first day on the job, the three worked a total of 21.5 hours and earned a total of $602.50. If the carpenter worked 2.5 more hours than the plumber did, then how many hours did the electrician work?

17. Find the equilibrium point for the demand and supply functions
    $D(p) = 133 - 10p$ and $S(p) = 65 + 7p$

18. Unethicon Inc. produces cloned dogs. For the first year, the fixed set-up costs are $51,300. The variable costs for cloning each dog are $51. The revenue from each dog is $108. Find the following.
    a) The total cost $C(x)$ of producing x dogs.
    b) The total revenue $R(x)$ from the sale of x dogs.
    c) The total profit $P(x)$ from the production and sale of x dogs.
    d) The profit or loss from production and sale of 750 dogs; of 1000 dogs.
    e) The break-even point.

19. The graph of the function $f(x) = mx + b$ contains the points
    $(3, -14)$ and $(1, 2)$. Find m and b.

20. At a county fair, an adult's ticket sold for $5.00, a senior citizen's ticket for $4.00, and a child's ticket for $2.50. On Tuesday, the number of adults' and senior citizens' tickets sold was 310 more than the number of children's tickets sold. The number of adults' tickets sold was 20 more than three times the number of senior citizens' tickets sold. Total receipts from the ticket sales were $2565.00. How many of each type of ticket were sold?

CHAPTER 3

TEST FORM E

NAME_____

CLASS_____ SCORE_____ GRADE_____

ANSWERS

Solve, if possible, using the substitution method.

1. $x + 8y = 16$,
   $2x + 7y = 10$

2. $-2x - 4y = 13$
   $y = x + 5$

1._____

Solve, if possible, using the elimination method.

2._____

3. $2x - 4y = 32$
   $4x + 5y = 25$

4. $4y + 5x = -1$
   $2x + 2y = -\frac{3}{10}$

3._____

5. The perimeter of a rectangle is 94. The length of the rectangle is three less than four times the width. Find the dimensions of the rectangle.

4._____

6. Between her home mortgage, car loan, and credit card bill, Helda is $28,000 in debt. Each month, Helda's credit card accumulates 1.8% interest, her car loan 1.2% interest, and her mortgage 1% interest. After one month, her total accumulated interest is $352. The interest on Helda's credit card was $8 more than the interest on her mortgage. Find the amount of each of these three debts.

5._____

6._____

7._____

Solve. If the system's equations are dependent or if there is no solution, then state this.

8._____

7. $3x + 2y + 3z = 10$,
   $-2x - 3y + 2z = -6$,
   $-3x + 2y + 2z = 3$

8. $8x - 4y + 2z = 18$,
   $-4x + 2y + z = -7$,
   $x + 2y - 8z = -1$

9._____

9. $x + 6z = -3$,
   $4x + 2z = 10$,
   $6x + 4y + 2z = 16$

10. $-4x - 5y = -4$,
    $5y + 4z = -4$,
    $-5x + 4z = -9$

10._____

11._____

Solve using matrices.

11. $2x - 3y = -6$,
    $-5x + 4y = 1$

12. $-x + y - z = -2$,
    $2x - 2y - 2z = 0$,
    $-3x - 3y + 3z = -12$

12._____

61

CHAPTER 3

TEST FORM E

NAME_____

ANSWERS

13._____

14._____

15._____

16._____

17._____

18. a._____

b._____

c._____

d._____

e._____

19._____

20._____

Evaluate.

13. $\begin{vmatrix} 2 & 4 \\ 4 & 8 \end{vmatrix}$

14. $\begin{vmatrix} 5 & 1 & -4 \\ 2 & 3 & -2 \\ 1 & 4 & -5 \end{vmatrix}$

Solve using Cramer's rule.

15. $4x - 3y = -7$,
    $6x + 2y = -4$

16. An electrician, a carpenter, and a plumber are hired to work on a house. The electrician earns $25 per hour, the carpenter $27 per hour, and the plumber $33 per hour. On the first day on the job, the three worked a total of 23.75 hours and earned a total of $689.25. If the plumber worked 4 more hours than the electrician did, then how many hours did the carpenter work?

17. Find the equilibrium point for the demand and supply functions
    $D(p) = 108 - 11p$ and $S(p) = 70 + 8p$

18. Unethicon Inc. produces cloned cats. For the first year, the fixed set-up costs are $52,290. The variable costs for cloning each cat are $47. The revenue from each cat is $130. Find the following.
    a) The total cost $C(x)$ of producing x cats.
    b) The total revenue $R(x)$ from the sale of x cats.
    c) The total profit $P(x)$ from the production and sale of x cats.
    d) The profit or loss from production and sale of 500 cats; of 700 cats.
    e) The break-even point.

19. The graph of the function $f(x) = mx + b$ contains the points
    $(-6, 7)$ and $(6, 5)$. Find m and b.

20. At a county fair, an adult's ticket sold for $5.00, a senior citizen's ticket for $4.00, and a child's ticket for $2.50. On Tuesday, the number of adults' and senior citizens' tickets sold was 410 more than the number of children's tickets sold. The number of adults' tickets sold was 90 more than three times the number of senior citizens' tickets sold. Total receipts from the ticket sales were $3420.00. How many of each type of ticket were sold?

CHAPTER 3

TEST FORM F

NAME_____

CLASS_____ SCORE_____ GRADE_____

ANSWERS

Solve, if possible, using the substitution method.

1. $3x + 4y = 5,$
   $x - 3y = -12$

2. $-4x + 2y = -7,$
   $y = -4x - 1$

Solve, if possible, using the elimination method.

3. $4x + 5y = 34$
   $3x - 10y = 53$

4. $5y + 4x = 0$
   $4x + 3y = -\dfrac{6}{5}$

5. The perimeter of a rectangle is 54. The length of the rectangle is three more than three times the width. Find the dimensions of the rectangle.

6. Between his home mortgage, car loan, and credit card bill, Leon is $212,000 in debt. Each month, Leon's credit card accumulates 1.8% interest, his car loan 1.2% interest, and his mortgage 1% interest. After one month, his total accumulated interest is $2174. The interest on Leon's credit card was $6 more than the interest on his car loan. Find the amount of each of these three debts.

Solve. If the system's equations are dependent or if there is no solution, then state this.

7. $x + 4y + 3z = 30,$
   $-4x + 2y + z = -30,$
   $3x - 4y + 2z = 10$

8. $3x - 5y + 7z = 49,$
   $-7x + 3y + 5z = 9,$
   $5x - 7y + 3z = 25$

9. $-4y + z = 19,$
   $-3x + 2y = -14,$
   $2x + 6y + 3z = -11$

10. $4x - 3y = 11,$
    $-3x + 4z = 2,$
    $4y + 3z = 2$

Solve using matrices.

11. $-x + 5y = 11,$
    $3x - 4y = 0$

12. $3x - y + 2z = -9,$
    $-3x + y + 2z = 5,$
    $3x + y - 2z = -9$

1._____

2._____

3._____

4._____

5._____

6._____

7._____

8._____

9._____

10._____

11._____

12._____

CHAPTER 3

TEST FORM F

NAME _____

ANSWERS

Evaluate.

13. $\begin{vmatrix} 2 & 8 \\ 4 & 4 \end{vmatrix}$

14. $\begin{vmatrix} -1 & 2 & 1 \\ 2 & 1 & 2 \\ -1 & -2 & -1 \end{vmatrix}$

Solve using Cramer's rule.

15. $-4x + 3y = -14$,
    $6x + 2y = 8$

16. An electrician, a carpenter, and a plumber are hired to work on a house. The electrician earns $25 per hour, the carpenter $27 per hour, and the plumber $33 per hour. On the first day on the job, the three worked a total of 23.75 hours and earned a total of $689.25. If the plumber worked 2.25 more hours than the carpenter did, then how many hours did the electrician work?

17. Find the equilibrium point for the demand and supply functions
    $D(p) = 164 - 12p$ and $S(p) = 80 + 9p$

18. Unethicon Inc. produces cloned goats. For the first year, the fixed set-up costs are $36,225. The variable costs for cloning each goat are $16. The revenue from each goat is $79. Find the following.
    a) The total cost $C(x)$ of producing x goats.
    b) The total revenue $R(x)$ from the sale of x goats.
    c) The total profit $P(x)$ from the production and sale of x goats.
    d) The profit or loss from production and sale of 400 goats; of 800 goats.
    e) The break-even point.

19. The graph of the function $f(x) = mx + b$ contains the points
    $(2, 3)$ and $(3, 8)$. Find m and b.

20. At a county fair, an adult's ticket sold for $5.00, a senior citizen's ticket for $4.00, and a child's ticket for $2.50. On Tuesday, the number of adults' and senior citizens' tickets sold was 160 more than the number of children's tickets sold. The number of adults' tickets sold was 20 more than three times the number of senior citizens' tickets sold. Total receipts from the ticket sales were $1490.00. How many of each type of ticket were sold?

CHAPTER 3

TEST FORM G

NAME_____

CLASS_____ SCORE_____ GRADE_____

ANSWERS

1. Solve, if possible, using the substitution method.

    $2x + 3y = -10,$
    $x - 2y = 7$

    What is the *x*-coordinate?

    a) $\frac{3}{7}$ b) $-\frac{10}{7}$ c) $\frac{1}{7}$ d) $-\frac{2}{7}$

1._____

2. Solve, if possible, using the substitution method.

    $5x - 2y = 15$
    $y = -3x + 1$

    What is the *y*-coordinate?

    a) $\frac{10}{11}$ b) 5   c) -15 d) $-\frac{40}{11}$

2._____

3. Solve, if possible, using the elimination method.

    $3x - 10y = 46$
    $x + 2y = 10$

    What is the *y* - coordinate?

    a) -1 b) 1   c) 0   d) not possible

3._____

4. Solve, if possible, using the elimination method.

    $5y - 6x = 0$
    $5x + y = \frac{31}{30}$

    What is the *x* - coordinate?

    a) $\frac{1}{6}$ b) $-\frac{5}{6}$ c) $\frac{1}{5}$ d) $-\frac{6}{5}$

4._____

5. The perimeter of a rectangle is 12. The length of the rectangle is five less than four times the width. Find the width of the rectangle.

    a) 4   b) 3   c) 6   d) 2

5._____

CHAPTER 3

TEST FORM G

NAME_____

ANSWERS

6. _____

7. _____

8. _____

9. _____

6. Between her home mortgage, car loan, and credit card bill, Deena is $117,000 in debt. Each month, Deena's credit card accumulates 1.5% interest, her car loan 1% interest, and her mortgage 0.8% interest. After one month, her total accumulated interest is $995. The interest on Deena's mortgage was $680 more than the interest on her car loan. How much does she owe on her car loan?

   a) $1200   b) $5000   c) $7500   d) $12,000

7. Solve.
$$3x - 5y + 2z = 19,$$
$$5x + 2y - 3z = -8,$$
$$-2x + 3y + 5z = 7$$

   What is the z-coordinate?

   a) -7 b) 7   c) 4   d) 3

8. Solve.
$$2x + 6y + 4z = -30,$$
$$6x + 4y + 2z = -32,$$
$$4x + 2y + 6z = -34$$

   What is the y-coordinate?

   a) 10   b) -10 c) -2 d) 6

9. Solve.
$$x - 2y = 2,$$
$$2x - z = -2,$$
$$x - y - 2z = 4$$

   What is the x-coordinate?

   a) 2   b) -2 c) -3 d) 5

CHAPTER 3

TEST FORM G

NAME_____

ANSWERS

10. Solve using matrices.

$$2x + 7y = 43,$$
$$-4x + 7y = 19$$

What is the y – coordinate?

a) 7   b) 5   c) 8   d) 3

10._____

11. Solve using matrices.

$$x + 2y + z = -13,$$
$$3x + 4y + 2z = -28,$$
$$x + 3y + z = -17$$

What is the x – coordinate?

a) – 2   b) – 4   c) – 1   d) – 5

11._____

12. Evaluate.

$$\begin{vmatrix} 9 & 2 \\ 5 & 1 \end{vmatrix}$$

a) – 13   b) 1   c) – 1   d) 42

12._____

13. Evaluate.

$$\begin{vmatrix} 3 & 4 & 3 \\ -1 & 0 & 5 \\ 3 & 2 & 3 \end{vmatrix}$$

a) – 30   b) – 36   c) 36   d) 24

13._____

14. Solve using Cramer's rule.

$$4x + 3y = 3,$$
$$6x - 2y = 24$$

What is the x – coordinate?

a) – 1   b) 3   c) 6   d) – 2

14._____

67

CHAPTER 3

TEST FORM G

NAME_____

ANSWERS

15._____

15. An electrician, a carpenter, and a plumber are hired to work on a house. The electrician earns $20 per hour, the carpenter $25 per hour, and the plumber $30 per hour. On the first day on the job, the three worked a total of 23.25 hours and earned a total of $571.25. If the electrician worked 4.75 more hours than the carpenter did, then how many hours did the plumber work?

 a) 8   b) 5.75   c) 7.25   d) 6

16._____

16. Find the equilibrium point for the demand and supply functions
$D(p) = 132 - 13p$ and $S(p) = 86 + 10p$

 a) ($5, 106)   b) ($2, 116)   c) ($3, 126)   d) ($2, 106)

17._____

17. Unethicon Inc. produces cloned pigs. For the first year, the fixed set-up costs are $75,000. The variable costs for cloning each pig are $25. The revenue from each pig is $75. Find the profit or loss from the production and sale of 900 pigs.

 a) $5000 loss   b) $500 loss   c) $300 profit   d) $30,000 loss

18._____

18. The graph of the function $f(x) = mx + b$ contains the points $(5, 15)$ and $(-3, -5)$. Find b.

 a) $\dfrac{5}{2}$ b) $-\dfrac{5}{2}$ c) $\dfrac{2}{5}$ d) 5

19._____

19. At a county fair, an adult's ticket sold for $6.00, a senior citizen's ticket for $4.50, and a child's ticket for $2.00. On Tuesday, the number of adults' and senior citizens' tickets sold was 350 more than the number of children's tickets sold. The number of adult' tickets sold was 50 more than three times the number of senior citizens' tickets sold. Total receipts from the ticket sales were $4275.00. How many children's tickets were sold?

 a) 200   b) 250   c) 300   d) 350

CHAPTER 3

TEST FORM H

NAME_____

CLASS_____ SCORE_____ GRADE_____

ANSWERS

1. Solve, if possible, using the substitution method.

$$4x + y = 5,$$
$$x - 4y = -8$$

What is the x - coordinate?

1._____

a) $\frac{3}{17}$ b) $-\frac{10}{7}$ c) $\frac{1}{7}$ d) $\frac{12}{17}$

2. Solve, if possible, using the substitution method.

$$2x + 5y = -14$$
$$y = -x - 8$$

What is the y - coordinate?

2._____

a) $\frac{2}{3}$ b) 5 c) -15 d) $-\frac{4}{7}$

3. Solve, if possible, using the elimination method.

$$x + 2y = 15$$
$$3x + 5y = 44$$

What is the y - coordinate?

3._____

a) 1 b) -1 c) 0 d) not possible

4. Solve, if possible, using the elimination method.

$$6y - 5x = 3$$
$$x + 4y = \frac{4}{15}$$

What is the x - coordinate?

4._____

a) $\frac{1}{6}$ b) $-\frac{2}{5}$ c) $\frac{1}{5}$ d) $-\frac{6}{5}$

5. The perimeter of a rectangle is 70. The length of the rectangle is five more than three times the width. Find the width of the rectangle.

5._____

a) 5 b) 13 c) 7 d) 10

CHAPTER 3
TEST FORM H

NAME_____

ANSWERS

6._____

6. Between his home mortgage, car loan, and credit card bill, Dino is $216,000 in debt. Each month, Dino's credit card accumulates 1.5% interest, his car loan 1% interest, and his mortgage 0.8% interest. After one month, his total accumulated interest is $1808. The interest on Dino's credit card was $8 more than the interest on his car loan. How much does he owe on his credit card?

   a) $1200   b) $5000   c) $7500   d) $6000

7._____

7. Solve.
$$-x + 2y - z = 2,$$
$$2x - y + 2z = -1,$$
$$x - 2y - z = -6$$

   What is the z – coordinate?

   a) –7   b) 2   c) 4   d) 3

8._____

8. Solve.
$$-x + 5y - z = 45,$$
$$-5x - 5y + 2z = -50,$$
$$2x + y - 2z = -25$$

   What is the y – coordinate?

   a) 10   b) 15   c) 5   d) 25

9._____

9. Solve.
$$5x + 4z = -9,$$
$$3x + 2y = -1,$$
$$x - y - 2z = 0$$

   What is the x – coordinate?

   a) –1   b) –2   c) –3   d) 5

CHAPTER 3

TEST FORM H

NAME_____

ANSWERS

10. Solve using matrices.

$$4x - y = 16,$$
$$4x + 3y = 32$$

What is the y – coordinate?

a) 7   b) 5   c) 4   d) 3

10._____

11. Solve using matrices.

$$-2x + 4y - 2z = 0,$$
$$4x + y + 5z = -2,$$
$$-2x + 5y - 2z = 0$$

What is the x – coordinate?

a) -2 b) 0   c) 2   d) -5

11._____

12. Evaluate.

$$\begin{vmatrix} -8 & 6 \\ -3 & 3 \end{vmatrix}$$

a) -13 b) 1   c) -1 d) -6

12._____

13. Evaluate.

$$\begin{vmatrix} 2 & 0 & 4 \\ -3 & 1 & 2 \\ 4 & -2 & 3 \end{vmatrix}$$

a) -30 b) 22   c) 36   d) 24

13._____

14. Solve using Cramer's rule.

$$4x + 3y = -4,$$
$$-6x + 2y = 32$$

What is the x – coordinate?

a) -4 b) 3   c) 6   d) -2

14._____

71

CHAPTER 3

TEST FORM H

NAME_____

ANSWERS

15. An electrician, a carpenter, and a plumber are hired to work on a house. The electrician earns $20 per hour, the carpenter $25 per hour, and the plumber $30 per hour. On the first day on the job, the three worked a total of 21.5 hours and earned a total of $532.50. If the carpenter worked 2.5 more hours than the plumber did, then how many hours did the electrician work?

15._____

a) 7.25   b) 5.5   c) 7   d) 6

16. Find the equilibrium point for the demand and supply functions
$D(p) = 165 - 14p$ and $S(p) = 90 + 11p$

16._____

a) ($5, 106)   b) ($2, 116)   c) ($3, 123)   d) ($2, 106)

17. Unethicon Inc. produces cloned pigs. For the first year, the fixed set-up costs are $75,000. The variable costs for cloning each pig are $25. The revenue from each pig is $75. Find the profit or loss from the production and sale of 2000 pigs.

17._____

a) $5000 loss   b) $500 loss   c) $300 profit   d) $25,000 profit

18. The graph of the function $f(x) = mx + b$ contains the points $(-2, 9)$ and $(4, -15)$. Find b.

18._____

a) -4 b) 1   c) 4   d) 5

19. At a county fair, an adult's ticket sold for $6.00, a senior citizen's ticket for $4.50, and a child's ticket for $2.00. On Tuesday, the number of adults' and senior citizens' tickets sold was 20 more than the number of children's tickets sold. The number of adult' tickets sold was 50 more than three times the number of senior citizens' tickets sold. Total receipts from the ticket sales were $3410.00. How many adults' tickets were sold?

19._____

a) 200   b) 250   c) 350   d) 450

CHAPTER 4

TEST FORM A

NAME_____

CLASS_____ SCORE_____ GRADE_____

Graph each inequality and write the solution set in both set-builder and interval notation.

1. $x - 4 < 2$

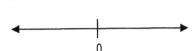

2. $-0.2y < 18$

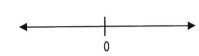

3. $-9y - 5 \geq 22$

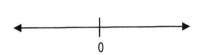

4. $7a - 1 \leq -2a + 3$

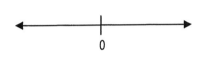

5. $9(2 - x) > 5x + 2$

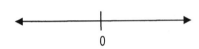

6. $-5(4x + 2) + 3(5 - 4x) \geq 4(2 - 6x) - 7(1 + 3x)$

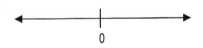

7. Let $f(x) = -2x + 8$ and $g(x) = -8x + 2$. Find all values of x for which $f(x) > g(x)$.

8. Greg can rent a van for either $50 per day with unlimited mileage or $35 per day with 75 free miles and an extra charge of 25¢ for each mile over 75. For what numbers of miles traveled would the unlimited mileage plan save Greg money?

ANSWERS

1._____
See graph.

2._____
See graph.

3._____
See graph.

4._____
See graph.

5._____
See graph.

6._____
See graph.

7._____

8._____

CHAPTER 4
TEST FORM A

NAME_____

ANSWERS

9._____

10._____

11._____

12._____

13._____
See graph.

14._____
See graph.

15._____
See graph.

16._____
See graph.

17._____
See graph.

18._____
See graph.

9. A refrigeration repair company charges $40 for the first half-hour of work and $30 for each additional hour. Sticky Mountain Resort has budgeted $200 to repair its walk-in cooler. For what lengths of service call will the budget not be exceeded?

10. Find the intersection: $\{1, 2, 3, 4, 5\} \cap \{0, 1, 2, 3\}$

11. Find the union: $\{1, 2, 3, 4, 5\} \cup \{0, 1, 2, 3\}$

12. Write the domain of f using interval notation if $f(x) = \sqrt{2-x}$

Solve and graph each solution set.

13. $-15 < x - 8 < -2$

14. $-7 \leq -4t - 5 < -3$

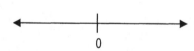

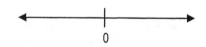

15. $9x - 5 < 1$ or $x - 5 > 2$

16. $-2x > 8$ or $3x > -9$

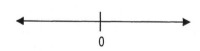

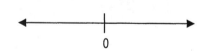

17. $-\dfrac{1}{5} \leq \dfrac{1}{10}x - 1 < \dfrac{1}{2}$

18. $|x| = 2$

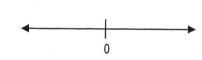

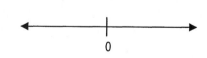

CHAPTER 4

TEST FORM A

NAME_____

Solve and graph each solution set.

19. $|a| > 9$

20. $|2x - 1| < 2.5$

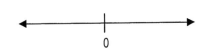
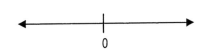

21. $|-6t - 1| \geq 8$

22. $|5 - 8x| = -7$

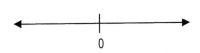
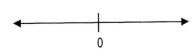

23. $g(x) < -6$ or $g(x) > 18$ where $g(x) = 6 - 12x$

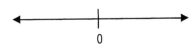

24. Let $f(x) = |x + 11|$ and $g(x) = |x - 13|$. Find all values of x for which $f(x) = g(x)$.

Graph the system of inequalities. Find the coordinates of any vertices formed.

25. $x + y \geq -6$
    $x - y \geq -2$

26. $3y - x \geq -3$
    $3y + 4x \leq 12$
    $y \leq 0$
    $x \leq 0$

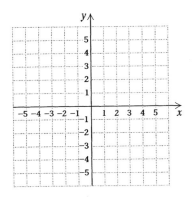
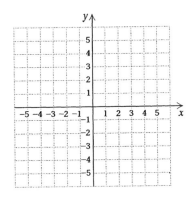

ANSWERS

19._____
See graph.

20._____
See graph.

21._____
See graph.

22._____
See graph.

23._____
See graph.

24._____

25. See graph.

26. See graph.

75

CHAPTER 4
TEST FORM A

NAME_____

ANSWERS

27._____

28._____

29._____

30._____

31._____

27. Find the maximum and minimum values of $F(x) = 8x + 5y$ subject to
$$x + y \leq 10$$
$$3 \leq x \leq 7$$
$$0 \leq y \leq 5$$

28. Integuments Salon makes $10 on each manicure and $17 on each cut and style. A manicure takes 30 minutes, a cut and style takes 60 minutes, and there are 8 stylists who each work 6 hours per day. If the salon can schedule 50 appointments per day, how many should be manicures and how many should be haircuts in order to maximize profit? What is the maximum profit?

Solve. Write the solution set using interval notation.

29. $|3x - 4| \leq 8$ and $|x - 3| \geq 3$

30. $9x < 11 - 5x < 8 + 9x$

31. Write an absolute-value inequality for which the interval shown is the solution.

CHAPTER 4
TEST FORM B

NAME_____

CLASS_____ SCORE_____ GRADE_____

Graph each inequality and write the solution set in both set-builder and interval notation.

ANSWERS

1. $x - 3 < 2$

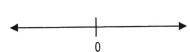

2. $-0.3y < 15$

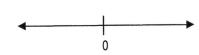

1._____
See graph.

3. $-8y - 6 \geq 26$

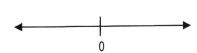

4. $3a - 2 \leq -4a + 4$

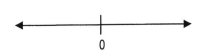

2._____
See graph.

5. $8(3 - x) > 6x + 3$

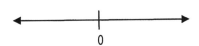

3._____
See graph.

4._____
See graph.

6. $-5(4x + 2) + 3(5 - 4x) \geq 4(2 - 6x) - 7(1 - 3x)$

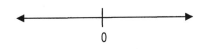

5._____
See graph.

7. Let $f(x) = -3x + 7$ and $g(x) = -5x + 2$. Find all values of x for which $f(x) > g(x)$.

6._____
See graph.

8. Dodie can rent a van for either $50 per day with unlimited mileage or $40 per day with 100 free miles and an extra charge of 25¢ for each mile over 100. For what numbers of miles traveled would the unlimited mileage plan save Dodie money?

7._____

8._____

77

CHAPTER 4

TEST FORM B

NAME_____

ANSWERS

9. _____

10. _____

11. _____

12. _____

13. _____
See graph.

14. _____
See graph.

15. _____
See graph.

16. _____
See graph.

17. _____
See graph.

18. _____
See graph.

9. An oven repair company charges $40 for the first half-hour of work and $25 for each additional hour. The Yeasty Spatula Bakery has budgeted $125 to repair its bread oven. For what lengths of service call will the budget not be exceeded?

10. Find the intersection: $\{2, 4, 6, 8, 10\} \cap \{1, 2, 3, 4\}$

11. Find the union: $\{2, 4, 6, 8, 10\} \cup \{1, 2, 3, 4\}$

12. Write the domain of f using interval notation if $f(x) = \sqrt{3 - 2x}$

Solve and graph each solution set.

13. $-12 < x - 6 < -1$

14. $-3 \leq -3t - 6 < 0$

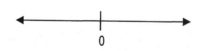

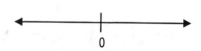

15. $8x - 6 < 2$ or $x - 4 > 2$

16. $-2x > 6$ or $3x > -6$

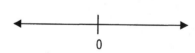

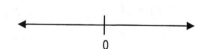

17. $-\dfrac{1}{6} \leq \dfrac{1}{12}x - 1 < \dfrac{1}{4}$

18. $|x| = 2.5$

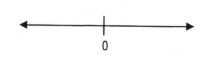

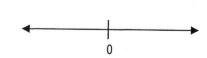

CHAPTER 4

NAME_____

TEST FORM B

Solve and graph each solution set.

19. $|a| > 8$

20. $|2x - 2| < 3$

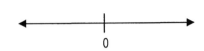
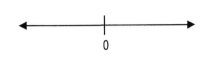

21. $|-7t - 9| \geq 10$

22. $|6 - 10x| = -6$

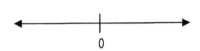
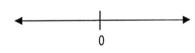

23. $g(x) < -5$ or $g(x) > 5$ where $g(x) = 12 - 9x$

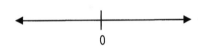

24. Let $f(x) = |x + 12|$ and $g(x) = |x - 8|$. Find all values of x for which $f(x) = g(x)$.

Graph the system of inequalities. Find the coordinates of any vertices formed.

25. $x + y \geq -4$
    $x - y \geq -1$

26. $4y - 3x \geq -6$
    $4y + 5x \leq 2$
    $y \leq 0$
    $x \leq 0$

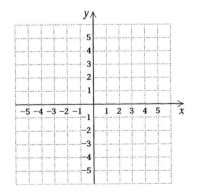
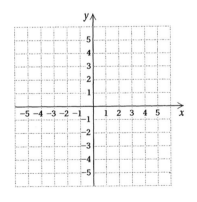

ANSWERS

19. _____
    See graph.

20. _____
    See graph.

21. _____
    See graph.

22. _____
    See graph.

23. _____
    See graph.

24. _____

25. See graph.

26. See graph.

CHAPTER 4
TEST FORM B

NAME_____

ANSWERS

27. Find the maximum and minimum values of $F(x) = 8x + 5y$ subject to
$$x + y \leq 12$$
$$5 \leq x \leq 10$$
$$0 \leq y \leq 10$$

27._____

28._____

28. Pampered Peds Spa makes $10 on each pedicure and $17 on each foot massage. A pedicure takes 15 minutes, a massage takes 45 minutes, and there are 3 employees who each work 8 hours per day. If the spa can schedule 50 appointments per day, how many should be pedicures and how many should be massages in order to maximize profit? What is the maximum profit?

29._____

Solve. Write the solution set using interval notation.

29. $|4x - 6| \leq 8$ and $|x - 4| \geq 4$

30._____

30. $8x < 12 - 4x < 7 + 8x$

31. Write an absolute-value inequality for which the interval shown is the solution.

31._____

CHAPTER 4

TEST FORM C

NAME_____

CLASS_____ SCORE_____ GRADE_____

Graph each inequality and write the solution set in both set-builder and interval notation.

ANSWERS

1. $x - 2 < 2$

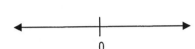

2. $-0.4y < 12$

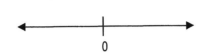

1._____
See graph.

3. $-7y - 7 \geq 14$

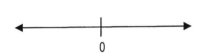

4. $5a - 3 \leq -6a + 5$

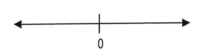

2._____
See graph.

5. $7(4 - x) > 7x + 4$

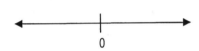

3._____
See graph.

6. $-5(4x + 2) + 3(5 - 4x) \geq 4(2 - 6x) - 7(1 + 2x)$

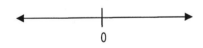

4._____
See graph.

5._____
See graph.

7. Let $f(x) = -4x + 6$ and $g(x) = -6x + 5$. Find all values of x for which $f(x) > g(x)$.

6._____
See graph.

8. Stan can rent a van for either $50 per day with unlimited mileage or $45 per day with 150 free miles and an extra charge of 25¢ for each mile over 150. For what numbers of miles traveled would the unlimited mileage plan save Stan money?

7._____

8._____

CHAPTER 4
TEST FORM C

NAME_____

ANSWERS

9._____

10._____

11._____

12._____

13._____
See graph.

14._____
See graph.

15._____
See graph.

16._____
See graph.

17._____
See graph.

18._____
See graph.

9. An oven repair company charges $40 for the first half-hour of work and $20 for each additional hour. The Dough Ball Bakery has budgeted $150 to repair its bread oven. For what lengths of service call will the budget not be exceeded?

10. Find the intersection: $\{1, 3, 5, 7, 9\} \cap \{1, 2, 3, 4\}$

11. Find the union: $\{1, 3, 5, 7, 9\} \cup \{1, 2, 3, 4\}$

12. Write the domain of f using interval notation if $f(x) = \sqrt{3-x}$

Solve and graph each solution set.

13. $-9 < x - 4 < 0$

14. $1 \le -2t - 7 < 2$

15. $7x - 2 < 3$ or $x - 3 > 2$

16. $-2x > 4$ or $3x > -9$

17. $-\dfrac{1}{4} \le \dfrac{1}{8}x - 1 < \dfrac{1}{2}$

18. $|x| = 3$

82

CHAPTER 4

TEST FORM C

NAME_____

Solve and graph each solution set.

19. $|a| > 7$

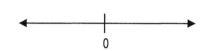

20. $|3x - 1| < 3.5$

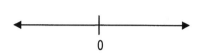

21. $|-8t - 1| \geq 5$

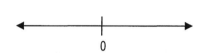

22. $|7 - 12x| = -5$

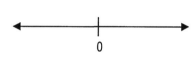

23. $g(x) < -4$ or $g(x) > 4$ where $g(x) = 10 - 6x$

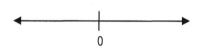

24. Let $f(x) = |x + 13|$ and $g(x) = |x - 3|$. Find all values of x for which $f(x) = g(x)$.

Graph the system of inequalities. Find the coordinates of any vertices formed.

25. $x + y \geq -2$
    $x - y \geq 0$

26. $5y - 2x \geq -5$
    $5y + 4x \leq 13$
    $y \leq 0$
    $x \leq 0$

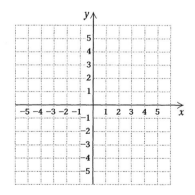

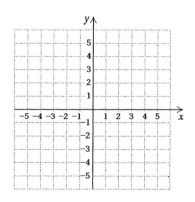

ANSWERS

19._____
See graph.

20._____
See graph.

21._____
See graph.

22._____
See graph.

23._____
See graph.

24._____

25. See graph.

26. See graph.

CHAPTER 4
TEST FORM C

NAME_____

ANSWERS

27._____

28._____

29._____

30._____

31._____

27. Find the maximum and minimum values of $F(x) = 8x + 5y$ subject to
$$x + y \leq 5$$
$$0 \leq x \leq 3$$
$$0 \leq y \leq 3$$

28. Pampered Peds Spa makes $10 on each pedicure and $17 on each foot massage. A pedicure takes 20 minutes, a massage takes 35 minutes, and there are 2 employees who each work 8 hours per day. If the spa can schedule 30 appointments per day, how many should be pedicures and how many should be massages in order to maximize profit? What is the maximum profit?

Solve. Write the solution set using interval notation.

29. $|5x - 8| \leq 6$ and $|x - 5| \geq 5$

30. $7x < 13 - 3x < 6 + 7x$

31. Write an absolute-value inequality for which the interval shown is the solution.

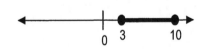

84

CHAPTER 4
TEST FORM D

NAME_____

CLASS_____ SCORE_____ GRADE_____

Graph each inequality and write the solution set in both set-builder and interval notation.

1. $x - 1 < 2$

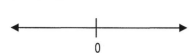

2. $-0.5y > 20$

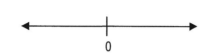

3. $-6y - 8 \leq 4$

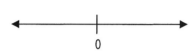

4. $8a - 4 \geq -8a + 6$

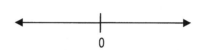

5. $6(5 - x) < 8x + 5$

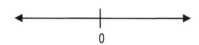

6. $-5(4x + 2) + 3(5 - 4x) \geq 4(2 - 6x) - 7(1 - 2x)$

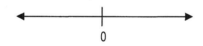

7. Let $f(x) = -5x + 9$ and $g(x) = -8x + 13$. Find all values of x for which $f(x) > g(x)$.

8. Jan can rent a van for either $75 per day with unlimited mileage or $50 per day with 75 free miles and an extra charge of 20¢ for each mile over 75. For what numbers of miles traveled would the unlimited mileage plan save Jan money?

ANSWERS

1._____
See graph.

2._____
See graph.

3._____
See graph.

4._____
See graph.

5._____
See graph.

6._____
See graph.

7._____

8._____

CHAPTER 4

TEST FORM D

NAME_____

ANSWERS

9._____

10._____

11._____

12._____

13._____
See graph.

14._____
See graph.

15._____
See graph.

16._____
See graph.

17._____
See graph.

18._____
See graph.

9. An mixer repair company charges $50 for the first half-hour of work and $35 for each additional hour. The Flakey Biscuit Bakery has budgeted $250 to repair its dough mixer. For what lengths of service call will the budget not be exceeded?

10. Find the intersection: $\{0, 1, 2, 3, 5\} \cap \{0, 2, 4, 6\}$

11. Find the union: $\{0, 1, 2, 3, 5\} \cup \{0, 2, 4, 6\}$

12. Write the domain of f using interval notation if $f(x) = \sqrt{4-3x}$

Solve and graph each solution set.

13. $-6 < x - 2 < 1$

14. $5 \leq -t - 8 < 8$

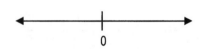

15. $6x - 4 < 10$ or $x - 2 > 2$

16. $-3x > 15$ or $5x > -15$

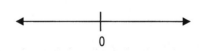
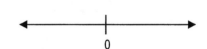

17. $-\dfrac{1}{3} \leq \dfrac{1}{6}x - 1 < \dfrac{1}{4}$

18. $|x| = 3.5$

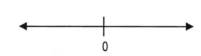
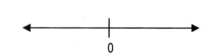

86

CHAPTER 4   NAME_____

TEST FORM D

Solve and graph each solution set.

19. $|a| > 6$

20. $|3x - 2| < 4$

21. $|-9t - 2| \geq 6$

22. $|8 - 14x| = -4$

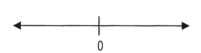
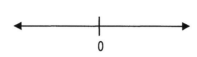

23. $g(x) < -3$ or $g(x) > 3$ where $g(x) = 8 - 3x$

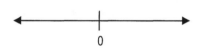

24. Let $f(x) = |x + 14|$ and $g(x) = |x - 2|$. Find all values of x for which $f(x) = g(x)$.

Graph the system of inequalities. Find the coordinates of any vertices formed.

25. $x + y \geq 0$
    $x - y \geq 1$

26. $5y - 4x \geq -10$
    $5y + 7x \leq 12$
    $y \geq 0$
    $x \geq 0$

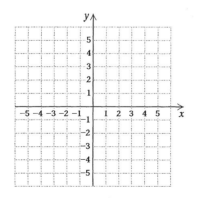
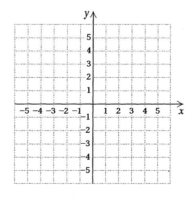

ANSWERS

19._____
See graph.

20._____
See graph.

21._____
See graph.

22._____
See graph.

23._____
See graph.

24._____

25. See graph._____

26. See graph._____

87

CHAPTER 4

TEST FORM D

NAME_____

ANSWERS

27. _____

28. _____

29. _____

30. _____

31. _____

27. Find the maximum and minimum values of $F(x) = 5x - 8y$ subject to
$$x + y \leq 10$$
$$0 \leq x \leq 5$$
$$0 \leq y \leq 5$$

28. Dirty Dog Grooming makes $15 on each brushing and $22 on each full shampoo wash. A good brushing takes 30 minutes, a full wash takes 75 minutes, and there are 8 employees who each work 4 hours per day. If the groomers can schedule 40 appointments per day, how many should be brushes and how many should be full washes in order to maximize profit? What is the maximum profit?

Solve. Write the solution set using interval notation.

29. $|6x - 2| \leq 6$ and $|x - 6| \geq 6$

30. $6x < 14 - 2x < 5 + 6x$

31. Write an absolute-value inequality for which the interval shown is the solution.

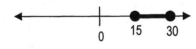

88

CHAPTER 4 NAME_____

TEST FORM E   CLASS_____ SCORE_____ GRADE_____

Graph each inequality and write the solution set in both set-builder and interval notation.

ANSWERS

1. $x + 1 < 10$

2. $-0.6y > 18$

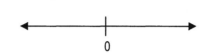

1._____
See graph.

3. $-5y - 9 \le 1$

4. $2a - 5 \ge -7a + 7$

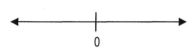
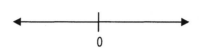

2._____
See graph.

5. $5(6 - x) < 9x + 6$

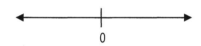

3._____
See graph.

6. $-5(4x + 2) + 3(5 - 4x) \ge 4(2 - 6x) - 7(1 + 4x)$

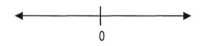

4._____
See graph.

5._____
See graph.

7. Let $f(x) = -6x + 1$ and $g(x) = -9x + 7$. Find all values of x for which $f(x) > g(x)$.

6._____
See graph.

8. Dan can rent a van for either $75 per day with unlimited mileage or $55 per day with 100 free miles and an extra charge of 20¢ for each mile over 100. For what numbers of miles traveled would the unlimited mileage plan save Dan money?

7._____

8._____

CHAPTER 4   NAME_____

TEST FORM E

ANSWERS

9._____

10._____

11._____

12._____

13._____
See graph.

14._____
See graph.

15._____
See graph.

16._____
See graph.

17._____
See graph.

18._____
See graph.

9. An computer repair company charges $50 for the first half-hour of work and $30 for each additional hour. Memleak Inc. has budgeted $225 to repair its intranet server. For what lengths of service call will the budget not be exceeded?

10. Find the intersection: $\{0, 5, 10, 15, 20\} \cap \{10, 20, 30, 40\}$

11. Find the union: $\{0, 5, 10, 15, 20\} \cup \{10, 20, 30, 40\}$

12. Write the domain of f using interval notation if $f(x) = \sqrt{-5-x}$

Solve and graph each solution set.

13. $-3 < x + 2 < 2$

14. $9 \leq -2t - 9 < 10$

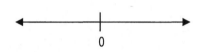
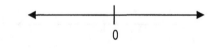

15. $5x - 3 < 7$ or $x - 1 > 2$

16. $-3x > 12$ or $5x > -20$

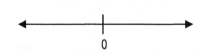

17. $-\dfrac{1}{2} \leq \dfrac{1}{4}x - 1 < \dfrac{1}{2}$

18. $|x| = 4$

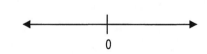
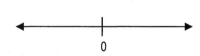

90

CHAPTER 4

TEST FORM E

NAME_____

Solve and graph each solution set.

19. $|a| > 5$

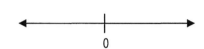

20. $|5x - 1| < 5.5$

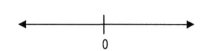

21. $|-t - 3| \geq 7$

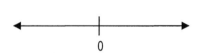

22. $|9 - 16x| = -3$

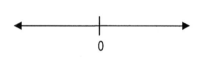

23. $g(x) < -2$ or $g(x) > 2$ where $g(x) = 6 - x$

24. Let $f(x) = |x + 15|$ and $g(x) = |x - 7|$. Find all values of x for which $f(x) = g(x)$.

Graph the system of inequalities. Find the coordinates of any vertices formed.

25. $x + y \geq 2$
    $x - y \geq 2$

26. $4y - 3x \geq -8$
    $4y + 6x \leq 8$
    $y \geq 0$
    $x \geq 0$

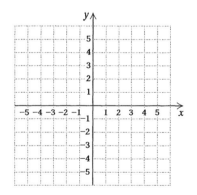

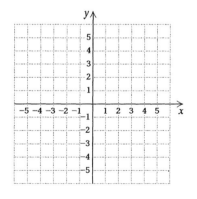

ANSWERS

19._____
   See graph.

20._____
   See graph.

21._____
   See graph.

22._____
   See graph.

23._____
   See graph.

24._____

25. See graph._____

26. See graph._____

91

CHAPTER 4

TEST FORM E

NAME_____

ANSWERS

27._____

27. Find the maximum and minimum values of $F(x) = 5x - 8y$ subject to
$$x + y \leq 8$$
$$2 \leq x \leq 6$$
$$1 \leq y \leq 5$$

28._____

29._____

28. Dirty Dog Grooming makes $15 on each brushing and $22 on each full shampoo wash. A good brushing takes 30 minutes, a full wash takes 50 minutes, and there are 4 employees who each work 4 hours per day. If the groomers can schedule 20 appointments per day, how many should be brushes and how many should be full washes in order to maximize profit? What is the maximum profit?

Solve. Write the solution set using interval notation.

29. $|7x - 12| \leq 14$ and $|x - 7| \geq 7$

30._____

30. $5x < 16 - x < 4 + 5x$

31. Write an absolute-value inequality for which the interval shown is the solution.

31._____

●━━●─────┼────▶
−18 −14 0

CHAPTER 4   NAME_____

TEST FORM F   CLASS_____ SCORE_____ GRADE_____

Graph each inequality and write the solution set in both set-builder and interval notation.

ANSWERS

1. $x + 2 < 9$

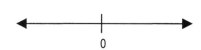

2. $-0.7y > 21$

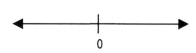

1._____
See graph.

3. $-4y - 1 \leq 7$

4. $9a - 6 \geq -5a + 8$

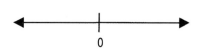

2._____
See graph.

5. $4(7 - x) < 2x + 7$

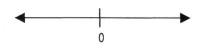

3._____
See graph.

6. $-5(4x + 2) + 3(5 - 4x) \geq 4(2 - 6x) - 7(1 - 4x)$

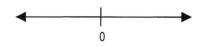

4._____
See graph.

5._____
See graph.

7. Let $f(x) = -7x + 3$ and $g(x) = -11x + 11$. Find all values of x for which $f(x) > g(x)$.

6._____
See graph.

8. Fran can rent a van for either $75 per day with unlimited mileage or $60 per day with 150 free miles and an extra charge of 15¢ for each mile over 150. For what numbers of miles traveled would the unlimited mileage plan save Fran money?

7._____

8._____

93

CHAPTER 4
TEST FORM F

NAME_____

ANSWERS

9._____

10._____

11._____

12._____

13._____
See graph.

14._____
See graph.

15._____
See graph.

16._____
See graph.

17._____
See graph.

18._____
See graph.

9. An computer repair company charges $50 for the first half-hour of work and $25 for each additional hour. Dataglut Inc. has budgeted $225 to repair its SQL server. For what lengths of service call will the budget not be exceeded?

10. Find the intersection: $\{3, 6, 9, 12, 15\} \cap \{9, 18, 27, 36\}$

11. Find the union: $\{3, 6, 9, 12, 15\} \cup \{9, 18, 27, 36\}$

12. Write the domain of f using interval notation if $f(x) = \sqrt{-5 - 2x}$

Solve and graph each solution set.

13. $0 < x + 4 < 3$

14. $5 \leq -3t - 4 < 6$

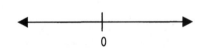
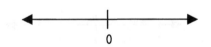

15. $4x - 1 < 7$ or $x - 6 > -4$

16. $-3x > 9$ or $5x > -10$

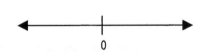
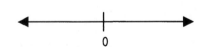

17. $-\dfrac{1}{4} \leq \dfrac{1}{8}x - 1 < \dfrac{1}{3}$

18. $|x| = 4.5$

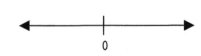
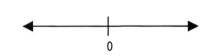

CHAPTER 4

TEST FORM F

NAME_____

Solve and graph each solution set.

19. $|a| > 4$

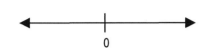

20. $|5x - 1| < 6$

21. $|-2t - 4| \geq 8$

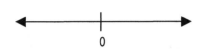

22. $|1 - 7x| = -2$

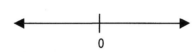

23. $g(x) < -1$ or $g(x) > 1$ where $g(x) = 4 - 2x$

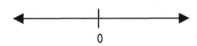

24. Let $f(x) = |x + 16|$ and $g(x) = |x - 12|$. Find all values of x for which $f(x) = g(x)$.

Graph the system of inequalities. Find the coordinates of any vertices formed.

25. $x + y \geq 4$
    $x - y \geq 3$

26. $3y - 2x \geq -9$
    $3y + 5x \leq 5$
    $y \geq 0$
    $x \geq 0$

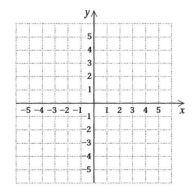

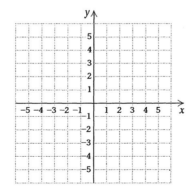

ANSWERS

19. _____
    See graph.

20. _____
    See graph.

21. _____
    See graph.

22. _____
    See graph.

23. _____
    See graph.

24. _____

25. See graph.

26. See graph.

CHAPTER 4
TEST FORM F

NAME_____

ANSWERS

27._____

27. Find the maximum and minimum values of $F(x) = 5x - 8y$ subject to
$$x + y \leq 6$$
$$-2 \leq x \leq 8$$
$$-2 \leq y \leq 2$$

28._____

28. Dirty Dog Grooming makes $15 on each brushing and $22 on each full shampoo wash. A good brushing takes 40 minutes, a full wash takes 60 minutes, and there are 10 employees who each work 6 hours per day. If the groomers can schedule 75 appointments per day, how many should be brushes and how many should be full washes in order to maximize profit? What is the maximum profit?

29._____

Solve. Write the solution set using interval notation.

29. $|8x - 14| \leq 4$ and $|x - 8| \geq 8$

30._____

30. $4x < 18 - x < 3 + 4x$

31. Write an absolute-value inequality for which the interval shown is the solution.

31._____

CHAPTER 4   NAME_____

TEST FORM G   CLASS_____ SCORE_____ GRADE_____

ANSWERS

1. Solve. $-0.8y > 24$

    a) $\{y \mid y > 30\}$ b) $\{y \mid y < 3\}$
    c) $\{y \mid y > -3\}$ d) $\{y \mid y < -30\}$

    1._____

2. Solve. $-3y - 2 \leq 4$

    a) $(-\infty, 2]$ b) $[-3, \infty)$ c) $[2, \infty)$ d) $(-\infty, -3]$

    2._____

3. Solve. $4a - 7 \geq -3a + 9$

    a) $\{a \mid a \leq \frac{7}{16}\}$ b) $\{a \mid a \leq \frac{16}{7}\}$
    c) $\{a \mid a \geq \frac{16}{7}\}$ d) $\{a \mid a \geq \frac{7}{16}\}$

    3._____

4. Solve. $3(8 - x) < 3x + 8$

    a) $\left(\frac{8}{3}, \infty\right)$ b) $\left(\frac{3}{8}, \infty\right)$ c) $\left(-\frac{8}{3}, \infty\right)$ d) $\left(-\infty, \frac{3}{8}\right)$

    4._____

5. Let $f(x) = -8x + 10$ and $g(x) = -13x + 15$. Find all values of x for which $f(x) > g(x)$.

    a) $\{x \mid x < 1\}$ b) $\{x \mid x > 1\}$ c) $\{x \mid x > -1\}$ d) $\{x \mid x > 2\}$

    5._____

6. Nan can rent a van for either $50 per day with unlimited mileage or $35 per day with 75 free miles and an extra charge of 20¢ for each mile over 75. For what numbers of miles traveled would the unlimited mileage plan save Nan money?

    a) at least 150   b) at most 150   c) at least 250   d) at most 350

    6._____

97

CHAPTER 4

TEST FORM G

NAME_____

ANSWERS

7._____

8._____

9._____

10._____

11._____

12._____

7. Find the intersection. $\{1, 2, 3, 5, 7\} \cap \{0, 2, 4, 6\}$

　a) $\{2\}$　　　　　　　　　　b) $\{0, 1, 2, 3, 4, 5, 6, 7\}$
　c) $\{0, 2, 4, 6\}$　　　　　　d) $\{1, 2, 3, 5, 7\}$

8. Find the domain of $f(x) = \sqrt{-4-x}$

　a) $(-\infty, -4)$　　b) $(-\infty, -4\,]$　　c) $(-\infty, 4)$　　d) $[-4, \infty)$

9. Identify the graph of the solution set for $1 \le -4t - 7 < 3$

　a) 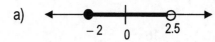　　　　b)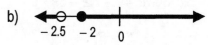

　c) 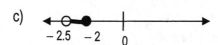　　　　d)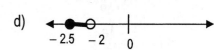

10. Identify the graph of the solution set for $3x - 5 < 1$ or $x - 7 > -6$

　a) 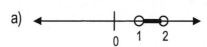　　　　b)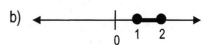

　c) 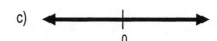　　　　d)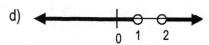

11. Solve. $-4x > 12$ or $7x > -14$

　a) $(-\infty, -2) \cup (-3, \infty)$　　　b) $(-\infty, -2) \cup (3, \infty)$
　c) $(-\infty, -3) \cup (2, \infty)$　　　d) $(-\infty, -3) \cup (-2, \infty)$

12. Solve. $-\dfrac{1}{8} \le \dfrac{1}{16}x - 1 < \dfrac{1}{4}$

　a) $\{14 \le x \le 20\}$　　　　b) $\{14 \le x < 20\}$
　c) $\{14 < x \le 20\}$　　　　d) $\{14 < x < 20\}$

CHAPTER 4

TEST FORM G

13. Solve. $|2x - 2| < 1$

a) $\left(\dfrac{1}{2}, \dfrac{5}{2}\right)$ b) $\left(-\dfrac{1}{2}, \dfrac{3}{2}\right)$ c) $\left(\dfrac{1}{2}, -\dfrac{3}{2}\right)$ d) $\left(\dfrac{1}{2}, \dfrac{3}{2}\right)$

14. Solve. $|-3t - 5| \geq 9$

a) $\{t \mid t \leq -\dfrac{14}{3} \text{ or } t \geq \dfrac{4}{3}\}$ b) $\{t \mid t \leq -\dfrac{14}{3} \text{ and } t \geq \dfrac{4}{3}\}$

c) $\{t \mid t \geq -\dfrac{14}{3} \text{ or } t \leq \dfrac{4}{3}\}$ d) $\{t \mid t \leq \dfrac{14}{3} \text{ or } t \geq -\dfrac{4}{3}\}$

15. Solve. $|2 - 6x| = -1$

a) $\pm \dfrac{1}{2}$ b) $\mathbb{R}$ c) $\varnothing$ d) $-\dfrac{1}{6}, \dfrac{1}{2}$

16. Graph. $x + y \geq 6$
            $x - y \geq 4$

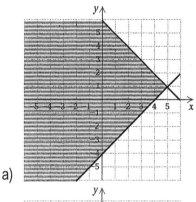
a)

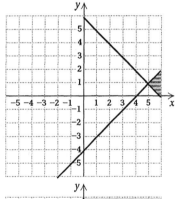
b)

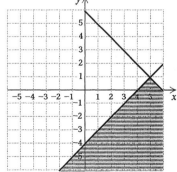
c)

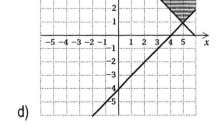
d)

ANSWERS

13. _____

14. _____

15. _____

16. _____

CHAPTER 4

TEST FORM G

NAME_____

ANSWERS

17._____

17. Find the maximum value of $F(x) = 2x + 3y$ subject to
$$x + y \leq 0$$
$$-5 \leq x \leq 0$$
$$0 \leq y \leq 5$$

    a) 25   b) 5   c) – 5   d) – 15

Pretentions Spa makes $10 on each pedicure and $17 on each foot massage. A pedicure takes 25 minutes, a massage takes 75 minutes, and there are 10 employees who each work 6 hours per day. The spa can schedule 60 appointments per day.

18._____

18. How many appointments in one day should be pedicures in order to maximize profit?

    a) 18   b) 42   c) 25   d) 35

19._____

19. What is the maximum profit?

    a) $805   b) $880   c) $948   d) $894

20._____

20. Solve. $|7x - 6| \leq 20$ and $|x - 9| \geq 9$

    a) $(-2, 1)$ b) $[2, 9]$ c) $[-2, 2]$ d) $[-2, 0]$

21. Solve. $3x < 20 + 2x < 2 + 3x$

    a) $(0, 18)$ b) $(0, 20)$ c) $(-18, 20)$ d) $(18, 20)$

21._____

22. Identify the graph corresponding to the solution for $|2x - 18| \leq 8$.

a)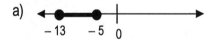
b)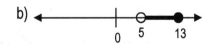

22._____

c)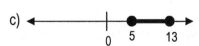
d)

CHAPTER 4   NAME_____

TEST FORM H   CLASS_____ SCORE_____ GRADE_____

1. Solve. $-0.9y < 18$ ANSWERS

   a) $\{y \mid y > 20\}$ b) $\{y \mid y < 2\}$
   c) $\{y \mid y > -2\}$ d) $\{y \mid y > -20\}$ 1._____

2. Solve. $-2y - 3 \geq 5$

   a) $(-\infty, -4]$ b) $[-4, \infty)$ c) $[4, \infty)$ d) $(-\infty, -4]$

                                                                      2._____

3. Solve. $6a - 8 \leq -a + 1$

   a) $\{a \mid a \leq \frac{9}{7}\}$ b) $\{a \mid a \leq \frac{7}{9}\}$
   c) $\{a \mid a \geq \frac{9}{7}\}$ d) $\{a \mid a \geq \frac{7}{9}\}$ 3._____

4. Solve. $2(9 - x) > 4x + 9$

   a) $\left(-\infty, \frac{3}{2}\right)$ b) $\left(\frac{3}{2}, \infty\right)$ c) $\left(-\frac{2}{3}, \infty\right)$ d) $\left(-\infty, \frac{9}{4}\right)$ 4._____

5. Let $f(x) = -9x + 12$ and $g(x) = -14x + 2$. Find all values of x for which $f(x) > g(x)$.

                                                                      5._____

   a) $\{x \mid x < 1\}$ b) $\{x \mid x > -2\}$ c) $\{x \mid x > -1\}$ d) $\{x \mid x > 2\}$

6. Hans can rent a van for either $75 per day with unlimited mileage or $45 per day with 100 free miles and an extra charge of 15¢ for each mile over 100. For what numbers of miles traveled would the unlimited mileage plan save Hans money?   6._____

   a) at least 350   b) at most 300   c) at least 300   d) at most 350

CHAPTER 4

TEST FORM H

NAME_____

ANSWERS

7._____

8._____

9._____

10._____

11._____

12._____

7. Find the intersection. $\{4, 6, 8, 9, 10\} \cap \{0, 2, 4, 6\}$

a) $\{2, 4\}$ b) $\{0, 2, 4, 6, 8, 9, 10\}$
c) $\{4, 6\}$ d) $\{0, 2, 4, 6\}$

8. Find the domain of $f(x) = \sqrt{6 - 2x}$

a) $(-\infty, -3)$ b) $(-\infty, 3]$ c) $(-\infty, 3)$ d) $[-3, \infty)$

9. Identify the graph of the solution set for $-3 \le -5t - 8 < 2$

a) 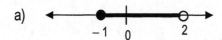   b)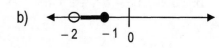

c) 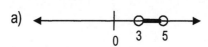   d)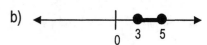

10. Identify the graph of the solution set for $2x - 3 < 3$ or $x - 8 > -3$

a) (number line with open circles at 3 and 5, segment between)
b) (number line with closed circles at 3 and 5, segment between)
c) (number line shaded entirely)
d) (number line with open circles at 3 and 5)

11. Solve. $-4x > 8$ or $7x > -7$

a) $(-\infty, -2) \cup (-3, \infty)$ b) $(-\infty, -2) \cup (1, \infty)$
c) $(-\infty, -2) \cup (-1, \infty)$ d) $(-\infty, -1) \cup (2, \infty)$

12. Solve. $-\dfrac{1}{5} \le \dfrac{1}{10}x - 1 < \dfrac{1}{5}$

a) $\{8 \le x \le 12\}$ b) $\{-12 \le x < 8\}$
c) $\{8 \le x < 12\}$ d) $\{8 < x < 12\}$

CHAPTER 4

TEST FORM H

13. Solve. $|3x - 2| < 2$

   a) $\left(\frac{1}{2}, \frac{5}{2}\right)$ b) $\left(0, \frac{3}{2}\right)$ c) $\left(0, \frac{4}{3}\right)$ d) $\left(\frac{1}{2}, \frac{3}{2}\right)$

13._____

14. Solve. $|-4t - 6| \geq 10$

   a) $\{t \mid t \leq -1 \text{ or } t \geq 4\}$ b) $\{t \mid t \leq 1 \text{ and } t \geq 4\}$
   c) $\{t \mid t \leq -4 \text{ or } t \geq 1\}$ d) $\{t \mid t \leq -6 \text{ or } t \geq -4\}$

15. Solve. $|3 - 5x| = -8$

   a) ± 1 b) $\mathbb{R}$ c) $-1, \frac{11}{5}$ d) $\varnothing$

14._____

16. Graph. $x + y \geq 2$
             $x - y \geq 5$

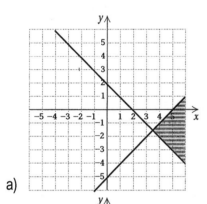
a)

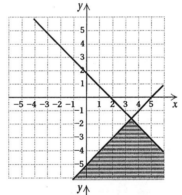
b)

15._____

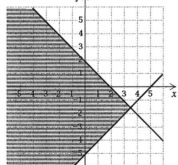
c)

d)

16._____

CHAPTER 4
TEST FORM H

NAME_____

ANSWERS

17._____

17. Find the maximum value of $F(x) = 2x + 3y$ subject to
$$x + y \leq 5$$
$$-5 \leq x \leq 5$$
$$0 \leq y \leq 3$$

a) 25   b) 13   c) −5   d) −10

Pretensions Spa makes $15 on each pedicure and $22 on each foot massage. A pedicure takes 30 minutes, a massage takes 45 minutes, and there are 4 employees who each work 8 hours per day. The spa can schedule 40 appointments per day.

18._____

18. How many appointments in one day should be pedicures in order to maximize profit?

a) 0   b) 40   c) 25   d) 15

19._____

19. What is the maximum profit?

a) $805   b) $880   c) $948   d) $894

20._____

20. Solve. $|8x - 10| \leq 18$ and $|x - 2| \geq 2$

a) (−1, 1)   b) [1, 9]   c) [−2, 1]   d) [−1, 0]

21. Solve. $2x < 22 + 3x < 1 + 2x$

a) (22, 21)   b) (0, 21)   c) (−22, −21)   d) (−22, 21)

21._____

22. Identify the graph corresponding to the solution for $|2x + 20| \leq 12$.

22._____

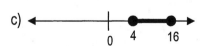

CHAPTER 5

TEST FORM A

NAME_____

CLASS_____ SCORE_____ GRADE_____

Given the polynomial $5xy^3 - 7x^5y + 9x^2y^3 - 2x^3y,$

ANSWERS

1. Determine the degree of the polynomial.
2. Arrange in descending powers of x.

1._____

2._____

3. Determine the leading term of the polynomial $6a - 7 + 10a^2 - a^3$.

4. Given $P(x) = 5x^3 + 3x^2 - 2x + 7$, find $P(0)$ and $P(-2)$.

3._____

5. Given $P(x) = x^2 - 3x$, find and simplify $P(a+h) - P(a)$.

4._____

6. Combine like terms: $7xy - 6xy^2 - 6xy + 7xy^2$.

5._____

Add.

6._____

7. $(5x^3 + 7x^2 - 9y) + (x^3 - 3y - 5y^2)$

7._____

8. $(5m^3 - 8m^2n - 3mn^2 - 6n^3) + (mn^2 - 4n^3 + 9m^3 + 2m^2n)$

8._____

Subtract.

9._____

9. $(a - 5b) - (6a + 3b)$

10._____

10. $(5y^2 - 2y - 9y^3) - (2y^2 - y - 7y^3)$

11._____

Multiply.

12._____

11. $(-15x^2y)(-3xy^2)$

13._____

12. $(5a - 3b)(2a + 4b)$

14._____

13. $(x - y)(x^2 + xy + y^2)$

14. $(2x^3 + 3)^2$

15._____

15. $(3y - 8)^2$

16._____

16. $(5x - 8y)(5x + 8y)$

CHAPTER 5

TEST FORM A

NAME_____

ANSWERS

Factor.

17. $14x - 7x^3$

18. $y^3 - 6y^2 + 3y - 18$

19. $p^2 - 16p - 17$

20. $6m^2 + 7m + 2$

21. $81y^2 - 1$

22. $5r^3 - 5$

23. $9x^2 + 1 - 6x$

24. $x^4 - y^4$

25. $y^2 + 10y + 25 - 144t^2$

26. $432a^2 - 3b^2$

27. $16x^2 - 28x + 6$

28. $16a^5b + 52ab^5$

29. $3y^4 x + 27yx^2 + 6y^2 x^2 - 12xy$

Solve.

30. $x^2 - 24 = 2x$

31. $2y^2 = 8$

32. $2x^3 + 3x = -5x^2$

33. $4x^2 + 2x = 0$

34. Let $f(x) = 4x^2 - 24x + 306$. Find a such that $f(a) = 306$.

35. Find the domain of the function f given by $f(x) = \dfrac{8-x}{x^2 + 4x + 4}$.

36. A photograph is 2 cm longer than it is wide. Its area is 120 cm^2. Find its length and its width.

37. To celebrate Shakeytown's centennial, fireworks are launched over a lake from atop a dam 128 ft above the water. The height of a display, t seconds after it is launched, is given by $h(t) = -16t^2 + 32t + 128$. After how long will the shell from the fireworks reach the water?

38. a) Multiply: $(x^2 + x + 1)(-x^3 - x^2 + 1)$
    b) Factor: $-x^5 - 2x^4 - 2x^3 + x + 1$

39. Factor: $6x^{2n} - x^n - 12$

CHAPTER 5

TEST FORM B

NAME_____

CLASS_____ SCORE_____ GRADE_____

Given the polynomial $6xy^2 - 8x^3y + x^4y^2 - 3x^2y$,

1. Determine the degree of the polynomial.
2. Arrange in descending powers of x.

3. Determine the leading term of the polynomial $7a - 6 + a^2 - 10a^3$.

4. Given $P(x) = 2x^3 + 3x^2 - 5x + 6$, find $P(0)$ and $P(-2)$.

5. Given $P(x) = x^2 - 4x$, find and simplify $P(a+h) - P(a)$.

6. Combine like terms: $8xy - 9xy^2 - 9xy + 8xy^2$.

Add.

7. $(4x^3 + 8x^2 - 8y) + (2x^3 - 2y - 6y^2)$

8. $(6m^3 - 7m^2n - 4mn^2 - 5n^3) + (2mn^2 - 3n^3 + 8m^3 + m^2n)$

Subtract.

9. $(2a - 6b) - (7a + 4b)$

10. $(3y^2 - 4y - 8y^3) - (8y^2 - 2y - 9y^3)$

Multiply.

11. $(-13x^2y)(-5xy^2)$

12. $(a - 4b)(3a + 5b)$

13. $(x - y)(-x^2 + xy + y^2)$

14. $(3x^3 + 4)^2$

15. $(5y - 2)^2$

16. $(6x - 5y)(6x + 5y)$

ANSWERS

1._____

2._____

3._____

4._____

5._____

6._____

7._____

8._____

9._____

10._____

11._____

12._____

13._____

14._____

15._____

16._____

CHAPTER 5

TEST FORM B

NAME_____

ANSWERS

Factor.

17. $24x^2 - 8x^4$

18. $y^3 - 4y^2 + 4y - 16$

19. $p^2 - 16p - 36$

20. $8m^2 + 10m + 3$

21. $100y^2 - 121$

22. $8r^3 - 1$

23. $9x^2 + 4 - 12x$

24. $5x^8 - 5y^8$

25. $y^2 + 12y + 36 - 121t^2$

26. $500a^2 - 5b^2$

27. $30x^2 - 51x + 9$

28. $15a^5b + 35ab^5$

29. $8y^4x + 4yx^2 + 20y^2x^2 - 12xy$

Solve.

30. $x^2 - 28 = 3x$

31. $3y^2 = 27$

32. $2x^3 + 2x = -5x^2$

33. $16x^2 + 4x = 0$

34. Let $f(x) = 3x^2 - 21x + 17$. Find a such that $f(a) = 17$.

35. Find the domain of the function f given by $f(x) = \dfrac{9-x}{x^2 + 6x + 9}$.

36. A photograph is 3 cm longer than it is wide. Its area is 88 cm^2. Find its length and its width.

37. To celebrate Asphalton's centennial, fireworks are launched over a lake from atop a dam 64 ft above the water. The height of a display, t seconds after it is launched, is given by $h(t) = -16t^2 + 48t + 64$. After how long will the shell from the fireworks reach the water?

38. a) Multiply: $(x^2 + x + 1)(-x^3 + x^2 - 1)$
    b) Factor: $-x^5 - x - 1$

39. Factor: $6x^{2n} + x^n - 12$

CHAPTER 5

TEST FORM C

NAME_____

CLASS_____ SCORE_____ GRADE_____

Given the polynomial $7xy^4 - 9x^2y + 2x^3y^3 - 4x^4y$,

1. Determine the degree of the polynomial.
2. Arrange in descending powers of x.

3. Determine the leading term of the polynomial $a - 10 + 6a^2 - 7a^3$.

4. Given $P(x) = 3x^3 + 2x^2 - 5x + 5$, find $P(0)$ and $P(-2)$.

5. Given $P(x) = x^2 - 41x$, find and simplify $P(a+h) - P(a)$.

6. Combine like terms: $9xy - 10xy^2 + 10xy - 9xy^2$.

Add.

7. $(3x^3 + 9x^2 - 7y) + (3x^3 - y - 7y^2)$

8. $(7m^3 - 6m^2n - 5mn^2 - 4n^3) + (3mn^2 - 2n^3 + m^3 + 8m^2n)$

Subtract.

9. $(3a - 7b) - (8a + 5b)$

10. $(y^2 - 6y - 7y^3) - (6y^2 - 3y - y^3)$

Multiply.

11. $(-11x^2y)(-7xy^2)$

12. $(2a - 5b)(4a + b)$

13. $(x - y)(-x^2 - xy + y^2)$

14. $(4x^3 + 5)^2$

15. $(7y - 4)^2$

16. $(7x - 3y)(7x + 3y)$

ANSWERS

1._____

2._____

3._____

4._____

5._____

6._____

7._____

8._____

9._____

10._____

11._____

12._____

13._____

14._____

15._____

16._____

CHAPTER 5

TEST FORM C

NAME_____

ANSWERS

17._____

18._____

19._____

20._____

21._____

22._____

23._____

24._____

25._____

26._____

27._____

28._____

29._____

30._____

31._____

32._____

33._____

34._____

35._____

36._____

37._____

38._____

39._____

Factor.

17. $36x^3 - 9x^5$

18. $y^3 - 2y^2 + 5y - 10$

19. $p^2 - 13p - 48$

20. $10m^2 + 15m + 5$

21. $121y^2 - 64$

22. $6r^3 - 6$

23. $9x^2 + 16 - 24x$

24. $x^4 - 81y^4$

25. $y^2 + 14y + 49 - 100t^2$

26. $448a^2 - 7b^2$

27. $60x^2 - 55x + 15$

28. $54a^5b + 12ab^5$

29. $25y^4x + 15yx^2 + 5y^2x^2 - 10xy$

Solve.

30. $x^2 - 21 = 4x$

31. $4y^2 = 64$

32. $2x^3 + 9x = -9x^2$

33. $25x^2 + 5x = 0$

34. Let $f(x) = 5x^2 - 40x + 135$. Find a such that $f(a) = 135$.

35. Find the domain of the function f given by $f(x) = \dfrac{10 - x}{x^2 + 8x + 16}$.

36. A photograph is 4 cm longer than it is wide. Its area is 60 cm². Find its length and its width.

37. To celebrate Slagville's centennial, fireworks are launched over a lake from atop a dam 144 ft above the water. The height of a display, t seconds after it is launched, is given by $h(t) = -16t^2 + 96t + 144$. After how long will the shell from the fireworks reach the water?

38. a) Multiply: $(x^2 + x - 1)(-x^3 + x^2 + 1)$
    b) Factor: $-x^5 + 2x^3 + x - 1$

39. Factor: $6x^{2n} - 6x^n - 12$

CHAPTER 5

TEST FORM D

NAME_____

CLASS_____ SCORE_____ GRADE_____

Given the polynomial $8xy^3 - x^5y + 3x^3y^4 - 5x^2y$,

ANSWERS

1. Determine the degree of the polynomial.
2. Arrange in descending powers of x.

1._____

2._____

3. Determine the leading term of the polynomial $10a - 1 + 7a^2 - 6a^3$.

4. Given $P(x) = 5x^3 + 2x^2 - 3x + 4$, find $P(0)$ and $P(-2)$.

3._____

5. Given $P(x) = x^2 + 7x$, find and simplify $P(a-h) + P(-a)$.

4._____

6. Combine like terms: $2xy + 7xy^2 - 7xy + 2xy^2$.

5._____

Add.

6._____

7. $(2x^3 + x^2 - 6y) + (4x^3 - 9y - 8y^2)$

7._____

8. $(5m^3 - 8m^2n - 3mn^2 - 6n^3) + (m^2n - 4m^3 + 9n^3 + 2mn^2)$

8._____

Subtract.

9._____

9. $(4a - 8b) - (9a + 6b)$

10._____

10. $(9y^2 - 8y - 6y^3) - (4y^2 - 4y - 3y^3)$

11._____

Multiply.

12._____

11. $(-20x^2y^3)(-4x^2y)$

13._____

12. $(3a - b)(5a + 2b)$

13. $(x + y)(x^2 + xy + y^2)$

14._____

14. $(2x^4 + 3)^2$

15._____

15. $(y - 6)^2$

16._____

16. $(8x - 3y)(8x + 3y)$

111

CHAPTER 5

TEST FORM D

NAME_____

ANSWERS

17._____
18._____
19._____
20._____
21._____
22._____
23._____
24._____
25._____
26._____
27._____
28._____
29._____
30._____
31._____
32._____
33._____
34._____
35._____
36._____
37._____
38._____
39._____

Factor.

17. $50x - 10x^3$

18. $y^3 + 6y^2 - 3y - 18$

19. $p^2 + 15p - 16$

20. $10m^2 + 12m + 2$

21. $64y^2 - 25$

22. $r^3 - 8$

23. $4x^2 + 1 - 4x$

24. $16x^4 - y^4$

25. $y^2 + 16y + 64 - 81t^2$

26. $324a^2 - 9b^2$

27. $30x^2 - 26x + 4$

28. $21a^4b^2 + 28ab^5$

29. $15y^4x + 9yx^2 + 3y^2x^2 - 6xy$

Solve.

30. $x^2 - 14 = 5x$

31. $6y^2 = 216$

32. $3x^3 + 4x = -13x^2$

33. $36x^2 + 6x = 0$

34. Let $f(x) = 2x^2 + 18x + 19$. Find a such that $f(a) = 19$.

35. Find the domain of the function f given by $f(x) = \dfrac{11-x}{x^2 - 10x + 25}$.

36. A photograph is 0.5 in longer than it is wide. Its area is 5 in^2. Find its length and its width.

37. To celebrate Puttyville's centennial, fireworks are launched over a lake from atop a dam 96 ft above the water. The height of a display, t seconds after it is launched, is given by $h(t) = -16t^2 + 80t + 96$. After how long will the shell from the fireworks reach the water?

38. a) Multiply: $(x^2 - x + 1)(-x^3 + x^2 + 1)$
    b) Factor: $-x^5 + 2x^4 - 2x^3 + 2x^2 - x + 1$

39. Factor: $6x^{2n} + 6x^n - 12$

CHAPTER 5

TEST FORM E

NAME_____

CLASS_____ SCORE_____ GRADE_____

Given the polynomial $9xy^2 - 2x^3y + 4x^4y^3 - 6x^2y$,

1. Determine the degree of the polynomial.
2. Arrange in descending powers of x.

3. Determine the leading term of the polynomial $10a - 7 + a^2 - 6a^3$.

4. Given $P(x) = 2x^3 + 5x^2 - 3x + 3$, find $P(0)$ and $P(-2)$.

5. Given $P(x) = x^2 + 8x$, find and simplify $P(a-h) + P(-a)$.

6. Combine like terms: $3xy + 11xy^2 + 11xy - 2xy^2$.

Add.

7. $(x^3 + 2x^2 - 5y) + (5x^3 - 8y - 9y^2)$

8. $(6m^3 - 7m^2n - 4mn^2 - 5n^3) + (2m^2n - 3m^3 + 8n^3 + mn^2)$

Subtract.

9. $(5a - 9b) - (a + 7b)$

10. $(7y^2 - 2y - 5y^3) - (2y^2 - 5y - 5y^3)$

Multiply.

11. $(-16x^2y^3)(-6x^2y)$

12. $(4a - 2b)(a + 3b)$

13. $(x + y)(-x^2 + xy + y^2)$

14. $(3x^4 + 4)^2$

15. $(2y - 5)^2$

16. $(9x - 5y)(9x + 5y)$

ANSWERS

1._____

2._____

3._____

4._____

5._____

6._____

7._____

8._____

9._____

10._____

11._____

12._____

13._____

14._____

15._____

16._____

CHAPTER 5

TEST FORM E

NAME_____

ANSWERS

17._____

18._____

19._____

20._____

21._____

22._____

23._____

24._____

25._____

26._____

27._____

28._____

29._____

30._____

31._____

32._____

33._____

34._____

35._____

36._____

37._____

38._____

39._____

Factor.

17. $44x^2 - 11x^4$

18. $y^3 + 2y^2 - 4y - 8$

19. $p^2 + 16p - 36$

20. $8m^2 + 14m + 3$

21. $144y^2 - 49$

22. $r^3 - 27$

23. $4x^2 + 25 - 20x$

24. $4x^4 - 64y^4$

25. $y^2 + 18y + 81 - 64t^2$

26. $176a^2 - 11b^2$

27. $30x^2 - 33x + 6$

28. $10a^4b^2 + 12ab^5$

29. $4y^4 x + 36yx^2 + 24y^2x^2 - 16xy$

Solve.

30. $x^2 - 24 = 5x$

31. $10y^2 = 1000$

32. $3x^3 + 10x = -17x^2$

33. $49x^2 + 7x = 0$

34. Let $f(x) = 105x^2 + 210x + 1087$. Find a such that $f(a) = 1087$.

35. Find the domain of the function f given by $f(x) = \dfrac{12 - x}{x^2 - 12x + 36}$.

36. A photograph is 3 in longer than it is wide. Its area is 40 in². Find its length and its width.

37. To celebrate Groutburgh's centennial, fireworks are launched over a lake from atop a dam 160 ft above the water. The height of a display, t seconds after it is launched, is given by $h(t) = -16t^2 + 48t + 160$. After how long will the shell from the fireworks reach the water?

38. a) Multiply: $(x^2 - x - 1)(-x^3 + x^2 + 1)$
    b) Factor: $-x^5 + 2x^4 + 2x^3 - x - 1$

39. Factor: $12x^{2n} + 3x^n - 15$

CHAPTER 5

TEST FORM F

NAME _____

CLASS_____ SCORE_____ GRADE_____

Given the polynomial $xy^4 - 3x^2y + 5x^4y^5 - 7x^3y$,

1. Determine the degree of the polynomial.
2. Arrange in descending powers of x.

3. Determine the leading term of the polynomial $6a - 11 + 7a^2 - a^3$.

4. Given $P(x) = 3x^3 + 5x^2 - 2x + 2$, find $P(0)$ and $P(-2)$.

5. Given $P(x) = x^2 + 37x$, find and simplify $P(a-h) + P(-a)$.

6. Combine like terms: $4xy + 8xy^2 - 8xy - 4xy^2$.

Add.

7. $(9x^3 + 3x^2 - 4y) + (6x^3 - 7y - y^2)$

8. $(7m^3 - 6m^2n - 5mn^2 - 4n^3) + (3m^2n - 2m^3 + n^3 + 8mn^2)$

Subtract.

9. $(6a - b) - (2a + 8b)$

10. $(5y^2 - 4y - 4y^3) - (8y^2 - 6y - 7y^3)$

Multiply.

11. $(-12x^2y^3)(-8x^2y)$

12. $(5a - 3b)(5a + 2b)$

13. $(x + y)(-x^2 - xy + y^2)$

14. $(4x^4 + 5)^2$

15. $(4y - 3)^2$

16. $(2x - 8y)(2x + 8y)$

ANSWERS

1. _____

2. _____

3. _____

4. _____

5. _____

6. _____

7. _____

8. _____

9. _____

10. _____

11. _____

12. _____

13. _____

14. _____

15. _____

16. _____

CHAPTER 5 NAME_____

TEST FORM F

ANSWERS

Factor.

17. _____ 17. $36x^3 - 12x^5$ 18. $y^3 + 4y^2 - 5y - 20$

18. _____ 19. $p^2 + 14p - 51$ 20. $6m^2 + 17m + 5$

19. _____ 21. $36y^2 - 25$ 22. $7r^3 - 7$

20. _____ 23. $4x^2 + 9 - 12x$ 24. $x^8 - 256y^8$

21. _____ 25. $y^2 + 20y + 100 - 49t^2$ 26. $52a^2 - 13b^2$

22. _____ 27. $40x^2 - 50x + 15$ 28. $30a^4b^2 + 9ab^5$

23. _____ 29. $10y^4x + 5yx^2 + 25y^2x^2 - 15xy$

24. _____ Solve.

25. _____

26. _____ 30. $x^2 - 60 = 4x$ 31. $\frac{3}{2}y^2 = \frac{27}{8}$

27. _____

28. _____ 32. $3x^3 + 12x = -20x^2$ 33. $64x^2 + 8x = 0$

29. _____ 34. Let $f(x) = 15x^2 + 45x + 75$. Find a such that $f(a) = 75$.

30. _____

31. _____ 35. Find the domain of the function f given by $f(x) = \frac{13 - x}{x^2 - 14x + 49}$.

32. _____ 36. A photograph is 5 in longer than it is wide. Its area is 126 in^2. Find its length and its width.

33. _____

34. _____ 37. To celebrate Slurry Falls' centennial, fireworks are launched over a lake from atop a dam 112 ft above the water. The height of a display, t seconds after it is launched, is given by $h(t) = -16t^2 + 96t + 112$. After how long will the shell from the fireworks reach the water?

35. _____

36. _____

37. _____

38. _____ 38. a) Multiply: $(x^2 - x - 1)(x^3 + x^2 + 1)$
   b) Factor: $x^5 - 2x^3 - x - 1$

39. _____ 39. Factor: $12x^{2n} - 3x^n - 15$

116

CHAPTER 5

TEST FORM G

NAME_____

CLASS_____ SCORE_____ GRADE_____

ANSWERS

1. Determine the leading term of the polynomial $6a - 1 + 10a^2 - 7a^3$.

   a) -1 b) 10 c) -7 d) 6

   1._____

2. Let $P(x) = 3x^3 + 5x^2 - 2x + 8$. Find $P(-2)$.

   a) -24 b) 81 c) -8 d) 8

   2._____

3. Find and simplify $P(a - h) - P(-a)$ given $P(x) = x^2 + 6x$.

   a) $h^2 - 6h + a^2$
   b) $h^2 - 6h - 2ah + 2a^2$
   c) $h^2 - 2h - 6ah + 2a^2$
   d) $h^2 + 2ah + 2a^2$

4. Add. $(8x^3 + 4x^2 - 3y) + (7x^3 - 6y - 2y^2)$

   a) $18x^3 + 3x^2 + 9y - 2y^2$
   b) $15x^3 + 4x^2 - 9y - 2y^2$
   c) $15x^3 + 4x^2 - 2y^2$
   d) $5x^3 + 14x^2 + 9y - 3y^2$

   3._____

5. Add. $(8m^3 - 5m^2n - 6mn^2 - 3n^3) + (4mn^2 - n^3 + 2m^3 + 9m^2n)$

   a) $10m^3 + 4m^2n - 2mn^2 - 4n^3$
   b) $8m^3 - 5m^2n - 6mn^2 - 3n^3$
   c) $10m^3 + 2m^2n - mn^2 - 4n^3$
   d) $8m^3 + 4m^2n - 4mn^2 - n^3$

   4._____

6. Subtract. $(7a - 2b) - (3a + 9b)$

   a) $4a + 11b$ b) $4a - 4b$ c) $4a - 11b$ d) $11a - 11b$

   5._____

   6._____

117

CHAPTER 5

TEST FORM G

NAME_____

ANSWERS

7. Subtract. $(3y^2 - 6y - 3y^3) - (6y^2 - 7y - 9y^3)$

   a) $3y^2 - 6y - 3y^3$ b) $6y^2 - 7y - 9y^3$
   c) $3y^2 + 3y^3$ d) $-3y^2 + y + 6y^3$

7._____

8. Multiply. $(-9x^2y)(-8xy^2)$

   a) $-72x^3y^3$ b) $17x^3y^3$ c) $72x^2y^2$ d) $72x^3y^3$

8._____

9. Multiply. $(a - 4b)(4a + b)$

   a) $4a^2 - 15ab + 4b^2$ b) $4a^2 - 15ab - 4b^2$
   c) $4a^2 + 15ab - 4b^2$ d) $4a^2 + 17ab + 4b^2$

9._____

10. Multiply. $(x - y)(-x^2 - xy - y^2)$

    a) $-x^3 + 2x^2y - y^3$ b) $-x^3 - 2x^2y + y^3$
    c) $-x^3 - x^2y + y^3$ d) $x^3 - 2x^2y - y^3$

10._____

11. Multiply. $(2x^3 + 5)^2$

    a) $4x^5 + 20x^3 + 25$ b) $4x^6 + 20x^3 + 25$
    c) $4x^6 + 10x^3 + 25$ d) $4x^6 + 20x^3 + 10$

11._____

12. Multiply. $(3x - 5y)(3x + 5y)$

    a) $6x^2 - 10y^2$ b) $9x^2 + 25y^2$
    c) $9x^2 - 25y^2$ d) $9x^2 - 30xy - 25y^2$

12._____

13. Find one factor of $65x - 13x^3$.

    a) $(5 - x^2)$ b) $5x$ c) $12x$ d) $(13 - x^2)$

13._____

CHAPTER 5

TEST FORM G

NAME_____

ANSWERS

14. Find one factor of $y^3 - 5y^2 + 2y - 10$.

   a) $(y^2 - 5)$ b) $(y + 5)$ c) $(y^2 + 2)$ d) $(y - 2)$

14._____

15. Find one factor of $25y^2 - 144$.

   a) $(y + 12)$ b) $(5y + 12)$ c) $(12y + 5)$ d) $(5y + 14)$

15._____

16. Find one factor of $16x^2 + 9 - 24x$.

   a) $(3x - 4)$ b) $(4x + 3)$ c) $(3x - 3)$ d) $(4x - 3)$

16._____

17. Find one factor of $y^2 + 22y + 121 - 36t^2$.

   a) $(y + 11 - 6t)$ b) $(y - 11 - 6t)$
   c) $(y + 11 - 18t)$ d) $(y - 11 + 6t)$

17._____

18. Find one factor of $20x^2 - 22x + 6$.

   a) $3x$ b) 2 c) $(2x + 1)$ d) $(5x + 3)$

18._____

19. Find one factor of $6y^4x + 3yx^2 + 15y^2x^2 - 9xy$.

   a) $3xy^2$ b) $2xy$ c) $3xy$ d) $3x^2y$

19._____

20. Solve. $x^2 - 10 = 3x$

   a) $-1, 7$ b) $-2, -5$ c) $2, 5$ d) $-2, 5$

20._____

CHAPTER 5　　　　　　　　　　　　　　　　NAME_____

TEST FORM G

ANSWERS

21. Solve. $2x^3 + 21x = -17x^2$

    a) $7, \frac{2}{3}, 0$　　b) $\frac{3}{2}, 0$　　c) $0, 7$　　d) $-7, -\frac{3}{2}, 0$

21._____

22. Let $f(x) = 8x^2 - 32x + 2$. Find a such that $f(a) = 2$

    a) $-2, 4$　　b) $-4, 0$　　c) $-4, 4$　　d) $0, 4$

22._____

23. Find the domain of the function f given by $f(x) = \dfrac{14 - x}{x^2 - 6x + 9}$

    a) $\{x \in \mathbb{R}, x \neq 3\}$　　　　b) $\{x \in \mathbb{R}, x \neq -3\}$
    c) $\{x \in \mathbb{R}, x \neq -2\}$　　　d) $\{x \in \mathbb{R}, x \neq 2\}$

23._____

24. To celebrate Plasticburgh's centennial, fireworks are launched over a lake from atop a dam 48 ft above the water. The height of a display, t seconds after it is launched, is given by $h(t) = -16t^2 + 32t + 48$. After how long will the shell from the fireworks reach the water?

    a) 2.5 sec　　b) 3 sec　　c) 3.2 sec　　d) 4 sec

24._____

25. Multiply. $(x^2 - x - 1)(x^3 + x^2 - 1)$

    a) $x^5 - 1$　　　　　　　b) $x^5 - 2x^4 - x - 1$
    c) $x^5 - x^4 + x^2 - x - 1$　　d) $x^5 - 2x^3 - 2x^2 - x + 1$

25._____

26. Find one factor of $12x^{2n} - 11x^n - 15$.

    a) $(4x^{2n} - 3)$　b) $(3x^n - 4)$　c) $(4x^n + 3)$　d) $(3x^n + 3)$

26._____

CHAPTER 5

TEST FORM H

NAME_____

CLASS_____ SCORE_____ GRADE_____

ANSWERS

1. Determine the leading term of the polynomial $a - 6 + 7a^2 - 10a^3$.

   a) -1 b) 10   c) -7 d) -10

1._____

2. Let $P(x) = 2x^3 + 5x^2 - 3x + 9$. Find $P(-2)$.

   a) 19   b) 81   c) -8 d) 8

2._____

3. Find and simplify $P(a + h) - P(a)$ given $P(x) = x^2 - 2x$.

   a) $h^2 - 6h + a^2$ b) $h^2 - 2h + 2ah$
   c) $h^2 - 2h - 6ah + 2a^2$ d) $h^2 + 2ah + 2a^2$

3._____

4. Add. $(7x^3 + 5x^2 - 2y) + (8x^3 - 5y - 3y^2)$

   a) $18x^3 + 3x^2 + 9y - 2y^2$ b) $7x^3 + 4x^2 - 9y - 2y^2$
   c) $7x^3 + 13x^2 - 5y - 5y^2$ d) $5x^3 + 13x^2 + 9y - 3y^2$

4._____

5. Add. $(8m^3 - 5m^2n - 6mn^2 - 3n^3) + (9mn^2 + 2n^3 - m^3 + 4m^2n)$

   a) $7m^3 + 4m^2n - 2mn^2 - 4n^3$
   b) $8m^3 - 5m^2n - 6mn^2 - 3n^3$
   c) $7m^3 - m^2n + 3mn^2 - n^3$
   d) $8m^3 + 4m^2n - 4mn^2 - n^3$

5._____

6. Subtract. $(8a - 3b) - (4a + b)$

   a) $4a - 4b$ b) $4a + 4b$ c) $4a - 11b$ d) $11a - 11b$

6._____

CHAPTER 5

TEST FORM H

NAME_____

ANSWERS

7. Subtract. $(y^2 - 8y - 2y^3) - (4y^2 - 8y - y^3)$

   a) $3y^2 - 6y - 3y^3$ b) $-3y^2 - y^3$
   c) $3y^2 + 3y^3$ d) $-3y^2 + y + 6y^3$

7._____

8. Multiply. $(-8x^2y^3)(-10x^2y)$

   a) $-80x^3y^3$ b) $80x^4y^3$ c) $80x^4y^4$ d) $80x^3y^3$

8._____

9. Multiply. $(2a - 5b)(3a + 5b)$

   a) $6a^2 - 5ab + 25b^2$ b) $6a^2 - 15ab - 25b^2$
   c) $6a^2 + 15ab - 25b^2$ d) $6a^2 - 5ab - 25b^2$

9._____

10. Multiply. $(x + y)(-x^2 - xy - y^2)$

    a) $-x^3 - 2x^2y - 2xy^2 - y^3$ b) $-x^3 - 2x^2y + y^3$
    c) $-x^3 - x^2y + y^3$ d) $-x^3 + 2x^2y + 2xy^2 - y^3$

10._____

11. Multiply. $(3x^4 + 3)^2$

    a) $9x^5 + 20x^3 + 9$ b) $9x^6 - 18x^4 + 9$
    c) $9x^8 + 18x^4 + 9$ d) $9x^6 + 18x^4 + 9$

11._____

12. Multiply. $(4x - 8y)(4x + 8y)$

    a) $8x^2 - 16y^2$ b) $16x^2 + 64y^2$
    c) $16x^2 - 64y^2$ d) $16x^2 - 64xy - 64y^2$

12._____

13. Find one factor of $28x^2 - 14x^4$.

    a) $(1 - 2x^2)$ b) $(2 - x^2)$ c) $14x^3$ d) $(4 - x^2)$

13._____

CHAPTER 5
TEST FORM H

NAME_____

ANSWERS

14. Find one factor of $y^3 + 3y^2 - 6y - 18$.

    a) $(y^2 - 5)$ b) $(y + 6)$ c) $(y^2 + 3)$ d) $(y + 3)$

14._____

15. Find one factor of $49y^2 - 16$.

    a) $(7y + 4)$ b) $(4y + 7)$ c) $(12y + 5)$ d) $(4y - 7)$

15._____

16. Find one factor of $16x^2 + 1 - 10x$.

    a) $(3x - 4)$ b) $(4x + 3)$ c) $(3x - 3)$ d) $(4x - 1)$

16._____

17. Find one factor of $y^2 + 24y + 144 - 25t^2$.

    a) $(y + 12 - 6t)$ b) $(y - 12 - 5t)$
    c) $(y + 12 - 5t)$ d) $(y - 12 + 6t)$

17._____

18. Find one factor of $36x^2 - 30x + 6$.

    a) $3x$ b) 4 c) $(3x - 1)$ d) $(5x + 3)$

18._____

19. Find one factor of $20y^4x + 12yx^2 + 4y^2x^2 - 8xy$.

    a) $3xy^2$ b) $4xy$ c) $3xy$ d) $3x^2y$

19._____

20. Solve. $x^2 - 63 = 2x$

    a) $-7, 9$ b) $-9, -7$ c) $7, 9$ d) $-9, 7$

20._____

CHAPTER 5

TEST FORM H

NAME_____

ANSWERS

21._____

21. Solve. $3x^3 + 8x = -25x^2$

a) $-8, -\frac{1}{3}, 0$ b) $-8, 0$ c) $-8, -\frac{1}{3}$ d) $0, \frac{1}{3}, 8$

22._____

22. Let $f(x) = 13x^2 + 65x + 3$. Find a such that $f(a) = 3$

a) $-3, 5$ b) $-5, 5$ c) $-5, 0$ d) $0, 5$

23._____

23. Find the domain of the function f given by $f(x) = \dfrac{15 - x}{x^2 + 10x + 25}$

a) $\{x \in \mathbb{R}, x \neq 15\}$ b) $\{x \in \mathbb{R}, x \neq -5\}$
c) $\{x \in \mathbb{R}, x \neq -3\}$ d) $\{x \in \mathbb{R}, x \neq 5\}$

24._____

24. To celebrate Glueburgh's centennial, fireworks are launched over a lake from atop a dam 32 ft above the water. The height of a display, t seconds after it is launched, is given by $h(t) = -16t^2 + 16t + 32$. After how long will the shell from the fireworks reach the water?

a) 2.5 sec   b) 2 sec   c) 3.2 sec   d) 4 sec

25._____

25. Multiply. $(x^2 - x - 1)(x^3 - x^2 + 1)$

a) $x^5 - 1$ b) $x^5 - 2x^4 - x - 1$
c) $x^5 - 2x^4 + 2x^2 - x - 1$ d) $x^5 - x^4 + x^2 - x - 1$

26._____

26. Find one factor of $12x^{2n} + 11x^n - 15$.

a) $(4x^n - 3)$ b) $(3x^n + 4)$ c) $(4x^n - 5)$ d) $(4x^n + 3)$

CHAPTER 6

TEST FORM A

NAME_____

CLASS_____ SCORE_____ GRADE_____

ANSWERS

Simplify.

1. $\dfrac{t-2}{t+1} \cdot \dfrac{5t+5}{6t^2-24}$

2. $\dfrac{x^3+1}{x^2-9} \div \dfrac{x^2+5x+4}{x^2+x-12}$

3. Find the LCD: $\dfrac{2x}{x^2+x-6}$, $\dfrac{x+5}{x^2+2x-8}$

Perform the indicated operation and simplify when possible.

4. $\dfrac{16x}{x+4} + \dfrac{x^3}{x+4}$

5. $\dfrac{5a^2}{a-b} - \dfrac{5b^2-10ab}{b-a}$

6. $\dfrac{ab}{a^2-b^2} + \dfrac{a^2+b^2}{a+b}$

7. $\dfrac{3}{x^3-1000} - \dfrac{1}{x^2-100}$

8. $\dfrac{2}{y+1} - \dfrac{y}{y-7} + \dfrac{y^2+2}{y^2-6y-7}$

Simplify.

9. $\dfrac{\dfrac{5}{a}+\dfrac{6}{b}}{\dfrac{11}{ab}+\dfrac{1}{a^2}}$

10. $\dfrac{\dfrac{x^2+3x-10}{x^2-100}}{\dfrac{x^2-3x-40}{x^2-20x+100}}$

11. $\dfrac{\dfrac{1}{x+5}-\dfrac{2}{x^2-5x+6}}{\dfrac{3}{x-3}+\dfrac{4}{x^2+3x-10}}$

Solve.

12. $\dfrac{1}{7x-8} = \dfrac{3}{5x+8}$

13. $\dfrac{t+10}{t^2-t-30} + \dfrac{5}{t-6} = \dfrac{9}{t+5}$

Let $f(x) = \dfrac{x+8}{x-5}$

14. Find $f(2)$ and $f(-3)$.
15. Find all a for which $f(a) = 9$.

1._____

2._____

3._____

4._____

5._____

6._____

7._____

8._____

9._____

10._____

11._____

12._____

13._____

14._____

15._____

CHAPTER 6 NAME_____

TEST FORM A

ANSWERS

16._____

17._____

18._____

19._____

20._____

21._____

22._____

23._____

24._____

25._____

26._____

27._____

28._____

16. David can paint a bedroom in 0.75 hr. Jay can paint the same room in 1.25 hr. How long will it take them, working together, to paint the bedroom?

Divide.

17. $(18ab^3c - 5ab^2c^2 + 12a^2b^2c) \div (3a^2b)$

18. $(y^2 - 10y + 5) \div (y - 5)$

19. $(3x^4 + 8x^2 + 4x + 1) \div (x^2 + 1)$

20. Divide using synthetic division: $(x^3 + 2x^2 + 3x - 10) \div (x - 4)$

21. If $f(x) = 5x^4 - 7x^3 + 3x - 9$, then use synthetic division to find $f(3)$.

22. Solve $\dfrac{1}{R} = \dfrac{1}{R_1} + \dfrac{1}{R_2}$ for R.

23. The product of the reciprocals of two consecutive integers is $\dfrac{1}{20}$. Find the integers.

24. Lance bicycles 38 km/h with no wind. Against the wind he bikes 28 km in the same time it takes to bike 48 km with the wind. What is the speed of the wind?

25. Let $f(x) = \dfrac{1}{x+2} + \dfrac{5}{x-3}$. Find all a for which $f(a) = f(a+5)$.

26. Solve: $\dfrac{2}{x-8} - \dfrac{2}{x} = \dfrac{16}{x^2 - 8x}$

27. Find the x- and y-intercepts for the function given by

$$f(x) = \dfrac{\dfrac{1}{x+4} - \dfrac{2}{x-3}}{\dfrac{3}{x-2} + \dfrac{4}{x+1}}$$

28. One summer, Tawana mowed 5 lawns for every 3 lawns mowed by Jeff. Together they mowed 120 lawns. How many did each mow?

CHAPTER 6

TEST FORM B

NAME_____

CLASS_____ SCORE_____ GRADE_____

ANSWERS

Simplify.

1. $\dfrac{t-3}{t+2} \cdot \dfrac{2t+4}{3t^2-27}$

2. $\dfrac{x^3-1}{x^2-9} \div \dfrac{x^2-3x+2}{x^2+x-6}$

3. Find the LCD: $\dfrac{3x}{x^2-2x-8}$, $\dfrac{x+4}{x^2-x-12}$

Perform the indicated operation and simplify when possible.

4. $\dfrac{9x}{x+3} + \dfrac{x^3}{x+3}$

5. $\dfrac{4a^2}{a-b} - \dfrac{4b^2+8ab}{b-a}$

6. $\dfrac{2ab}{a^2-b^2} + \dfrac{a^2-b^2}{a+b}$

7. $\dfrac{6}{x^3-125} - \dfrac{2}{x^2-25}$

8. $\dfrac{3}{y+2} - \dfrac{y}{y-6} + \dfrac{y^2+1}{y^2-4y-12}$

Simplify.

9. $\dfrac{\dfrac{4}{a}+\dfrac{7}{b}}{\dfrac{11}{ab}+\dfrac{3}{a^2}}$

10. $\dfrac{\dfrac{x^2+2x-15}{x^2-81}}{\dfrac{x^2-2x-35}{x^2-18x+81}}$

11. $\dfrac{\dfrac{2}{x+5} - \dfrac{1}{x^2-5x+6}}{\dfrac{3}{x-2} + \dfrac{4}{x^2+2x-15}}$

Solve.

12. $\dfrac{9}{6x-9} = \dfrac{2}{4x+7}$

13. $\dfrac{t+12}{t^2-t-20} + \dfrac{4}{t-5} = \dfrac{1}{t+4}$

Let $f(x) = \dfrac{x+8}{x-5}$

14. Find $f(3)$ and $f(-4)$.
15. Find all a for which $f(a) = 2$.

1._____

2._____

3._____

4._____

5._____

6._____

7._____

8._____

9._____

10._____

11._____

12._____

13._____

14._____

15._____

127

CHAPTER 6

TEST FORM B

NAME_____

ANSWERS

16. _____

17. _____

18. _____

19. _____

20. _____

21. _____

22. _____

23. _____

24. _____

25. _____

26. _____

27. _____

28. _____

16. Joe can paint a bedroom in 1.25 hr. Sara can paint the same room in 1.75 hr. How long will it take them, working together, to paint the bedroom?

Divide.

17. $(20ab^3c - 3ab^2c^2 + 32a^2b^2c) \div (4a^2b)$

18. $(y^2 - 22y + 30) \div (y + 4)$

19. $(4x^4 + 7x^2 + 3x + 2) \div (x^2 + 2)$

20. Divide using synthetic division: $(x^3 + 3x^2 + 5x - 9) \div (x - 3)$

21. If $f(x) = 5x^4 - 7x^3 + 3x - 9$, then use synthetic division to find $f(4)$.

22. Solve $\dfrac{a}{b} = \dfrac{c}{d} + \dfrac{e}{g}$ for g.

23. The product of the reciprocals of two consecutive integers is $\dfrac{1}{72}$. Find the integers.

24. Catherine bicycles 35 km/h with no wind. Against the wind she bikes 52 km in the same time it takes to bike 88 km with the wind. What is the speed of the wind?

25. Let $f(x) = \dfrac{1}{x+3} + \dfrac{7}{x-4}$. Find all a for which $f(a) = f(a+7)$.

26. Solve: $\dfrac{3}{x-9} - \dfrac{3}{x} = \dfrac{27}{x^2 - 9x}$

27. Find the x- and y-intercepts for the function given by

$$f(x) = \dfrac{\dfrac{2}{x+4} - \dfrac{3}{x-3}}{\dfrac{4}{x-2} + \dfrac{1}{x+1}}$$

28. One summer, Xavier mowed 5 lawns for every 4 lawns mowed by Phyllis. Together they mowed 126 lawns. How many did each mow?

CHAPTER 6

TEST FORM C

NAME_____

CLASS_____ SCORE_____ GRADE_____

Simplify.

ANSWERS

1. $\dfrac{t-4}{t+3} \cdot \dfrac{3t+9}{4t^2-64}$

2. $\dfrac{x^3+8}{x^2-25} \div \dfrac{x^2+4x+4}{x^2-3x-10}$

3. Find the LCD: $\dfrac{4x}{x^2-x-6}$, $\dfrac{x+3}{x^2+x-12}$

Perform the indicated operation and simplify when possible.

4. $\dfrac{64x}{x+8} + \dfrac{x^3}{x+8}$

5. $\dfrac{6a^2}{a-b} - \dfrac{6b^2-12ab}{b-a}$

6. $\dfrac{3ab}{a^2-b^2} + \dfrac{a^2+b^2}{a+b}$

7. $\dfrac{9}{x^3+8} - \dfrac{3}{x^2-4}$

8. $\dfrac{4}{y+3} - \dfrac{y}{y-5} + \dfrac{y^2+9}{y^2-2y-15}$

Simplify.

9. $\dfrac{\dfrac{3}{a}+\dfrac{8}{b}}{\dfrac{11}{ab}+\dfrac{5}{a^2}}$

10. $\dfrac{\dfrac{x^2-16}{x^2-64}}{\dfrac{x^2-2x-24}{x^2-16x+64}}$

11. $\dfrac{\dfrac{1}{x-3}-\dfrac{2}{x^2+3x-10}}{\dfrac{3}{x+5}+\dfrac{4}{x^2-5x+6}}$

Solve.

12. $\dfrac{8}{5x-1} = \dfrac{1}{3x+6}$

13. $\dfrac{t+14}{t^2-t-12} + \dfrac{3}{t-4} = \dfrac{2}{t+3}$

Let $f(x) = \dfrac{x+8}{x-5}$

14. Find $f(4)$ and $f(-5)$.

15. Find all a for which $f(a) = 3$.

1._____

2._____

3._____

4._____

5._____

6._____

7._____

8._____

9._____

10._____

11._____

12._____

13._____

14._____

15._____

CHAPTER 6

TEST FORM C

NAME _____

ANSWERS

16. _____

17. _____

18. _____

19. _____

20. _____

21. _____

22. _____

23. _____

24. _____

25. _____

26. _____

27. _____

28. _____

16. Jodi can paint a bedroom in 1.5 hr. Sara can paint the same room in 2.75 hr. How long will it take them, working together, to paint the bedroom?

Divide.

17. $(35ab^3c - 7ab^2c^2 + 10a^2b^2c) \div (5a^2b)$

18. $(y^2 + 18y + 12) \div (y - 4)$

19. $(5x^4 + 6x^2 + 2x + 3) \div (x^2 + 3)$

20. Divide using synthetic division: $(x^3 + 4x^2 + 7x - 8) \div (x - 2)$

21. If $f(x) = 5x^4 - 7x^3 + 3x - 9$, then use synthetic division to find $f(5)$.

22. Solve $F = \dfrac{kq_0q_2}{d_1} + \dfrac{kq_0q_2}{d_2}$ for q_0.

23. The product of the reciprocals of two consecutive integers is $\dfrac{1}{42}$. Find the integers.

24. Jan bicycles 36 km/h with no wind. Against the wind he bikes 42 km in the same time it takes to bike 66 km with the wind. What is the speed of the wind?

25. Let $f(x) = \dfrac{1}{x+4} + \dfrac{9}{x-5}$. Find all a for which $f(a) = f(a+9)$.

26. Solve: $\dfrac{4}{x-10} - \dfrac{4}{x} = \dfrac{40}{x^2 - 10x}$

27. Find the x- and y-intercepts for the function given by

$$f(x) = \dfrac{\dfrac{3}{x+4} - \dfrac{4}{x-3}}{\dfrac{1}{x-2} + \dfrac{2}{x+1}}$$

28. One summer, Hosni mowed 5 lawns for every 6 lawns mowed by Oumy. Together they mowed 121 lawns. How many did each mow?

CHAPTER 6

TEST FORM D

Simplify.

1. $\dfrac{t-5}{t+4} \cdot \dfrac{4t+16}{5t^2-125}$

2. $\dfrac{x^3-8}{x^2-25} \div \dfrac{x^2+x-6}{x^2+8x+15}$

3. Find the LCD: $\dfrac{5x}{x^2-5x+6}$, $\dfrac{x+2}{x^2+x-12}$

Perform the indicated operation and simplify when possible.

4. $\dfrac{49x}{x+7} + \dfrac{x^3}{x+7}$

5. $\dfrac{3a^2}{a-b} - \dfrac{3b^2+6ab}{b-a}$

6. $\dfrac{4ab}{a^2-b^2} + \dfrac{a^2-b^2}{a+b}$

7. $\dfrac{2}{x^3+27} - \dfrac{4}{x^2-9}$

8. $\dfrac{5}{y+4} - \dfrac{y}{y-4} + \dfrac{y^2+8}{y^2-16}$

Simplify.

9. $\dfrac{\dfrac{2}{a}+\dfrac{9}{b}}{\dfrac{11}{ab}+\dfrac{7}{a^2}}$

10. $\dfrac{\dfrac{x^2-x-20}{x^2-49}}{\dfrac{x^2-16}{x^2-14x+49}}$

11. $\dfrac{\dfrac{1}{x-3} - \dfrac{2}{x^2+3x-10}}{\dfrac{3}{x-2} + \dfrac{4}{x^2+2x-15}}$

Solve.

12. $\dfrac{7}{4x-2} = \dfrac{9}{2x+5}$

13. $\dfrac{t+16}{t^2-t-6} + \dfrac{2}{t-3} = \dfrac{4}{t+2}$

Let $f(x) = \dfrac{x+8}{x-5}$

14. Find $f(6)$ and $f(-7)$.
15. Find all a for which $f(a) = 4$.

ANSWERS

1._____

2._____

3._____

4._____

5._____

6._____

7._____

8._____

9._____

10._____

11._____

12._____

13._____

14._____

15._____

CHAPTER 6

TEST FORM D

NAME_____

ANSWERS

16._____

17._____

18._____

19._____

20._____

21._____

22._____

23._____

24._____

25._____

26._____

27._____

28._____

16. Erin can paint a bedroom in 1.75 hr. Tom can paint the same room in 2.5 hr. How long will it take them, working together, to paint the bedroom?

Divide.

17. $(42ab^3c - 9ab^2c^2 + 12a^2b^2c) \div (6a^2b)$

18. $(y^2 + 15y + 10) \div (y + 3)$

19. $(3x^4 + 7x^2 + x + 4) \div (x^2 + 1)$

20. Divide using synthetic division: $(x^3 + 5x^2 + 9x - 7) \div (x - 4)$

21. If $f(x) = 5x^4 - 7x^3 + 3x - 9$, then use synthetic division to find $f(2)$.

22. Solve $p = m\left(\dfrac{x_1 - x_0}{t_1 - t_0}\right)$ for t_1.

23. The product of the reciprocals of two consecutive integers is $\dfrac{1}{132}$. Find the integers.

24. Jeanie bicycles 34 km/h with no wind. Against the wind she bikes 27 km in the same time it takes to bike 41 km with the wind. What is the speed of the wind?

25. Let $f(x) = \dfrac{1}{x+1} + \dfrac{3}{x-2}$. Find all a for which $f(a) = f(a+3)$.

26. Solve: $\dfrac{5}{x-11} - \dfrac{5}{x} = \dfrac{55}{x^2 - 11x}$

27. Find the x- and y-intercepts for the function given by

$$f(x) = \dfrac{\dfrac{4}{x+4} - \dfrac{1}{x-3}}{\dfrac{2}{x-2} + \dfrac{3}{x+1}}$$

28. One summer, Erin mowed 5 lawns for every 7 lawns mowed by Julia. Together they mowed 144 lawns. How many did each mow?

CHAPTER 6   NAME_____

TEST FORM E   CLASS_____ SCORE_____ GRADE_____

Simplify.

ANSWERS

1. $\dfrac{t-5}{t+1} \cdot \dfrac{6t+6}{7t^2-175}$

2. $\dfrac{x^3+27}{x^2-4} \div \dfrac{x^2-2x-15}{x^2-7x+10}$

1._____

2._____

3. Find the LCD: $\dfrac{2x}{x^2-6x+8}$, $\dfrac{x+3}{x^2-x-12}$

3._____

Perform the indicated operation and simplify when possible.

4. $\dfrac{4x}{x+2} + \dfrac{x^3}{x+2}$

5. $\dfrac{7a^2}{a-b} - \dfrac{7b^2-14ab}{b-a}$

4._____

6. $\dfrac{4ab}{a^2-b^2} + \dfrac{a^2+b^2}{a+b}$

7. $\dfrac{6}{x^3-8} - \dfrac{5}{x^2-4}$

5._____

8. $\dfrac{6}{y+5} - \dfrac{y}{y-3} + \dfrac{y^2+7}{y^2+2y-15}$

6._____

7._____

Simplify.

8._____

9. $\dfrac{\dfrac{5}{a}+\dfrac{9}{b}}{\dfrac{14}{ab}+\dfrac{4}{a^2}}$

10. $\dfrac{\dfrac{x^2-x-12}{x^2-36}}{\dfrac{x^2-2x-15}{x^2-12x+36}}$

9._____

10._____

11. $\dfrac{\dfrac{2}{x-2}-\dfrac{1}{x^2+2x-15}}{\dfrac{3}{x-3}+\dfrac{4}{x^2+3x-10}}$

11._____

Solve.

12._____

12. $\dfrac{6}{3x-3} = \dfrac{8}{x+4}$

13. $\dfrac{t+18}{t^2-t-2} + \dfrac{1}{t-2} = \dfrac{4}{t+1}$

13._____

Let $f(x) = \dfrac{x+8}{x-5}$

14._____

14. Find $f(7)$ and $f(-8)$.
15. Find all a for which $f(a) = 5$.

15._____

133

CHAPTER 6

TEST FORM E

NAME_____

ANSWERS

16._____

17._____

18._____

19._____

20._____

21._____

22._____

23._____

24._____

25._____

26._____

27._____

28._____

16. Enid can paint a bedroom in 2 hr. Jacob can paint the same room in 3.25 hr. How long will it take them, working together, to paint the bedroom?

Divide.

17. $(14ab^3c - 5ab^2c^2 + 7a^2b^2c) \div (7a^2b)$

18. $(y^2 - 26y + 9) \div (y - 3)$

19. $(4x^4 + 8x^2 + x + 5) \div (x^2 + 2)$

20. Divide using synthetic division: $(x^3 + 6x^2 + 11x - 6) \div (x - 3)$

21. If $f(x) = 2x^4 - 9x^3 + 6x - 5$, then use synthetic division to find $f(3)$.

22. Solve $W = \dfrac{Fg}{a}(h_1 - h_0)$ for h_0.

23. The product of the reciprocals of two consecutive integers is $\dfrac{1}{90}$. Find the integers.

24. Simon bicycles 37 km/h with no wind. Against the wind he bikes 75 km in the same time it takes to bike 110 km with the wind. What is the speed of the wind?

25. Let $f(x) = \dfrac{1}{x+5} + \dfrac{11}{x-6}$. Find all a for which $f(a) = f(a+11)$.

26. Solve: $\dfrac{2}{x-11} - \dfrac{2}{x} = \dfrac{22}{x^2 - 11x}$

27. Find the x- and y-intercepts for the function given by

$$f(x) = \dfrac{\dfrac{1}{x+5} - \dfrac{2}{x-4}}{\dfrac{3}{x-3} + \dfrac{4}{x+2}}$$

28. One summer, Robin mowed 4 lawns for every 3 lawns mowed by Segal. Together they mowed 119 lawns. How many did each mow?

134

CHAPTER 6   NAME_____

TEST FORM F   CLASS_____ SCORE_____ GRADE_____

Simplify.   ANSWERS

1. $\dfrac{t-4}{t+2} \cdot \dfrac{2t+4}{4t^2-64}$ 2. $\dfrac{x^3-27}{x^2-4} \div \dfrac{x^2-9x+18}{x^2-2x-8}$ 1._____

2._____

3. Find the LCD: $\dfrac{3x}{x^2-7x+12}$, $\dfrac{x+5}{x^2-2x-8}$

3._____

Perform the indicated operation and simplify when possible.

4. $\dfrac{25x}{x+5} + \dfrac{x^3}{x+5}$ 5. $\dfrac{2a^2}{a-b} - \dfrac{2b^2+2ab}{b-a}$ 4._____

6. $\dfrac{3ab}{a^2-b^2} + \dfrac{a^2-b^2}{a+b}$ 7. $\dfrac{10}{x^3+1000} - \dfrac{6}{x^2-100}$ 5._____

8. $\dfrac{7}{y+6} - \dfrac{y}{y-2} + \dfrac{y^2+6}{y^2+4y-12}$ 6._____

7._____

Simplify.

9. $\dfrac{\dfrac{4}{a}+\dfrac{8}{b}}{\dfrac{12}{ab}+\dfrac{4}{a^2}}$ 10. $\dfrac{\dfrac{x^2-9}{x^2-25}}{\dfrac{x^2-3x-18}{x^2-10x+25}}$ 8._____

9._____

10._____

11. $\dfrac{\dfrac{1}{x-2}-\dfrac{2}{x^2+2x-15}}{\dfrac{3}{x+5}+\dfrac{4}{x^2-5x+6}}$

11._____

Solve.   12._____

12. $\dfrac{5}{2x-4} = \dfrac{7}{9x+3}$ 13. $\dfrac{t+20}{t^2+8t-9} + \dfrac{9}{t-1} = \dfrac{5}{t+9}$

13._____

Let $f(x) = \dfrac{x+8}{x-5}$ 14._____

14. Find $f(8)$ and $f(-2)$.
15. Find all a for which $f(a) = 6$.   15._____

135

CHAPTER 6

TEST FORM F

NAME_____

ANSWERS

16. Enos can paint a bedroom in 2.25 hr. Zoe can paint the same room in 3 hr. How long will it take them, working together, to paint the bedroom?

Divide.

17. $(8ab^3c - 6ab^2c^2 + 40a^2b^2c) \div (8a^2b)$

18. $(y^2 - 36y + 12) \div (y + 8)$

19. $(5x^4 + 6x^2 + 4x + 6) \div (x^2 + 3)$

20. Divide using synthetic division: $(x^3 + 7x^2 + 13x - 5) \div (x - 2)$

21. If $f(x) = 2x^4 - 9x^3 + 6x - 5$, then use synthetic division to find $f(4)$.

22. Solve $F = m\left(\dfrac{v_1 - v_0}{t_1 - t_0}\right)$ for v_1.

23. The product of the reciprocals of two consecutive integers is $\dfrac{1}{12}$. Find the integers.

24. Leontine bicycles 33 km/h with no wind. Against the wind she bikes 75 km in the same time it takes to bike 123 km with the wind. What is the speed of the wind?

25. Let $f(x) = \dfrac{2}{x} + \dfrac{1}{x-1}$. Find all a for which $f(a) = f(a+1)$.

26. Solve: $\dfrac{3}{x-10} - \dfrac{3}{x} = \dfrac{30}{x^2 - 10x}$

27. Find the x- and y-intercepts for the function given by

$$f(x) = \dfrac{\dfrac{2}{x+5} - \dfrac{3}{x-4}}{\dfrac{4}{x-3} + \dfrac{1}{x+2}}$$

28. One summer, Oscar mowed 4 lawns for every 5 lawns mowed by Luis. Together they mowed 144 lawns. How many did each mow?

CHAPTER 6   NAME_____

TEST FORM G   CLASS_____ SCORE_____ GRADE_____

ANSWERS

1. Simplify. $\dfrac{t-3}{t+3} \cdot \dfrac{3t+9}{3t^2-27}$

   a) $\dfrac{1}{t+5}$ b) $\dfrac{1}{t+3}$ c) $\dfrac{2}{t+3}$ d) $\dfrac{1}{t-3}$

1._____

2. Simplify. $\dfrac{x^3+64}{x^2-16} \div \dfrac{x^2+3x-4}{x^2-5x+4}$

   a) $\dfrac{x^2-8x+16}{x+4}$ b) $\dfrac{x^2-4x+16}{x+6}$

   c) $\dfrac{x^2-4x+16}{x+4}$ d) $\dfrac{x^2-4x+16}{2x+2}$

2._____

3. Find the LCD: $\dfrac{4x}{x^2-6x+8}$, $\dfrac{x+2}{x^2-5x+6}$

   a) $(x-6)(x-3)(x-2)$ b) $(x-4)(x-3)(x-1)$
   c) $(x-4)(x-3)(x+2)$ d) $(x-4)(x-3)(x-2)$

3._____

4. Add and simplify. $\dfrac{36x}{x+6} + \dfrac{x^3}{x+6}$

   a) $\dfrac{x^3+36x}{x+6}$ b) $\dfrac{x^3+6x}{x+6}$ c) $\dfrac{x^2+36x}{x+6}$ d) $\dfrac{x^3+36x}{x-6}$

4._____

5. Subtract and simplify. $\dfrac{5a^2}{a-b} - \dfrac{5b^2+10ab}{b-a}$

   a) $\dfrac{5(a-b)^2}{a-b}$ b) $\dfrac{5(a+b)^2}{a-b}$ c) $\dfrac{4(a+b)^2}{a-b}$ d) $\dfrac{5(a+2b)^2}{a-b}$

5._____

137

CHAPTER 6

TEST FORM G

NAME_____

ANSWERS

6. Simplify. $\dfrac{8}{y+7} - \dfrac{y}{y-1} + \dfrac{y^2+5}{y^2+6y-7}$

    a) $\dfrac{y-3}{y-1}$ b) $\dfrac{y-3}{y+7}$

    c) $\dfrac{y-3}{(y-1)(y+7)}$ d) $\dfrac{y-3}{(y+1)(y-7)}$

6._____

7. Simplify. $\dfrac{\dfrac{3}{a}+\dfrac{7}{b}}{\dfrac{10}{ab}+\dfrac{4}{a^2}}$

    a) $\dfrac{a^2(3a+7b)}{10a+4b}$ b) $\dfrac{a^2(7a+3b)}{4a+10b}$

    c) $\dfrac{a^2(3a+3b)}{4a+4b}$ d) $\dfrac{a^2(7a+3b)}{10a+4b}$

7._____

8. Simplify. $\dfrac{\dfrac{x^2-4}{x^2-16}}{\dfrac{x^2-5x-14}{x^2-8x+16}}$

    a) $\dfrac{(x+4)(x+2)}{(x-7)(x+4)}$ b) $\dfrac{(x-4)(x-2)}{(x-7)(x+4)}$

    c) $\dfrac{(x-4)(x-2)}{(x+7)(x-4)}$ d) $\dfrac{(x-4)(x-7)}{(x-2)(x+4)}$

8._____

9. Solve. $\dfrac{4}{x-5} = \dfrac{6}{8x+2}$

    a) $-\dfrac{13}{19}$ b) $\dfrac{19}{13}$ c) $-\dfrac{1}{13}$ d) $-\dfrac{19}{13}$

9._____

CHAPTER 6

TEST FORM G

NAME_____

ANSWERS

10. Solve. $\dfrac{t+22}{t^2-t-72} + \dfrac{8}{t-9} = \dfrac{4}{t+8}$

    a) $-\dfrac{12}{5}$ b) $\dfrac{122}{5}$ c) $-\dfrac{122}{5}$ d) $-\dfrac{22}{5}$

10._____

11. Edward can paint a bedroom in 2.5 hr. Jeb can paint the same room in 4 hr. How long will it take them, working together, to paint the bedroom?

    a) $\dfrac{20}{13}$ hr   b) 2 hr   c) $\dfrac{7}{3}$ hr   d) 1.5 hr

11._____

12. Divide. $(y^2 + 6y + 4) \div (y - 8)$

    a) $y + 24 + \dfrac{116}{y-8}$ b) $y + 116$

    c) $y^2 + 14 + \dfrac{116}{y-8}$ d) $y + 14 + \dfrac{116}{y-8}$

12._____

13. Divide. $(3x^4 + 6x^2 + 3x + 7) \div (x^2 + 2)$

    a) $3x^2 + \dfrac{4x+7}{x^2+2}$ b) $3x + \dfrac{3x+7}{x^2+2}$

    c) $3x^2 + \dfrac{3x+7}{x^2+2}$ d) $3x^2 - \dfrac{3x+7}{x^2+2}$

13._____

14. If $f(x) = 2x^4 - 9x^3 + 6x - 5$, then use synthetic division to find $f(5)$.

    a) 150   b) 159   c) -304 d) 10

14._____

15. Solve $K = \dfrac{v^2(m_0 + m_1 + m_2)}{2}$ for m_2.

    a) $m_2 = 2K - m_0v^2 - m_1v^2$ b) $m_2 = \dfrac{2K - m_0v^2 - m_1v^2}{v^2}$

    c) $m_2 = \dfrac{2K + m_0v^2 + m_1v^2}{v^2}$ d) $m_2 = \dfrac{2K}{v^2}$

15._____

CHAPTER 6

TEST FORM G

NAME_____

ANSWERS

16._____

16. The product of the reciprocals of two consecutive integers is $\frac{1}{56}$. Find the integers.

 a) 6, 7   b) –7, –8   c) 7, 8   d) –8, –7 or 7, 8

17._____

17. Erik bicycles 36 km/h with no wind. Against the wind he bikes 27 km in the same time it takes to bike 45 km with the wind. What is the speed of the wind?

 a) 11 km/h   b) 8 km/h   c) 9 km/h   d) 10 km/h

18._____

18. Let $f(x) = \frac{1}{x+6} + \frac{13}{x-7}$. Find all a for which $f(a) = f(a+13)$.

 a) $-\frac{120}{7}$ b) 14   c) $\frac{12}{7}$ d) $\frac{7}{2}$

19._____

19. Solve: $\frac{4}{x-9} - \frac{4}{x} = \frac{36}{x^2-9x}$

 a) 4   b) $\{x \mid x \in \mathbb{R}\}$ c) $\varnothing$ d) –4, 9

20._____

20. Find the y-intercepts for the function given by
$$f(x) = \frac{\frac{3}{x+5} - \frac{4}{x-4}}{\frac{1}{x-3} + \frac{2}{x+2}}$$

 a) $\left(0, \frac{12}{5}\right)$ b) $\left(0, \frac{2}{5}\right)$ c) $\left(0, -\frac{12}{5}\right)$ d) $\left(\frac{12}{5}, 0\right)$

21._____

21. One summer, Van mowed 4 lawns for every 6 lawns mowed by Fiona. Together they mowed 130 lawns. How many did Fiona mow?

 a) 74   b) 82   c) 52   d) 78

CHAPTER 6　　　　　　　　　　　　　　　　　NAME_____

TEST FORM H　　　　　　　　　　　　CLASS_____ SCORE_____ GRADE_____

ANSWERS

1. Simplify. $\dfrac{t-2}{t+4} \cdot \dfrac{4t+16}{2t^2-8}$

   a) $\dfrac{1}{t+2}$　　b) $3t$　　c) $\dfrac{1}{t+3}$　　d) $\dfrac{2}{t+2}$

1._____

2. Simplify. $\dfrac{x^3-64}{x^2-16} \div \dfrac{x^2-2x-8}{x^2+6x+8}$

   a) $x^2 - 4x + 16$　　　　b) $x^2 + 4x + 16$

   c) $\dfrac{x^2-4x+16}{x+4}$　　d) $\dfrac{x^2+4x+16}{x-4}$

2._____

3. Find the LCD: $\dfrac{5x}{x^2+7x+12}$, $\dfrac{x+4}{x^2+5x+6}$

   a) $(x-4)(x-3)$　　　　b) $(x-4)(x-3)(x-2)$
   c) $(x+2)(x+3)(x+4)$　　d) $(x-4(x-2)$

3._____

4. Add and simplify. $\dfrac{100x}{x+10} + \dfrac{x^3}{x+10}$

   a) $\dfrac{x^2+10x}{x+20}$　b) $\dfrac{x^3+10x}{x+10}$　c) $\dfrac{x^3+100x}{x+10}$　d) $\dfrac{x^3+110x}{x+20}$

4._____

5. Subtract and simplify. $\dfrac{4a^2}{a-b} - \dfrac{4b^2-8ab}{b-a}$

   a) $4(a+b)$　b) $\dfrac{4(a+b)^2}{a-b}$　c) $4(a-b)$　d) $\dfrac{4(a-b)^2}{a-b}$

5._____

CHAPTER 6　　　　　　　　　　　　　　　NAME_____

TEST FORM H

ANSWERS

6. Simplify. $\dfrac{9}{y+8} - \dfrac{y}{y-2} + \dfrac{y^2+4}{y^2+6y-16}$

   a) $\dfrac{y-14}{(y-2)(y+8)}$ 　　　　b) $\dfrac{y-3}{y-1}$

   c) $\dfrac{y-3}{(y-1)(y+7)}$ 　　　　d) $\dfrac{y-3}{y+1}$

6._____

7. Simplify. $\dfrac{\dfrac{2}{a}+\dfrac{6}{b}}{\dfrac{8}{ab}+\dfrac{4}{a^2}}$

   a) $\dfrac{a^2(9a+3b)}{8a+4b}$ 　　　　b) $\dfrac{a^2(3a+b)}{4a+2b}$

   c) $\dfrac{a^2(7a+3b)}{10a+4b}$ 　　　　d) $\dfrac{a^2(7a+3b)}{12a+b}$

7._____

8. Simplify. $\dfrac{\dfrac{x^2+x-2}{x^2-9}}{\dfrac{x^2-6x-16}{x^2-6x+9}}$

   a) $\dfrac{(x-4)(x-2)}{(x-7)(x+4)}$ 　　　　b) $\dfrac{(x-3)(x-1)}{(x-8)(x+3)}$

   c) $\dfrac{(x-4)(x-2)}{(x+8)(x-3)}$ 　　　　d) $\dfrac{(x-3)(x-1)}{(x-7)(x+4)}$

8._____

9. Solve. $\dfrac{3}{9x-6} = \dfrac{5}{7x+1}$

   a) $\dfrac{11}{8}$ 　　b) $-\dfrac{19}{13}$ 　　c) $-\dfrac{11}{8}$ 　　d) $\dfrac{9}{13}$

9._____

CHAPTER 6

TEST FORM H

NAME_____

ANSWERS

10. Solve. $\dfrac{t+24}{t^2-t-56} + \dfrac{7}{t-8} = \dfrac{3}{t+7}$

    a) 13   b) $\dfrac{13}{5}$ c) $-\dfrac{97}{5}$ d) $-\dfrac{122}{5}$

10._____

11. Edie can paint a bedroom in 3 hr. Zeb can paint the same room in 3.75 hr. How long will it take them, working together, to paint the bedroom?

    a) $\dfrac{2}{13}$ hr   b) $\dfrac{20}{13}$ hr   c) $\dfrac{20}{3}$ hr   d) $\dfrac{5}{3}$ hr

11._____

12. Divide. $(y^2 + 20y + 30) \div (y+5)$

    a) $y+15$ b) $y + 15 - \dfrac{116}{y+5}$

    c) $y + 15 + \dfrac{45}{y+5}$ d) $y + 15 + \dfrac{16}{y+5}$

12._____

13. Divide. $(4x^4 + 8x^2 + 2x + 9) \div (x^2 + 1)$

    a) $4x^2 + \dfrac{3x+7}{x^2+1}$ b) $4x^2 + \dfrac{2x+5}{x^2+1}$

    c) $4x^2 + 4 + \dfrac{2x+5}{x^2+1}$ d) $4x^2 + 4x - \dfrac{3x+7}{x^2+1}$

13._____

14. If $f(x) = 2x^4 - 9x^3 + 6x - 5$, then use synthetic division to find $f(6)$.

    a) 679   b) 150   c) -467 d) 650

14._____

15. Solve $\dfrac{1}{C} = \dfrac{1}{C_1} + \dfrac{1}{C_2}$ for C_1.

    a) $C_1 = \dfrac{CC_2}{C_2 - C}$ b) $C_1 = \dfrac{C - C_2}{C}$

    c) $C_1 = \dfrac{C + C_2}{CC_2}$ d) $C_1 = \dfrac{C + C_2}{C_2}$

15._____

CHAPTER 6　　　　　　　　　　　　　　　　　　NAME_____

TEST FORM H

ANSWERS

16. The product of the reciprocals of two consecutive integers is $\frac{1}{182}$. Find the integers.

16._____

    a) $-14, -13$ b) $-14, -13$ c) $-13, -12$ d) $13, 14$
                           $13, 14$ $12, 13$

17. Alyson bicycles 32 km/h with no wind. Against the wind she bikes 33km in the same time it takes to bike 63 km with the wind. What is the speed of the wind?

17._____

    a) 10 km/h   b) 9.5 km/h   c) 8.5 km/h   d) 8 km/h

18. Let $f(x) = \dfrac{1}{x+7} + \dfrac{15}{x-8}$. Find all a for which $f(a) = f(a+15)$.

18._____

    a) -17 b) $\dfrac{161}{7}$ c) $\dfrac{120}{7}$ d) $-\dfrac{161}{8}$

19. Solve: $\dfrac{5}{x-8} - \dfrac{5}{x} = \dfrac{40}{x^2 - 8x}$

    a) 40   b) $\varnothing$ c) $\{x \mid x \in \mathbb{R}\}$ d) $-5, 8$

19._____

20. Find the y-intercepts for the function given by

$$f(x) = \dfrac{\dfrac{4}{x+5} - \dfrac{1}{x-4}}{\dfrac{2}{x-3} + \dfrac{3}{x+2}}$$

20._____

    a) $\left(0, \dfrac{12}{5}\right)$ b) $\left(0, \dfrac{63}{50}\right)$ c) $\left(0, -\dfrac{12}{5}\right)$ d) $\left(0, \dfrac{6}{5}\right)$

21. One summer, Kael mowed 4 lawns for every 7 lawns mowed by Sian. Together they mowed 132 lawns. How many did Kael mow?

21._____

    a) 48   b) 78   c) 84   d) 52

CHAPTER 7

TEST FORM A

Simplify. Assume that variables can represent any real number.

1. $\sqrt{32}$

2. $\sqrt[3]{-\dfrac{27}{x^{12}}}$

3. $\sqrt{25a^2}$

4. $\sqrt{x^2 - 14x + 49}$

5. $\sqrt[5]{x^6 y}$

6. $\sqrt{\dfrac{16x^6}{25y^2}}$

7. $\sqrt[3]{5x}\,\sqrt[3]{3y}$

8. $\dfrac{\sqrt[5]{x^4 y^6}}{\sqrt[5]{x^3 y^2}}$

9. $\sqrt[4]{x^2 y^3}\,\sqrt{xy}$

10. $\dfrac{\sqrt[5]{a^4}}{\sqrt[4]{a^3}}$

11. $6\sqrt{3} - 2\sqrt{3}$

12. $\sqrt{x^5 y} + \sqrt{16xy^3}$

13. $(2 + \sqrt{x})(8 - 6\sqrt{x})$

14. Rewrite using radical notation: $(3ab)^{4/5}$

15. Rewrite using exponential notation: $\sqrt{5x^3 y}$

16. If $f(x) = \sqrt{10 - 5x}$, determine the domain of f.

17. If $f(x) = x^2$, find $f(3 + \sqrt{5})$.

18. Rationalize the denominator: $\dfrac{\sqrt{10}}{2 + \sqrt{7}}$

1._____

2._____

3._____

4._____

5._____

6._____

7._____

8._____

9._____

10._____

11._____

12._____

13._____

14._____

15._____

16._____

17._____

18._____

CHAPTER 7 NAME_____

TEST FORM A

ANSWERS

19._____

20._____

21._____

22._____

23._____

24._____

25._____

26._____

27._____

28._____

29._____

30._____

Solve.

19. $x = \sqrt{2x-1} + 2$

20. $\sqrt{x} = \sqrt{x+2} + 3$

Solve. Give exact answers and, where appropriate, approximation to three decimal points.

21. One leg of an isosceles right triangle is 5 cm long. Find the lengths of the other sides.

22. A pedestrian crosses diagonally from one corner of a 2 km by 5 km city park to the far corner. How far does the pedestrian walk?

23. Express in terms of i and simplify: $\sqrt{-162}$

24. Subtract: $(5 + 2i) - (-6 + 7i)$

25. Multiply: $\sqrt{-9} \sqrt{-64}$

26. Multiply. Write the answer in the form $a + bi$. $(2 - i)^2$

27. Divide and simplify to the form $a + bi$. $\dfrac{-6 + i}{3 - 4i}$

28. Simplify: i^{19}

29. Solve: $\sqrt{2x-7} + \sqrt{6x+1} = \sqrt{10x-4}$

30. Simplify: $\dfrac{1 - 2i}{2i(1 + 2i)^{-1}}$

146

CHAPTER 7
TEST FORM B

NAME_____

CLASS_____ SCORE_____ GRADE_____

Simplify. Assume that variables can represent any real number.

1. $\sqrt{75}$

2. $\sqrt[3]{-\dfrac{1}{x^9}}$

3. $\sqrt{36a^2}$

4. $\sqrt{x^2 - 16x + 64}$

5. $\sqrt[5]{x^7 y^2}$

6. $\sqrt{\dfrac{81x^2}{25y^4}}$

7. $\sqrt[3]{3x^2} \; \sqrt[3]{5y}$

8. $\dfrac{\sqrt[5]{x^5 y^4}}{\sqrt[5]{x^3 y^2}}$

9. $\sqrt[4]{xy^2} \sqrt{xy}$

10. $\dfrac{\sqrt[5]{a^3}}{\sqrt[4]{a^2}}$

11. $7\sqrt{2} - 3\sqrt{2}$

12. $\sqrt{xy^2} + \sqrt{4x^3}$

13. $(3 + \sqrt{x})(7 - 5\sqrt{x})$

14. Rewrite using radical notation: $(4a^3 b^2)^{5/6}$

15. Rewrite using exponential notation: $\sqrt{3xy^2}$

16. If $f(x) = \sqrt{9 - 3x}$, determine the domain of f.

17. If $f(x) = x^2$, find $f(4 + \sqrt{6})$.

18. Rationalize the denominator: $\dfrac{\sqrt{7}}{3 - \sqrt{6}}$

1._____
2._____
3._____
4._____
5._____
6._____
7._____
8._____
9._____
10._____
11._____
12._____
13._____
14._____
15._____
16._____
17._____
18._____

CHAPTER 7
TEST FORM B

NAME_____

ANSWERS

19. _____

20. _____

21. _____

22. _____

23. _____

24. _____

25. _____

26. _____

27. _____

28. _____

29. _____

30. _____

Solve.

19. $x = \sqrt{15x - 9} - 3$

20. $\sqrt{x} = \sqrt{x - 2} - 4$

Solve. Give exact answers and, where appropriate, approximation to three decimal points.

21. One leg of an isosceles right triangle is 8 cm long. Find the lengths of the other sides.

22. A shopper crosses diagonally from one corner of a 750 m by 1200 m parking lot to the far corner. How far does the shopper walk?

23. Express in terms of i and simplify: $\sqrt{-128}$

24. Subtract: $(6 + i) - (-4 + 8i)$

25. Multiply: $\sqrt{-16} \sqrt{-9}$

26. Multiply. Write the answer in the form $a + bi$. $(3 + i)^2$

27. Divide and simplify to the form $a + bi$. $\dfrac{5 - 2i}{2 + 5i}$

28. Simplify: i^{25}

29. Solve: $\sqrt{5x - 6} + \sqrt{10x + 16} = \sqrt{34x - 4}$

30. Simplify: $\dfrac{2 - 3i}{3i(2 + 3i)^{-1}}$

CHAPTER 7 NAME_____

TEST FORM C CLASS_____ SCORE_____ GRADE_____

Simplify. Assume that variables can represent any real number.

1. $\sqrt{180}$

2. $\sqrt[3]{-\dfrac{1000}{x^6}}$

3. $\sqrt{49a^2}$

4. $\sqrt{x^2 - 18x + 81}$

5. $\sqrt[5]{x^8 y^3}$

6. $\sqrt{\dfrac{25x^6}{49y^2}}$

7. $\sqrt[3]{5x^2}\,\sqrt[3]{3y^2}$

8. $\dfrac{\sqrt[5]{x^6 y^3}}{\sqrt[5]{x^3 y^2}}$

9. $\sqrt[4]{x^3 y}\,\sqrt{xy}$

10. $\dfrac{\sqrt[5]{a^6}}{\sqrt[4]{a}}$

11. $8\sqrt{5} - 4\sqrt{5}$

12. $\sqrt{x^2 y} + \sqrt{25 x^2 y^3}$

13. $(4 + \sqrt{x})(6 - 4\sqrt{x})$

14. Rewrite using radical notation: $(5a^2 b)^{4/5}$

15. Rewrite using exponential notation: $\sqrt{2x^2 y}$

16. If $f(x) = \sqrt{15 - 3x}$, determine the domain of f.

17. If $f(x) = x^2$, find $f(5 + \sqrt{6})$.

18. Rationalize the denominator: $\dfrac{\sqrt{6}}{5 - \sqrt{5}}$

1._____

2._____

3._____

4._____

5._____

6._____

7._____

8._____

9._____

10._____

11._____

12._____

13._____

14._____

15._____

16._____

17._____

18._____

CHAPTER 7 NAME_____

TEST FORM C

ANSWERS

Solve.

19. $x = \sqrt{3x - 12} + 4$

19._____

20. $\sqrt{x} = \sqrt{x - 8} + 2$

20._____

Solve. Give exact answers and, where appropriate, approximation to three decimal points.

21._____

21. One leg of an isosceles right triangle is 11 cm long. Find the lengths of the other sides.

22._____

22. A housefly crosses diagonally from one corner of a 0.75 m by 0.6 m television screen to the far corner. How far does the fly walk?

23._____

24._____

23. Express in terms of i and simplify: $\sqrt{-98}$

25._____

24. Subtract: $(7 + 3i) - (-2 + 9i)$

26._____

25. Multiply: $\sqrt{-4} \sqrt{-81}$

26. Multiply. Write the answer in the form $a + bi$. $(3 + 2i)^2$

27._____

27. Divide and simplify to the form $a + bi$. $\dfrac{3 - 9i}{-1 + 4i}$

28._____

28. Simplify: i^{32}

29._____

29. Solve: $\sqrt{6x - 9} + \sqrt{5x + 1} = \sqrt{12x + 13}$

30._____

30. Simplify: $\dfrac{3 - 4i}{4i(3 + 4i)^{-1}}$

150

CHAPTER 7   NAME_____

TEST FORM D   CLASS_____ SCORE_____ GRADE_____

Simplify. Assume that variables can represent any real number.

1. $\sqrt{24}$

2. $\sqrt[3]{-\dfrac{8}{x^3}}$

3. $\sqrt{64a^2}$

4. $\sqrt{x^2 - 20x + 100}$

5. $\sqrt[5]{x^9 y^4}$

6. $\sqrt{\dfrac{4x^8}{36y^2}}$

7. $\sqrt[3]{5x}\, \sqrt[3]{3y^2}$

8. $\dfrac{\sqrt[5]{x^7 y^2}}{\sqrt[5]{x^3 y^2}}$

9. $\sqrt[4]{x^2 y}\, \sqrt{xy}$

10. $\dfrac{\sqrt[5]{a^6}}{\sqrt[4]{a^2}}$

11. $9\sqrt{6} - 5\sqrt{6}$

12. $\sqrt{xy^5} + \sqrt{9x^3 y}$

13. $(5 + \sqrt{x})(5 - 3\sqrt{x})$

14. Rewrite using radical notation: $(2ab^2)^{5/6}$

15. Rewrite using exponential notation: $\sqrt[3]{7x^2 y}$

16. If $f(x) = \sqrt{20 + 5x}$, determine the domain of f.

17. If $f(x) = x^2$, find $f(6 - \sqrt{5})$.

18. Rationalize the denominator: $\dfrac{\sqrt{5}}{6 + \sqrt{3}}$

1._____

2._____

3._____

4._____

5._____

6._____

7._____

8._____

9._____

10._____

11._____

12._____

13._____

14._____

15._____

16._____

17._____

18._____

CHAPTER 7

TEST FORM D

NAME_____

ANSWERS

19._____

20._____

21._____

22._____

23._____

24._____

25._____

26._____

27._____

28._____

29._____

30._____

Solve.

19. $x = \sqrt{25x - 31} - 3$

20. $\sqrt{x} = \sqrt{x+2} - 1$

Solve. Give exact answers and, where appropriate, approximation to three decimal points.

21. One leg of an isosceles right triangle is 14 cm long. Find the lengths of the other sides.

22. A disoriented driver crosses diagonally from one corner of a 100 m by 300 m lawn to the far corner. How far does the wayward driver drive across the lawn?

23. Express in terms of i and simplify: $\sqrt{-108}$

24. Subtract: $(8 - 2i) - (4 - 7i)$

25. Multiply: $\sqrt{-25} \sqrt{-36}$

26. Multiply. Write the answer in the form $a + bi$. $(4 - 3i)^2$

27. Divide and simplify to the form $a + bi$. $\dfrac{-4 + 2i}{7 - i}$

28. Simplify: i^7

29. Solve: $\sqrt{10x - 4} + \sqrt{16x + 17} = \sqrt{75x - 29}$

30. Simplify: $\dfrac{4 + 3i}{3i(4 - 3i)^{-1}}$

CHAPTER 7   NAME_____

TEST FORM E   CLASS_____ SCORE_____ GRADE_____

Simplify. Assume that variables can represent any real number.

1. $\sqrt{63}$

2. $\sqrt[3]{-\dfrac{216}{x^6}}$

3. $\sqrt{81a^2}$

4. $\sqrt{x^2 - 12x + 36}$

5. $\sqrt[5]{x^{10} y^5}$

6. $\sqrt{\dfrac{9x^4}{81 y^{10}}}$

7. $\sqrt[3]{3x} \, \sqrt[3]{5y^2}$

8. $\dfrac{\sqrt[5]{x^7 y^2}}{\sqrt[5]{x^4 y}}$

9. $\sqrt[4]{x^2 y} \, \sqrt[3]{x^2 y}$

10. $\dfrac{\sqrt[5]{a^4}}{\sqrt[4]{a^2}}$

11. $9\sqrt{3} - 6\sqrt{3}$

12. $\sqrt{x^2 y^3} + \sqrt{4 y^3}$

13. $(6 + \sqrt{x})(4 - 2\sqrt{x})$

14. Rewrite using radical notation: $(3a^3 b)^{4/5}$

15. Rewrite using exponential notation: $\sqrt[3]{2x\,y}$

16. If $f(x) = \sqrt{16 + 8x}$, determine the domain of f.

17. If $f(x) = x^2$, find $f(5 + \sqrt{7})$.

18. Rationalize the denominator: $\dfrac{\sqrt{3}}{7 - \sqrt{2}}$

1._____

2._____

3._____

4._____

5._____

6._____

7._____

8._____

9._____

10._____

11._____

12._____

13._____

14._____

15._____

16._____

17._____

18._____

CHAPTER 7

TEST FORM E

NAME_____

ANSWERS

Solve.

19. _____

19. $x = \sqrt{x-3} + 5$

20. _____

20. $\sqrt{x} = \sqrt{x+13} + 7$

21. _____

Solve. Give exact answers and, where appropriate, approximation to three decimal points.

21. One leg of an isosceles right triangle is 17 cm long. Find the lengths of the other sides.

22. _____

22. A shopper crosses diagonally from one corner of a 0.7 km by 0.6 km parking lot to the far corner. How far does the shopper walk?

23. _____

24. _____

23. Express in terms of i and simplify: $\sqrt{-75}$

25. _____

24. Subtract: $(9-i)-(6-8i)$

26. _____

25. Multiply: $\sqrt{-64} \sqrt{-16}$

26. Multiply. Write the answer in the form $a + bi$. $(2+3i)^2$

27. _____

27. Divide and simplify to the form $a + bi$. $\dfrac{9+5i}{-2+i}$

28. _____

28. Simplify: i^{15}

29. _____

29. Solve: $\sqrt{2x-13} + \sqrt{3x+4} = \sqrt{5x+1}$

30. _____

30. Simplify: $\dfrac{3+2i}{2i(3-2i)^{-1}}$

154

CHAPTER 7   NAME_____

TEST FORM F   CLASS_____ SCORE_____ GRADE_____

Simplify. Assume that variables can represent any real number.

1. $\sqrt{810}$

2. $\sqrt[3]{-\dfrac{64}{x^{12}}}$

3. $\sqrt{121a^2}$

4. $\sqrt{x^2 - 10x + 25}$

5. $\sqrt[5]{x^5 y^6}$

6. $\sqrt{\dfrac{9x^4}{49y^8}}$

7. $\sqrt[3]{3x^2} \, \sqrt[3]{5y^2}$

8. $\dfrac{\sqrt[5]{x^6 y^4}}{\sqrt[5]{x^5 y}}$

9. $\sqrt[4]{x^3 y} \, \sqrt[3]{xy^2}$

10. $\dfrac{\sqrt[5]{a^4}}{\sqrt[4]{a}}$

11. $8\sqrt{2} - 7\sqrt{2}$

12. $\sqrt{x} + \sqrt{36 x^3 y^2}$

13. $(5 + \sqrt{x})(3 - \sqrt{x})$

14. Rewrite using radical notation: $(4a^2 b^2)^{5/6}$

15. Rewrite using exponential notation: $\sqrt[3]{5x^2 y^2}$

16. If $f(x) = \sqrt{6 + 2x}$, determine the domain of f.

17. If $f(x) = x^2$, find $f(4 - \sqrt{3})$.

18. Rationalize the denominator: $\dfrac{\sqrt{2}}{10 + \sqrt{3}}$

1._____

2._____

3._____

4._____

5._____

6._____

7._____

8._____

9._____

10._____

11._____

12._____

13._____

14._____

15._____

16._____

17._____

18._____

CHAPTER 7

TEST FORM F

NAME_____

ANSWERS

19._____

20._____

21._____

22._____

23._____

24._____

25._____

26._____

27._____

28._____

29._____

30._____

Solve.

19. $x = \sqrt{18x - 8} - 4$

20. $\sqrt{x} = \sqrt{x - 5} + 2$

Solve. Give exact answers and, where appropriate, approximation to three decimal points.

21. One leg of an isosceles right triangle is 20 cm long. Find the lengths of the other sides.

22. A pedestrian crosses diagonally from one corner of a 5 km by 7 km parking lot to the far corner. How far does the shopper walk?

23. Express in terms of i and simplify: $\sqrt{-48}$

24. Subtract: $(10 - 3i) - (2 - 9i)$

25. Multiply: $\sqrt{-81} \sqrt{-25}$

26. Multiply. Write the answer in the form $a + bi$. $(4 + 2i)^2$

27. Divide and simplify to the form $a + bi$. $\dfrac{-10 - 4i}{1 + 3i}$

28. Simplify: i^{28}

29. Solve: $\sqrt{3x - 11} + \sqrt{8x - 4} = \sqrt{12x + 4}$

30. Simplify: $\dfrac{2 + i}{i(2 - i)^{-1}}$

156

CHAPTER 7

TEST FORM G

NAME_____

CLASS_____ SCORE_____ GRADE_____

Assume that variables can represent any real number.

ANSWERS

1. Simplify. $\sqrt{54}$
   a) $3\sqrt{6}$　　b) $6\sqrt{3}$　　c) $7\sqrt{3}$　　d) $2\sqrt{6}$

1._____

2. Simplify. $\sqrt[3]{-\dfrac{125}{x^9}}$
   a) $-\dfrac{5}{x^2}$　　b) $\dfrac{5}{x^3}$　　c) $-\dfrac{5}{x^3}$　　d) $-\dfrac{3}{x^2}$

2._____

3. Simplify. $\sqrt{16a^2}$
   a) $4a^2$　　b) $3a$　　c) $4a$　　d) $2a^2$

3._____

4. Simplify. $\sqrt{x^2 - 6x + 9}$
   a) $x - 2$　　b) $x + 3$　　c) $x + 2$　　d) $x - 3$

4._____

5. Simplify. $\sqrt[5]{x^4 y^7}$
   a) $y\sqrt[5]{x^3 y^3}$　　b) $\sqrt[5]{x^4 y^2}$　　c) $\sqrt[5]{x^3 y^3}$　　d) $y\sqrt[5]{x^4 y^2}$

5._____

6. Simplify. $\sqrt{\dfrac{36x^2}{4y^6}}$
   a) $\dfrac{3x^2}{y^3}$　　b) $\dfrac{x^2}{2y^2}$　　c) $\dfrac{3x}{y^2}$　　d) $\dfrac{3x}{y^3}$

6._____

7. Simplify. $\sqrt[3]{3x} \sqrt[3]{5y}$
   a) $\sqrt[3]{15x y^2}$　　b) $\sqrt[3]{15x y}$　　c) $\sqrt[3]{5x y}$　　d) $\sqrt[3]{15x^2 y}$

7._____

CHAPTER 7

TEST FORM G

NAME_____

ANSWERS

8. Simplify. $\dfrac{\sqrt[5]{x^5 y^6}}{\sqrt[5]{x^2 y^2}}$

8._____

a) $\sqrt{x^3 y^4}$ b) $\sqrt[5]{x^4 y^3}$ c) $\sqrt[5]{x^3 y^4}$ d) $\sqrt[5]{x^2 y^4}$

9. Simplify. $\sqrt[4]{x y^2} \sqrt[3]{x^2 y^2}$

9._____

a) $y \sqrt[12]{x^{11} y^2}$ b) $y \sqrt[12]{x^{10} y}$ c) $\sqrt[12]{x^{11} y^2}$ d) $\sqrt[12]{x^{10} y}$

10. Simplify. $\dfrac{\sqrt[5]{a^3}}{\sqrt[4]{a}}$

10._____

a) $\sqrt[20]{a^5}$ b) $\sqrt[20]{a^7}$ c) $\sqrt[9]{a^7}$ d) $\sqrt{a^5}$

11. Simplify. $7\sqrt{5} - 8\sqrt{5}$

11._____

a) $-\sqrt{5}$ b) $\sqrt{5}$ c) $-15\sqrt{5}$ d) $15\sqrt{5}$

12. Simplify. $\sqrt{x y^3} + \sqrt{4 x^5 y}$

a) $\sqrt{xy}(y + x^2)$ b) $\sqrt{x}(y + 2x^2)$

c) $\sqrt{xy}(y + 2x^2)$ d) $\sqrt{xy}(y^2 + 2x^2)$

12._____

13. Simplify. $(4 + \sqrt{x})(2 - 3\sqrt{x})$

a) $8 - \sqrt{10}\, x - 3x$ b) $3 - \sqrt{11}\, x - 4x$

c) $3 - 11\sqrt{x} - 4x$ d) $8 - 10\sqrt{x} - 3x$

13._____

CHAPTER 7

TEST FORM G

NAME_____

ANSWERS

14. Express using radical notation: $(5ab)^{4/3}$

a) $5ab\sqrt[3]{5a^2b}$ b) $5ab\sqrt[3]{5ab}$

c) $5\sqrt[3]{5ab}$ d) $ab\sqrt[3]{5ab}$

14._____

15. Express using exponential notation: $\sqrt{7xy^3}$

a) $(3xy^3)^{1/3}$ b) $(3xy^2)^{1/3}$

c) $(7xy^3)^{1/2}$ d) $(7xy^2)^{1/2}$

15._____

16. If $f(x) = \sqrt{8-2x}$, determine the domain of f.

a) $(-\infty, 4]$ b) $[-4, \infty)$ c) $(-\infty, 3]$ d) $(-\infty, -4]$

16._____

17. If $f(x) = x^2$, find $f(3-\sqrt{7})$.

a) $7 - 6\sqrt{3}$ b) $5 - 6\sqrt{7}$ c) $5 + 6\sqrt{7}$ d) $7 + 4\sqrt{3}$

17._____

18. Rationalize the denominator: $\dfrac{\sqrt{3}}{7+\sqrt{5}}$

a) $\dfrac{7\sqrt{5}+\sqrt{15}}{19}$ b) $\dfrac{7\sqrt{3}+\sqrt{15}}{42}$

c) $\dfrac{5\sqrt{5}-\sqrt{30}}{19}$ d) $\dfrac{5\sqrt{3}+\sqrt{30}}{42}$

18._____

19. Solve. $x = \sqrt{5x-21} + 3$

a) 5 b) 3 c) 4 d) 2

19._____

CHAPTER 7

TEST FORM G

NAME_____

ANSWERS

20. A housefly crosses diagonally from one corner of a 0.8 m by 0.5 m television screen to the far corner. How far does the fly walk? Give an approximate answer to three decimal places.

    a) 1.012 m   b) 2.876 m   c) 0.943 m   d) 0.765 m

20._____

21. Express in terms of i and simplify: $\sqrt{-45}$

    a) $5i\sqrt{5}$ b) $-3i\sqrt{5}$ c) $i\sqrt{5}$ d) $3i\sqrt{5}$

21._____

22. Subtract: $(4 + 3i) - (-5 + 4i)$

    a) $-3 + i$ b) $9 + i$ c) $3 - i$ d) $9 - i$

22._____

23. Multiply: $\sqrt{-36}\ \sqrt{-100}$

    a) -40 b) -60 c) -30 d) -20

23._____

24. Multiply. Write the answer in the form $a + bi$. $(1 - 4i)^2$

    a) $-21 - 20i$ b) $-15 + 8i$ c) $-15 - 8i$ d) $-21 + 20i$

24._____

25. Divide and simplify to the form $a + bi$: $\dfrac{7 - 3i}{5 + 4i}$

    a) $\dfrac{23}{41} - \dfrac{43}{41}i$ b) $\dfrac{23}{13} + \dfrac{43}{13}i$ c) $-\dfrac{19}{13} - \dfrac{22}{13}i$ d) $\dfrac{19}{41} + \dfrac{22}{41}i$

25._____

26. Solve: $\sqrt{4x - 7} + \sqrt{2x + 8} = \sqrt{9x + 13}$

    a) 3   b) -4 c) 5   d) 4

26._____

27. Simplify: $\dfrac{5 - 3i}{3i(5 + 3i)^{-1}}$

    a) $-\dfrac{10}{3}i$ b) $-\dfrac{34}{3}i$ c) $\dfrac{34}{3}i$ d) $-10i$

27._____

28._____

CHAPTER 7

TEST FORM H

Assume that variables can represent any real number.

1. Simplify. $\sqrt{147}$

   a) $3\sqrt{6}$ b) $6\sqrt{3}$ c) $7\sqrt{3}$ d) $2\sqrt{6}$

2. Simplify. $\sqrt[3]{-\dfrac{27}{x^6}}$

   a) $-\dfrac{5}{x^2}$ b) $\dfrac{5}{x^3}$ c) $-\dfrac{5}{x^3}$ d) $-\dfrac{3}{x^2}$

3. Simplify. $\sqrt{9a^2}$

   a) $4a^2$ b) $3a$ c) $4a$ d) $2a^2$

4. Simplify. $\sqrt{x^2-4x+4}$

   a) $x-2$ b) $x+3$ c) $x+2$ d) $x-3$

5. Simplify. $\sqrt[5]{x^3 y^8}$

   a) $y\sqrt[5]{x^3 y^3}$ b) $\sqrt[5]{x^4 y^2}$ c) $\sqrt[5]{x^3 y^3}$ d) $y\sqrt[5]{x^4 y^2}$

6. Simplify. $\sqrt{\dfrac{25x^4}{100y^4}}$

   a) $\dfrac{3x^2}{y^3}$ b) $\dfrac{x^2}{2y^2}$ c) $\dfrac{3x}{y^2}$ d) $\dfrac{3x}{y^3}$

7. Simplify. $\sqrt[3]{5x^2}\,\sqrt[3]{3y}$

   a) $\sqrt[3]{15xy^2}$ b) $\sqrt[3]{15xy}$ c) $\sqrt[3]{5xy}$ d) $\sqrt[3]{15x^2 y}$

ANSWERS

1. _____

2. _____

3. _____

4. _____

5. _____

6. _____

7. _____

CHAPTER 7

TEST FORM H

NAME_____

ANSWERS

8. Simplify. $\dfrac{\sqrt[5]{x^8 y^8}}{\sqrt[5]{x^4 y^6}}$

a) $\sqrt{x^3 y^4}$ b) $\sqrt[5]{x^4 y^3}$ c) $\sqrt[5]{x^4 y^2}$ d) $\sqrt[5]{x^2 y^4}$

8._____

9. Simplify. $\sqrt[4]{x^2 y^3} \sqrt[3]{xy}$

a) $y\sqrt[12]{x^{11} y^2}$ b) $y\sqrt[12]{x^{10} y}$ c) $\sqrt[12]{x^{11} y^2}$ d) $\sqrt[12]{x^{10} y}$

9._____

10. Simplify. $\dfrac{\sqrt[5]{a^6}}{\sqrt[4]{a^3}}$

a) $\sqrt[20]{a^5}$ b) $\sqrt[20]{a^9}$ c) $\sqrt[9]{a^7}$ d) $\sqrt{a^5}$

10._____

11. Simplify. $6\sqrt{6} - 9\sqrt{6}$

a) $-3\sqrt{6}$ b) $3\sqrt{6}$ c) $-15\sqrt{6}$ d) $15\sqrt{6}$

11._____

12. Simplify. $\sqrt{x^5 y^2} + \sqrt{9xy^2}$

a) $\sqrt{xy}\,(y + x^2)$ b) $\sqrt{x}\,(y + 2x^2)$
c) $\sqrt{x}\,(x^2 y + 3y)$ d) $\sqrt{xy}\,(y^2 + 2x^2)$

12._____

13. Simplify. $(3 + \sqrt{x})(1 - 4\sqrt{x})$

a) $8 - \sqrt{10}\,x - 3x$ b) $3 - \sqrt{11}\,x - 4x$
c) $3 - 11\sqrt{x} - 4x$ d) $8 - 10\sqrt{x} - 3x$

13._____

CHAPTER 7

TEST FORM H

ANSWERS

14. Express using radical notation: $(2a^3b^2)^{4/5}$

    a) $5ab\sqrt[3]{5a^2b}$ b) $a^2b\sqrt[5]{16a^2b^3}$

    c) $5\sqrt[5]{5ab}$ d) $ab\sqrt[3]{5ab}$

    14._____

15. Express using exponential notation: $\sqrt[3]{3xy^2}$

    a) $(3xy^3)^{1/3}$ b) $(3xy^2)^{1/3}$

    c) $(7xy^3)^{1/2}$ d) $(7xy^2)^{1/2}$

    15._____

16. If $f(x) = \sqrt{18+6x}$, determine the domain of f.

    a) $[-3, \infty)$ b) $[-4, \infty)$ c) $(-\infty, 3]$ d) $(-\infty, -4]$

    16._____

17. If $f(x) = x^2$, find $f(2+\sqrt{3})$.

    a) $7-6\sqrt{3}$ b) $5-6\sqrt{7}$ c) $5+6\sqrt{7}$ d) $7+4\sqrt{3}$

    17._____

18. Rationalize the denominator: $\dfrac{\sqrt{5}}{5+\sqrt{6}}$

    a) $\dfrac{7\sqrt{5}+\sqrt{15}}{19}$ b) $\dfrac{7\sqrt{3}+\sqrt{15}}{42}$

    c) $\dfrac{5\sqrt{5}-\sqrt{30}}{19}$ d) $\dfrac{5\sqrt{3}+\sqrt{30}}{42}$

    18._____

19. Solve. $x = \sqrt{10x-4} - 2$

    a) 5   b) 3   c) 4   d) 2

    19._____

CHAPTER 7

TEST FORM H

NAME_____

ANSWERS

20._____

21._____

22._____

23._____

24._____

25._____

26._____

27._____

28._____

20. A disoriented driver crosses diagonally from one corner of a 150 m by 275 m lawn to the far corner. How far does the wayward driver drive across the lawn? Give an approximate answer to three decimal places.

a) 269.981 m   b) 313.249 m   c) 297.121 m   d) 308.750 m

21. Express in terms of i and simplify: $\sqrt{-20}$

a) $4i\sqrt{5}$ b) $-2i\sqrt{5}$ c) $i\sqrt{5}$ d) $2i\sqrt{5}$

22. Subtract: $(3 - 2i) - (6 - 3i)$

a) $-3 + i$ b) $9 + i$ c) $3 - i$ d) $9 - i$

23. Multiply: $\sqrt{-100} \sqrt{-4}$

a) -40 b) -60 c) -30 d) -20

24. Multiply. Write the answer in the form $a + bi$. $(2 + 5i)^2$

a) $-21 - 20i$ b) $-15 + 8i$ c) $-15 - 8i$ d) $-21 + 20i$

25. Divide and simplify to the form $a + bi$: $\dfrac{-8 + i}{2 - 3i}$

a) $\dfrac{23}{41} - \dfrac{43}{41}i$ b) $\dfrac{23}{13} + \dfrac{43}{13}i$ c) $-\dfrac{19}{13} - \dfrac{22}{13}i$ d) $\dfrac{19}{41} + \dfrac{22}{41}i$

26. Solve: $\sqrt{7x - 5} + \sqrt{15x + 4} = \sqrt{30x + 31}$

a) 3   b) -4 c) 5   d) 4

27. Simplify: $\dfrac{4 + 2i}{2i(4 - 2i)^{-1}}$

a) $-\dfrac{10}{3}i$ b) $-\dfrac{34}{3}i$ c) $\dfrac{34}{3}i$ d) $-10i$

CHAPTER 8　　　　　　　　　　　　　　　　　　　NAME_____

TEST FORM A　　　　　　　　　　　CLASS_____ SCORE_____ GRADE_____

Solve.

ANSWERS

1. $2x^2 - 17 = 0$
2. $8x(x-4) - 7x(x+2) = -360$
3. $x^2 + x + 2 = 0$
4. $x + 3 = x^2$
5. $x^{-2} - x^{-1} = \dfrac{2}{3}$
6. $x^2 + x = 3$. Use a calculator to approximate the solutions with rational numbers.
7. Let $f(x) = 6x^2 + x - 15$. Find x such that $f(x) = 0$.

Complete the square. Then write the perfect-square trinomial in factored form.

8. $x^2 + 12x$
9. $x^2 - \dfrac{10}{9}x$

10. Solve by completing the square. Show your work.

$x^2 + 12x + 24 = 0$

Solve.

11. The Deerfield River flows at a rate of 2 km/h. In order for a boat to travel 38.4 km upriver and then return in a total of 8 hr, how fast must the boat travel in still water?

12. Elsworth and Priscilla can eat an entire chocolate cream pie in 42 minutes. Eating alone, it takes Priscilla 80 minutes longer than Elsworth to eat the same type of pie. How long would it take for Elsworth to eat the pie by himself?

13. Determine the type of number that the solutions of $x^2 + 12x + 27 = 0$.

14. Write a quadratic equation having solutions -5 and $\dfrac{1}{4}$.

15. Find all x-intercepts of the graph of $(x^2 + 5x)^2 - 8(x^2 + 5x) - 84$.

16. The surface area of a balloon varies directly as the square of its radius. The area is 314 cm² when the radius is 5 cm. What is the area when the radius is 6 cm?

1._____

2._____

3._____

4._____

5._____

6._____

7._____

8._____

9._____

10._____

11._____

12._____

13._____

14._____

15._____

16._____

CHAPTER 8

TEST FORM A

NAME_____

ANSWERS

17.a)-c) See graph.

d)_____

18._____

19._____

20._____

21._____

22.a)_____

b) See graph.

23._____

24._____

25._____

26._____

27._____

17. a) Graph: $f(x) = 2(x-1)^2 + 2$
    b) Label the vertex.
    c) Draw the axis of symmetry.
    d) Find the maximum or the minimum function value.

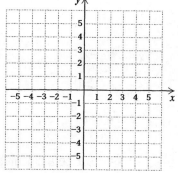

18. Find the x- and y-intercepts of $x^2 - x - 90$.

19. Rosie's Palm Pirates, a manufacturer of electronic organizers, estimates that when x hundred Palm Pirates are made, the average cost per unit is given by $C(x) = 0.3x^2 - 1.8x + 4.24$, where C is in hundreds of dollars. What is the minimum cost per unit, and how many units should be made to achieve that minimum?

20. Solve $A = \pi r^2$ for r.

21. Find the quadratic function that fits the data points $(1, 0)$, $(4, 0)$, and $(7, 3)$.

22. For the function $f(x) = 2x^2 + x + 1$
    a) find the vertex and the axis of symmetry;
    b) graph the function.

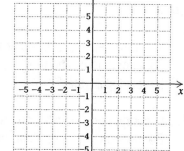

Solve

23. $x^2 + 2x \le 3$

24. $x - \dfrac{2}{x} > 0$

25. One solution of $kx^2 + 5x - k = 0$ is -2. Find the other solution.

26. Find a fourth-degree polynomial equation, with integer coefficients, for which $2 - \sqrt{2}$ and $2 - i$ are solutions.

27. Find a polynomial equation, with integer coefficients, for which 7 is a repeated root and $\sqrt{2}$ and $\sqrt{5}$ are solutions.

CHAPTER 8

TEST FORM B

Solve.

1. $3x^2 - 10 = 0$
2. $7x(x-3) - 6x(x+5) = -50$
3. $x^2 + 2x + 1 = 0$
4. $2x + 7 = x^2$
5. $x^{-2} - x^{-1} = \dfrac{3}{2}$
6. $x^2 + 3x = 7$. Use a calculator to approximate the solutions with rational numbers.
7. Let $f(x) = 10x^2 - 17x - 20$. Find x such that $f(x) = 0$.

Complete the square. Then write the perfect-square trinomial in factored form.

8. $x^2 - 8x$
9. $x^2 + \dfrac{8}{7}x$

10. Solve by completing the square. Show your work.
    $x^2 + 10x + 20 = 0$

Solve.

11. The Greenfield River flows at a rate of 3 km/h. In order for a boat to travel 36.4 km upriver and then return in a total of 8 hr, how fast must the boat travel in still water?

12. Jean Paul and Maurice can bake a dozen loaves of bread in 3 hours. Working alone, it takes Maurice 2.5 hours longer than Jean Paul to bake this much bread. How long would it take for Jean Paul to bake the bread by himself?

13. Determine the type of number that the solutions of $x^2 + 11x - 10 = 0$.

14. Write a quadratic equation having solutions -4 and $\dfrac{2}{3}$.

15. Find all x-intercepts of the graph of $(x^2 + 2x)^2 - 2(x^2 + 2x) - 3$.

16. The surface area of a balloon varies directly as the square of its radius. The area is 314 cm² when the radius is 5 cm. What is the area when the radius is 7 cm?

ANSWERS

1. _____
2. _____
3. _____
4. _____
5. _____
6. _____
7. _____
8. _____
9. _____
10. _____
11. _____
12. _____
13. _____
14. _____
15. _____
16. _____

CHAPTER 8

NAME_____

TEST FORM B

ANSWERS

17.a)-c) _See graph._

    d) _____

18. _____

19. _____

20. _____

21. _____

22.a) _____

  b) _See graph._

23. _____

24. _____

25. _____

26. _____

27. _____

---

17. a) Graph: $f(x) = 2(x-2)^2 - 3$
    b) Label the vertex.
    c) Draw the axis of symmetry.
    d) Find the maximum or the minimum function value.

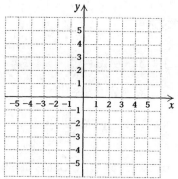

18. Find the x- and y-intercepts of $x^2 + x - 30$.

19. Rosie's Palm Pirates, a manufacturer of electronic organizers, estimates that when x hundred Palm Pirates are made, the average cost per unit is given by $C(x) = 0.4x^2 - 2.64x + 5.816$, where C is in hundreds of dollars. What is the minimum cost per unit, and how many units should be made to achieve that minimum?

20. Solve $K = \frac{1}{2}mv^2$ for v.

21. Find the quadratic function that fits the data points $(2, 0)$, $(3, 0)$, and $(4, 1)$.

22. For the function $f(x) = 2x^2 + x + 2$
    a) find the vertex and the axis of symmetry;
    b) graph the function.

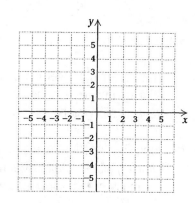

Solve

23. $x^2 + 3x \leq -2$

24. $x + \frac{3}{x} > 0$

---

25. One solution of $kx^2 - 5x + k = 0$ is 3. Find the other solution.

26. Find a fourth-degree polynomial equation, with integer coefficients, for which $2 - \sqrt{3}$ and $2 - i$ are solutions.

27. Find a polynomial equation, with integer coefficients, for which 2 is a repeated root and $\sqrt{2}$ and $\sqrt{5}$ are solutions.

CHAPTER 8

TEST FORM C

NAME_____

CLASS_____ SCORE_____ GRADE_____

Solve.

1. $4x^2 - 15 = 0$
2. $6x(x-2) - 5x(x+3) = -140$
3. $x^2 + 2x + 2 = 0$
4. $3x + 5 = x^2$
5. $x^{-2} - x^{-1} = \dfrac{4}{3}$
6. $x^2 + 5x = 5$. Use a calculator to approximate the solutions with rational numbers.
7. Let $f(x) = 12x^2 - 22x - 14$. Find x such that $f(x) = 0$.

Complete the square. Then write the perfect-square trinomial in factored form.

8. $x^2 + 20x$
9. $x^2 - \dfrac{6}{5}x$

10. Solve by completing the square. Show your work.

   $x^2 + 8x + 12 = 0$

Solve.

11. The White River flows at a rate of 4 km/h. In order for a boat to travel 76.8 km upriver and then return in a total of 8 hr, how fast must the boat travel in still water?

12. Gunther and Gerta can peel a bushel of potatoes in 48 minutes. Working alone, it takes Gerta 28 minutes longer than Gunther to peel the potatoes. How long would it take for Gunther to do this job by himself?

13. Determine the type of number that the solutions of $x^2 + 10x + 26 = 0$.

14. Write a quadratic equation having solutions -3 and $\dfrac{3}{4}$.

15. Find all x-intercepts of the graph of $(x^2 + 3x)^2 + 3(x^2 + 3x) - 4$.

16. The surface area of a balloon varies directly as the square of its radius. The area is 314 cm² when the radius is 5 cm. What is the area when the radius is 8 cm?

ANSWERS

1._____

2._____

3._____

4._____

5._____

6._____

7._____

8._____

9._____

10._____

11._____

12._____

13._____

14._____

15._____

16._____

CHAPTER 8

TEST FORM C

NAME_____

ANSWERS

17.a)-c) _See graph._

d) _____

18. _____

19. _____

20. _____

21. _____

22.a) _____

b) _See graph._

23. _____

24. _____

25. _____

26. _____

27. _____

17. a) Graph: $f(x) = 3(x-2)^2 + 2$
    b) Label the vertex.
    c) Draw the axis of symmetry.
    d) Find the maximum or the minimum function value.

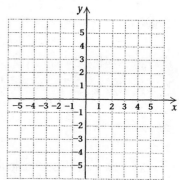

18. Find the x- and y-intercepts of $x^2 - x - 20$.

19. Rosie's Palm Pirates, a manufacturer of electronic organizers, estimates that when x hundred Palm Pirates are made, the average cost per unit is given by $C(x) = 0.25x^2 - 1.8x + 4.62$, where C is in hundreds of dollars. What is the minimum cost per unit, and how many units should be made to achieve that minimum?

20. Solve $F = \dfrac{GMm}{r^2}$ for r.

21. Find the quadratic function that fits the data points $(3, 0)$, $(5, 0)$, and $(6, 4)$.

22. For the function $f(x) = 2x^2 + x + 3$
    a) find the vertex and the axis of symmetry;
    b) graph the function.

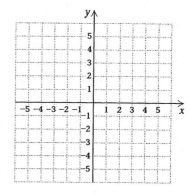

Solve

23. $x^2 + 4x \geq -2$

24. $x + \dfrac{4}{x} > 0$

25. One solution of $kx^2 + 4x - k = 0$ is -2. Find the other solution.

26. Find a fourth-degree polynomial equation, with integer coefficients, for which $2 - \sqrt{5}$ and $2 - i$ are solutions.

27. Find a polynomial equation, with integer coefficients, for which 3 is a repeated root and $\sqrt{2}$ and $\sqrt{5}$ are solutions.

CHAPTER 8

TEST FORM D

NAME_____

CLASS_____ SCORE_____ GRADE_____

Solve.

ANSWERS

1. $5x^2 - 8 = 0$
2. $5x(x-1) - 4x(x+4) = -110$
3. $x^2 + 3x + 1 = 0$
4. $4x + 1 = x^2$
5. $x^{-2} - x^{-1} = \frac{5}{3}$
6. $x^2 + 7x = 1$. Use a calculator to approximate the solutions with rational numbers.
7. Let $f(x) = 15x^2 + 8x - 12$. Find x such that $f(x) = 0$.

Complete the square. Then write the perfect-square trinomial in factored form.

8. $x^2 - 18x$
9. $x^2 + \frac{4}{3}x$

10. Solve by completing the square. Show your work.

   $x^2 + 6x + 3 = 0$

Solve.

11. The Rat River flows at a rate of 5 km/h. In order for a boat to travel 75 km upriver and then return in a total of 8 hr, how fast must the boat travel in still water?

12. Nisha and T'Mar can proofread a manuscript in 12 hours. Working alone, it takes T'Mar 10 hours longer than Nisha to proofread the same manuscript. How long would it take for Nisha to do this job by herself?

13. Determine the type of number that the solutions of $x^2 + 9x + 4 = 0$.

14. Write a quadratic equation having solutions -2 and $\frac{5}{7}$.

15. Find all x-intercepts of the graph of $(x^2 + 4x)^2 - 2(x^2 + 4x) - 15$.

16. The surface area of a balloon varies directly as the square of its radius. The area is 314 cm² when the radius is 5 cm. What is the area when the radius is 9 cm?

1._____

2._____

3._____

4._____

5._____

6._____

7._____

8._____

9._____

10._____

11._____

12._____

13._____

14._____

15._____

16._____

CHAPTER 8
TEST FORM D

NAME_____

ANSWERS

17.a)-c) See graph.

d)_____

18._____

19._____

20._____

21._____

22.a)_____

b) See graph.

23._____

24._____

25._____

26._____

27._____

17. a) Graph: $f(x) = 3(x-1)^2 - 3$
    b) Label the vertex.
    c) Draw the axis of symmetry.
    d) Find the maximum or the minimum function value.

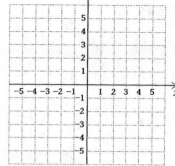

18. Find the x- and y-intercepts of $x^2 + x - 110$.

19. Rosie's Palm Pirates, a manufacturer of electronic organizers, estimates that when x hundred Palm Pirates are made, the average cost per unit is given by $C(x) = 0.5x^2 - 3.9x + 8.905$, where C is in hundreds of dollars. What is the minimum cost per unit, and how many units should be made to achieve that minimum?

20. Solve $c^2 = a^2 + b^2$ for b.

21. Find the quadratic function that fits the data points $(0, 0)$, $(6, 0)$, and $(8, 2)$.

22. For the function $f(x) = 2x^2 + 2x + 1$
    a) find the vertex and the axis of symmetry;
    b) graph the function.

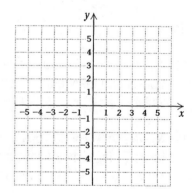

Solve

23. $x^2 + 5x \geq -4$

24. $x + \dfrac{5}{x} > 0$

25. One solution of $kx^2 - 4x + k = 0$ is 2. Find the other solution.

26. Find a fourth-degree polynomial equation, with integer coefficients, for which $2 - \sqrt{6}$ and $2 - i$ are solutions.

27. Find a polynomial equation, with integer coefficients, for which 4 is a repeated root and $\sqrt{2}$ and $\sqrt{5}$ are solutions.

CHAPTER 8

TEST FORM E

NAME_____

CLASS_____ SCORE_____ GRADE_____

Solve.

ANSWERS

1. $6x^2 - 13 = 0$
2. $4x(x-2) - 3x(x+2) = -40$
3. $x^2 + 3x + 2 = 0$
4. $3x + 3 = x^2$
5. $x^{-2} - x^{-1} = \dfrac{3}{5}$
6. $x^2 + x = 7$. Use a calculator to approximate the solutions with rational numbers.
7. Let $f(x) = 6x^2 - 9x - 42$. Find x such that $f(x) = 0$.

Complete the square. Then write the perfect-square trinomial in factored form.

8. $x^2 + 6x$
9. $x^2 - \dfrac{4}{9}x$

10. Solve by completing the square. Show your work.
    $x^2 + 4x + 1 = 0$

Solve.

11. The Spaz River flows at a rate of 5 km/h. In order for a boat to travel 30 km upriver and then return in a total of 8 hr, how fast must the boat travel in still water?

12. Lauren and Humphrey can scrub a floor in 3 hours. Working alone, it takes Humphrey 3.25 hours longer than Lauren to scrub the same floor. How long would it take for Lauren to do this job by herself?

13. Determine the type of number that the solutions of $x^2 + 8x - 9 = 0$.

14. Write a quadratic equation having solutions -2 and $\dfrac{1}{4}$.

15. Find all x-intercepts of the graph of $(x^2 + 4x)^2 - 3(x^2 + 4x) - 10$.

16. The surface area of a balloon varies directly as the square of its radius. The area is 314 cm² when the radius is 5 cm. What is the area when the radius is 10 cm?

1._____

2._____

3._____

4._____

5._____

6._____

7._____

8._____

9._____

10._____

11._____

12._____

13._____

14._____

15._____

16._____

CHAPTER 8

TEST FORM E

NAME_____

ANSWERS

17.a)-c) __See graph.__

d) _____

18. _____

19. _____

20. _____

21. _____

22.a) _____

b) __See graph.__

23. _____

24. _____

25. _____

26. _____

27. _____

17. a) Graph: $f(x) = 2(x+2)^2 + 1$
    b) Label the vertex.
    c) Draw the axis of symmetry.
    d) Find the maximum or the minimum function value.

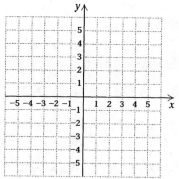

18. Find the x- and y-intercepts of $x^2 - x - 42$.

19. Rosie's Palm Pirates, a manufacturer of electronic organizers, estimates that when x hundred Palm Pirates are made, the average cost per unit is given by $C(x) = 0.6x^2 - 4.8x + 10.82$, where C is in hundreds of dollars. What is the minimum cost per unit, and how many units should be made to achieve that minimum?

20. Solve $\frac{1}{2}mv^2 = \frac{3}{2}kT$ for v.

21. Find the quadratic function that fits the data points $(-3, 0)$, $(4, 0)$, and $(2, 2)$.

22. For the function $f(x) = 2x^2 + 2x + 2$
    a) find the vertex and the axis of symmetry;
    b) graph the function.

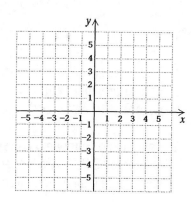

Solve

23. $x^2 + 6x \leq -8$

24. $x - \frac{5}{x} < 0$

25. One solution of $kx^2 + 7x - k = 0$ is -3. Find the other solution.

26. Find a fourth-degree polynomial equation, with integer coefficients, for which $3 - \sqrt{2}$ and $5 - i$ are solutions.

27. Find a polynomial equation, with integer coefficients, for which -4 is a repeated root and $\sqrt{2}$ and $\sqrt{5}$ are solutions.

CHAPTER 8 NAME_____

TEST FORM F CLASS_____ SCORE_____ GRADE_____

Solve.

ANSWERS

1. $7x^2 - 20 = 0$
2. $3x(x-3) - 2x(x+5) = -60$
3. $x^2 + 3x + 3 = 0$
4. $2x + 7 = x^2$
5. $x^{-2} - x^{-1} = \dfrac{4}{5}$
6. $x^2 + 3x = 3$. Use a calculator to approximate the solutions with rational numbers.
7. Let $f(x) = 10x^2 - 7x - 12$. Find x such that $f(x) = 0$.

Complete the square. Then write the perfect-square trinomial in factored form.

8. $x^2 - 2x$
9. $x^2 + \dfrac{6}{7}x$

10. Solve by completing the square. Show your work.
    $x^2 + 14x + 36 = 0$

Solve.

11. The Stemsell River flows at a rate of 4 km/h. In order for a boat to travel 33.6 km upriver and then return in a total of 8 hr, how fast must the boat travel in still water?

12. Katherine and Spencer can weed the garden in 35 minutes. Working alone, it takes Spencer 24 minutes longer than Katherine to weed the garden. How long would it take for Katherine to do this job by herself?

13. Determine the type of number that the solutions of $x^2 + 7x + 10 = 0$.

14. Write a quadratic equation having solutions -3 and $\dfrac{2}{3}$.

15. Find all x-intercepts of the graph of $(x^2 + 3x)^2 - 2(x^2 + 3x) - 8$.

16. The surface area of a balloon varies directly as the square of its radius. The area is 314 cm² when the radius is 5 cm. What is the area when the radius is 4 cm?

1._____

2._____

3._____

4._____

5._____

6._____

7._____

8._____

9._____

10._____

11._____

12._____

13._____

14._____

15._____

16._____

CHAPTER 8

TEST FORM F

NAME_____

ANSWERS

17. a)-c) __See graph.__

d) _____

18. _____

19. _____

20. _____

21. _____

22. a) _____

b) __See graph.__

23. _____

24. _____

25. _____

26. _____

27. _____

17. a) Graph: $f(x) = 2(x+3)^2 - 2$
    b) Label the vertex.
    c) Draw the axis of symmetry.
    d) Find the maximum or the minimum function value.

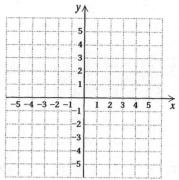

18. Find the x- and y-intercepts of $x^2 + x - 56$.

19. Rosie's Palm Pirates, a manufacturer of electronic organizers, estimates that when x hundred Palm Pirates are made, the average cost per unit is given by $C(x) = 0.3x^2 - 2.58x + 6.687$, where C is in hundreds of dollars. What is the minimum cost per unit, and how many units should be made to achieve that minimum?

20. Solve $K = \dfrac{v^2}{2}\left(\dfrac{I}{R^2} + m\right)$ for R.

21. Find the quadratic function that fits the data points $(-5, 0)$, $(2, 0)$, and $(3, 4)$.

22. For the function $f(x) = 2x^2 + 2x + 3$
    a) find the vertex and the axis of symmetry;
    b) graph the function.

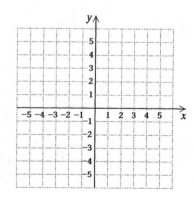

Solve

23. $x^2 + 7x \geq 8$

24. $x + \dfrac{4}{x} < 0$

25. One solution of $kx^2 - 7x + k = 0$ is 4. Find the other solution.

26. Find a fourth-degree polynomial equation, with integer coefficients, for which $3 - \sqrt{3}$ and $5 - i$ are solutions.

27. Find a polynomial equation, with integer coefficients, for which -3 is a repeated root and $\sqrt{2}$ and $\sqrt{5}$ are solutions.

CHAPTER 8   NAME_____

TEST FORM G   CLASS_____ SCORE_____ GRADE_____

ANSWERS

1. Solve. $4x^2 - 11 = 0$

   a) $\pm\sqrt{\dfrac{11}{4}}$ b) $\pm\sqrt{\dfrac{11}{3}}$ c) $\pm\sqrt{2}$ d) $\pm\sqrt{\dfrac{8}{3}}$

   1._____

2. Solve. $2x(x-4) - x(x+3) = -30$

   a) 5, 18   b) 5, 10   c) 5, 6   d) 4, 5

   2._____

3. Solve. $x^2 + 4x + 1$

   a) $2 \pm \sqrt{3}$ b) $2 \pm \sqrt{2}$ c) $-2 \pm \sqrt{2}$ d) $-2 \pm \sqrt{3}$

   3._____

4. Solve. $x + 5 = x^2$

   a) $\dfrac{1 \pm \sqrt{21}}{2}$ b) $\dfrac{1 \pm \sqrt{29}}{2}$ c) $\dfrac{5 \pm \sqrt{21}}{2}$ d) $\dfrac{5 \pm \sqrt{29}}{2}$

4._____

5. Solve. $x^{-2} - x^{-1} = \dfrac{5}{4}$

   a) $\dfrac{4 \pm \sqrt{39}}{5}$ b) $\dfrac{5 \pm \sqrt{29}}{2}$ c) $\dfrac{-3 \pm \sqrt{39}}{5}$ d) $\dfrac{-2 \pm 2\sqrt{6}}{5}$

5._____

6. Solve. $x^2 + 5x = 1$
   Use a calculator to approximate the solutions with rational numbers.

   a) 0.65331193,
      7.65331193

   b) −5.1925824,
      0.1925824

   c) −0.1925824,
      5.65331193

   d) −7.65331193,
      0.65331193

6._____

177

CHAPTER 8   NAME_____

TEST FORM G

ANSWERS

7. Let $f(x) = 12x^2 + 7x - 10$. Find x such that $f(x) = 0$

  a) $-\dfrac{4}{5}, \dfrac{5}{3}$ b) $-\dfrac{5}{4}, \dfrac{2}{3}$ c) $-\dfrac{2}{5}, \dfrac{5}{3}$ d) $-\dfrac{5}{3}, \dfrac{2}{3}$

7._____

8. Complete the square: $x^2 + 22x$. Which of the following is the correct perfect-square trinomial?

  a) $x^2 + 22x + 121$ b) $x^2 - 10x + 25$
  c) $x^2 - 20x + 44$ d) $x^2 + 11x + 484$

8._____

9. The Mad River flows at a rate of 3 km/h. In order for a boat to travel 78.2 km upriver and then return in a total of 8 hr, how fast must the boat travel in still water?

  a) 16.5 km/h   b) 10 km/h   c) 14 km/h   d) 20 km/h

9._____

10. Sabine and Hercule can assemble a table in 45 minutes. Working alone, it takes Hercule 48 minutes longer than Sabine to put this table together. How long would it take for Sabine to assemble the table by herself?

  a) 2 hours   b) 48 minutes   c) 72 minutes   d) 27 minutes

10._____

11. Determine the type of number that the solutions of $x^2 + 6x + 7 = 0$ will be.

  a) no solution   b) complex   c) rational   d) irrational

11._____

12. Which of the following quadratic equations has -4 and $\dfrac{3}{4}$ as solutions?

  a) $x^2 + \dfrac{13}{4}x - 3$ b) $x^2 + \dfrac{30}{7}x - \dfrac{25}{7}$
  c) $x^2 + \dfrac{30}{7}x + \dfrac{25}{7}$ d) $x^2 - \dfrac{13}{4}x + 3$

12._____

CHAPTER 8   NAME_____

TEST FORM G

13. Which of the following points is an *x*-intercept of the graph of
    $f(x) = (x^2 + 2x)^2 - 11(x^2 + 2x) + 24$?

    a) (4, 0)   b) (−6, 0)   c) (6, 0)   d) (−4, 0)

    13._____

14. Graph $2x^2 + 3x + 1$

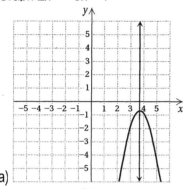

    14._____

15. What is the *y*–intercept of $f(x) = x^2 - x - 12$?

    a) (0, 12)   b) (−72, 0)   c) (0, −12)   d) (0, −72)

16. Solve $w = \dfrac{va}{r^2 + a^2}$ for *a*.

    a) $a = \sqrt{\dfrac{wr^2}{v}}$ b) $a = \dfrac{v \pm \sqrt{v^2 - 4w^2 r^2}}{2w}$

    c) $a = \dfrac{v \pm \sqrt{v^2 + 4w^2 r^2}}{w}$ d) $a = \sqrt{\dfrac{w(r^2 + v^2)}{v}}$

    15._____

    16._____

179

CHAPTER 8

TEST FORM G

ANSWERS

17. Rosie's Palm Pirates, a manufacturer of electronic organizers, estimates that when x hundred Palm Pirates are made, the average cost per unit is given by $C(x) = 0.4x^2 - 3.68x + 9.524$, where C is in hundreds of dollars. What is the minimum cost per unit?

17._____

a) $106   b) $98   c) $10.60   d) $0.98

18. Find the quadratic function that fits the data points $(-2, 0), (1, 0), (2, 3)$.

18._____

a) $\frac{3}{4}x^2 + \frac{3}{4}x - \frac{3}{2}$ b) $\frac{1}{4}x^2 - \frac{1}{2}x - \frac{3}{4}$

c) $\frac{1}{4}x^2 + \frac{1}{2}x + \frac{3}{4}$ d) $-\frac{3}{4}x^2 + \frac{3}{4}x + \frac{3}{2}$

19._____

19. Solve. $x^2 + 8x \leq 9$

a) $(-\infty, -5] \cup [-4, \infty)$ b) $[-5, -4]$

c) $[-9, 1]$ d) $(-\infty, -9] \cup [1, \infty)$

20._____

20. Solve. $x - \frac{3}{x} < 0$

a) $(-\infty, -\sqrt{3}) \cup (0, \sqrt{3})$ b) $(0, \sqrt{3})$

c) $(0, \infty)$ d) $(-\infty, 0)$

21._____

21. One solution of $kx^2 + 6x - k = 0$ is -3. Find the other solution.

a) $-\frac{1}{3}$ b) $\frac{2}{3}$ c) $\frac{1}{3}$ d) $-\frac{2}{3}$

22._____

22. Find a fourth-degree polynomial equation with integer coefficients for which $3 - \sqrt{5}$ and $5 - i$ are solutions.

a) $x^4 - 16x^3 + 92x^2 - 216x + 144$
b) $x^4 - 16x^3 + 91x^2 - 206x + 128$
c) $x^4 - 16x^3 + 90x^2 - 196x + 104$
d) $x^4 - 16x^3 + 89x^2 - 186x + 78$

CHAPTER 8   NAME_____

TEST FORM H   CLASS_____ SCORE_____ GRADE_____

ANSWERS

1. Solve. $3x^2 - 8 = 0$

    a) $\pm\sqrt{\dfrac{11}{4}}$ b) $\pm\sqrt{\dfrac{11}{3}}$ c) $\pm\sqrt{2}$ d) $\pm\sqrt{\dfrac{8}{3}}$

1._____

2. Solve. $3x(x-5) - 2x(x+4) = -90$

    a) 5, 18   b) 5, 10   c) 5, 6   d) 4, 5

2._____

3. Solve. $x^2 + 4x + 2$

    a) $2 \pm \sqrt{3}$ b) $2 \pm \sqrt{2}$ c) $-2 \pm \sqrt{2}$ d) $-2 \pm \sqrt{3}$

4. Solve. $5x + 1 = x^2$

    a) $\dfrac{1 \pm \sqrt{21}}{2}$ b) $\dfrac{1 \pm \sqrt{29}}{2}$ c) $\dfrac{5 \pm \sqrt{21}}{2}$ d) $\dfrac{5 \pm \sqrt{29}}{2}$

3._____

5. Solve. $x^{-2} - x^{-1} = \dfrac{5}{6}$

4._____

    a) $\dfrac{4 \pm \sqrt{39}}{5}$ b) $\dfrac{5 \pm \sqrt{29}}{2}$ c) $\dfrac{-3 \pm \sqrt{39}}{5}$ d) $\dfrac{-2 \pm 2\sqrt{6}}{5}$

6. Solve. $x^2 + 7x = 5$
Use a calculator to approximate the solutions with rational numbers.

5._____

    a) 0.65331193,   b) -5.1925824,
       7.65331193   0.1925824

    c) -0.1925824,   d) -7.65331193,
       5.65331193   0.65331193

6._____

CHAPTER 8   NAME_____

TEST FORM H

ANSWERS

7. Let $f(x) = 15x^2 - 19x - 10$. Find x such that $f(x) = 0$

   a) $-\dfrac{4}{5}, \dfrac{5}{3}$ b) $-\dfrac{5}{4}, \dfrac{2}{3}$ c) $-\dfrac{2}{5}, \dfrac{5}{3}$ d) $-\dfrac{5}{3}, \dfrac{2}{3}$

7._____

8. Complete the square : $x^2 - 10x$. Which of the following is the correct perfect-square trinomial?

   a) $x^2 + 22x + 121$ b) $x^2 - 10x + 25$

   c) $x^2 - 20x + 44$ d) $x^2 + 11x + 484$

8._____

9. The Mad River flows at a rate of 2 km/h. In order for a boat to travel 79.2 km upriver and then return in a total of 8 hr, how fast must the boat travel in still water?

   a) 16.5 km/h   b) 10 km/h   c) 20 km/h   d) 13 km/h

9._____

10. Loretta and Dolly can tile a bathroom in 8 hours. Working alone, it takes Dolly 4 hours and 40 minutes longer than Loretta to tile this bathroom. How long would it take for Loretta to tile the bathroom by herself?

    a) 13 hours   b) 6.5 hours   c) 7.75 hours   d) 20 hours

10._____

11. Determine the type of number that the solutions of $x^2 + 5x + 7 = 0$ will be.

    a) no solution   b) complex   c) rational   d) irrational

11._____

12. Which of the following quadratic equations has -5 and $\dfrac{5}{7}$ as solutions?

    a) $x^2 + \dfrac{13}{4}x - 3$ b) $x^2 + \dfrac{30}{7}x - \dfrac{25}{7}$

    c) $x^2 + \dfrac{30}{7}x + \dfrac{25}{7}$ d) $x^2 - \dfrac{13}{4}x + 3$

12._____

CHAPTER 8

TEST FORM H

NAME_____

ANSWERS

13. Which of the following points is an *x*-intercept of the graph of
$f(x) = (x^2 + 5x)^2 - 5(x^2 + 5x) - 6$?

a) $(4, 0)$ b) $(-6, 0)$ c) $(6, 0)$ d) $(-4, 0)$

13._____

14. Graph $2x^2 + 3x + 2$

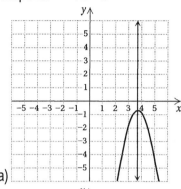
a)

b)

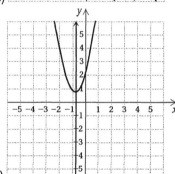
c)

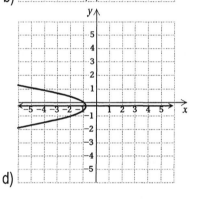
d)

14._____

15. What is the *y*-intercept of $f(x) = x^2 + x - 72$?

a) $(0, 12)$ b) $(-72, 0)$ c) $(0, -12)$ d) $(0, -72)$

15._____

16. Solve $\left(\dfrac{v_1}{v_2}\right)^2 = \dfrac{m_2}{m_1}$ for v_2.

a) $v_2 = v_1 \sqrt{\dfrac{m_1}{m_2}}$

b) $v_2 = v_1 \sqrt{\dfrac{m_2}{m_1}}$

c) $v_2 = \dfrac{v_1 m_2}{m_1}$

d) $v_2 = v_1 \sqrt[3]{\dfrac{m_1}{m_2}}$

16._____

183

CHAPTER 8

TEST FORM H

NAME_____

ANSWERS

17._____

17. Rosie's Palm Pirates, a manufacturer of electronic organizers, estimates that when x hundred Palm Pirates are made, the average cost per unit is given by $C(x) = 0.3x^2 - 2.94x + 8.183$, where C is in hundreds of dollars. What is the minimum cost per unit?

a) $106   b) $98   c) $10.60   d) $0.98

18._____

18. Find the quadratic function that fits the data points $(-1, 0), (3, 0), (5, 3)$.

a) $\frac{3}{4}x^2 + \frac{3}{4}x - \frac{3}{2}$ b) $\frac{1}{4}x^2 - \frac{1}{2}x - \frac{3}{4}$

c) $\frac{1}{4}x^2 + \frac{1}{2}x + \frac{3}{4}$ d) $-\frac{3}{4}x^2 + \frac{3}{4}x + \frac{3}{2}$

19._____

19. Solve. $x^2 + 9x \geq -20$

a) $(-\infty, -5] \cup [-4, \infty)$ b) $[-5, -4]$

c) $[-9, 1]$ d) $(-\infty, -9] \cup [1, \infty)$

20._____

20. Solve. $x - \frac{2}{x} < 0$

a) $(-\infty, -\sqrt{3}) \cup (0, \sqrt{3})$ b) $(0, \sqrt{3})$

c) $(0, \infty)$ d) $(-\infty, 0)$

21._____

21. One solution of $kx^2 - 6x + k = 0$ is 3. Find the other solution.

a) $-\frac{1}{3}$ b) $\frac{2}{3}$ c) $\frac{1}{3}$ d) $-\frac{2}{3}$

22._____

22. Find a fourth-degree polynomial equation with integer coefficients for which $3 - \sqrt{6}$ and $5 - i$ are solutions.

a) $x^4 - 16x^3 + 92x^2 - 216x + 144$
b) $x^4 - 16x^3 + 91x^2 - 206x + 128$
c) $x^4 - 16x^3 + 90x^2 - 196x + 104$
d) $x^4 - 16x^3 + 89x^2 - 186x + 78$

CHAPTER 9   NAME_____

TEST FORM A   CLASS_____ SCORE_____ GRADE_____

1. Find $(f \circ g)(x)$ and $(g \circ f)(x)$ if $f(x) = x + x^2$ and $g(x) = 3x + 2$.

2. Determine whether $f(x) = \sqrt{x}$ is one to one.

Find a formula for the inverse of each function.

3. $f(x) = 3x - 4$ 4. $g(x) = (x+4)^3$

Graph.

5. $f(x) = 2^x - 1$ 6. $g(x) = \log_2 x$

Simplify.

7. $\log_3 81$ 8. $\log_4 2$ 9. $6^{\log_6 25}$

Convert to logarithmic equations.

10. $5^{-3} = \dfrac{1}{125}$ 11. $121^{1/2} = 11$

Convert to exponential equations.

12. $m = \log_2 4$ 13. $\log_2 128 = 7$

14. Express in terms of logarithms of a, b, and c. $\log \dfrac{a^2 b^3}{c}$

15. Express as a single logarithm: $\dfrac{1}{4} \log_a x + 5 \log_a z$

Simplify.

16. $\log_p p$ 17. $\log_t t^5$ 18. $\log_a 1$

If $\log_a 8 = 2.079$, $\log_a 5 = 1.609$, and $\log_a 3 = 1.099$, find the following.

19. $\log_a \dfrac{8}{5}$ 20. $\log_a 40$ 21. $\log_a 45$

ANSWERS

1._____
2._____
3._____
4._____
5. See graph.
6. See graph.
7._____
8._____
9._____
10._____
11._____
12._____
13._____
14._____
15._____
16._____
17._____
18._____
19._____
20._____
21._____

CHAPTER 9

TEST FORM A

NAME_____

ANSWERS

22._____

23._____

24._____

25._____

26._____

27._____

28._____

29._____

30._____

31._____

32._____

33._____

34._____

35._____

36._____

State the domain and the range of each function.

22. $f(x) = e^x + 2$

23. $g(x) = \ln(x - 3)$

Solve. Where appropriate, include the approximation to the nearest ten-thousandth.

24. $3^x = \dfrac{1}{27}$

25. $\log_x 8 = 3$

26. $\log_{81} x = \dfrac{1}{2}$

27. $\log x = -4$

28. $9^{3-7x} = 81$

29. $6^x = 3.2$

30. $\ln x = \dfrac{1}{8}$

31. $\log(x - 2) + \log(x + 2) = \log 2$

32. The average walking speed R of people living in a city of population P, in thousands, is given by $R = 0.37 \ln P + 0.05$, where R is in feet per second.

 a) The population of Burlington, Vermont is 38,450. Find the average walking speed.

 b) A city's average walking speed is 2.10 ft / sec. Find the population.

33. On Monday, Lassiter began monitoring the size of a bacteria colony. He noted that the mass of the colony grew exponentially from 5 grams at 8AM to 6.5 grams at 2 PM.

 a) Find the value for k and write an exponential function that estimates the mass M of the bacteria colony t hours after 8AM Monday.

 b) What will the mass be at 6PM Monday?

 c) To the nearest hour, at what time will the mass of the growing colony reach 15 grams?

34. An investment with interest compounded continuously doubled itself in 22 years. What is the interest rate?

35. Solve. $\log_3 |2x - 11| = 5$

36. If $\log_a x = 2$, $\log_a y = 3$, and $\log_a z = 4$, then find $\log_a \dfrac{\sqrt[3]{x^2 z}}{\sqrt[3]{y z^2}}$.

CHAPTER 9
TEST FORM B

NAME_____ CLASS____ SCORE____ GRADE____

ANSWERS

1. Find $(f \circ g)(x)$ and $(g \circ f)(x)$ if $f(x) = x - x^2$ and $g(x) = 2x + 3$.

2. Determine whether $f(x) = x^2$ is one to one.

Find a formula for the inverse of each function.

3. $f(x) = 3x + 5$
4. $g(x) = (x - 2)^3$

Graph.

5. $f(x) = 2^x - 5$
6. $g(x) = \log_3 x$

Simplify.

7. $\log_4 256$
8. $\log_{64} 8$
9. $5^{\log_5 32}$

Convert to logarithmic equations.

10. $4^{-4} = \dfrac{1}{256}$
11. $64^{1/2} = 8$

Convert to exponential equations.

12. $m = \log_3 27$
13. $\log_3 9 = 2$

14. Express in terms of logarithms of a, b, and c. $\log \dfrac{a^3 b}{c^2}$

15. Express as a single logarithm: $\dfrac{1}{2} \log_a x + 4 \log_a z$

Simplify.

16. $\log_q q$
17. $\log_t t^{19}$
18. $\log_b 1$

If $\log_a 8 = 2.079$, $\log_a 5 = 1.609$, and $\log_a 3 = 1.099$, find the following.

19. $\log_a \dfrac{8}{3}$
20. $\log_a 24$
21. $\log_a 25$

1._____
2._____
3._____
4._____
5. See graph.
6. See graph.
7._____
8._____
9._____
10._____
11._____
12._____
13._____
14._____
15._____
16._____
17._____
18._____
19._____
20._____
21._____

CHAPTER 9

TEST FORM B

NAME _____

ANSWERS

State the domain and the range of each function.

22. $f(x) = e^x - 2$
23. $g(x) = \ln(x+3)$

Solve. Where appropriate, include the approximation to the nearest ten-thousandth.

24. $4^x = \dfrac{1}{16}$
25. $\log_x 64 = 2$
26. $\log_{64} x = \dfrac{1}{2}$

27. $\log x = -3$
28. $3^{4-8x} = \dfrac{1}{81}$
29. $7^x = 4.2$

30. $\ln x = \dfrac{3}{8}$
31. $\log(x-4) + \log(x+1) = \log 6$

32. The average walking speed R of people living in a city of population P, in thousands, is given by $R = 0.37 \ln P + 0.05$, where R is in feet per second.

   a) The population of Cleveland, Ohio is 495,820. Find the average walking speed.

   b) A city has an average walking speed of 2.30 ft / sec. Find the population.

33. On Monday, Matilde began monitoring the size of a bacteria colony. She noted that the mass of the colony grew exponentially from 1.5 grams at 8AM to 4 grams at 2 PM.

   a) Find the value for k and write an exponential function that estimates the mass M of the bacteria colony t hours after 8AM Monday.

   b) What will the mass be at 6PM Monday?

   c) To the nearest hour, at what time will the mass of the growing colony reach 15 grams?

34. An investment with interest compounded continuously doubled itself in 20 years. What is the interest rate?

35. Solve. $\log_4 |2x+8| = 2$

36. If $\log_a x = 2$, $\log_a y = 3$, and $\log_a z = 4$, then find $\log_a \dfrac{\sqrt{xz}}{\sqrt[3]{zy^2}}$.

22. _____
23. _____
24. _____
25. _____
26. _____
27. _____
28. _____
29. _____
30. _____
31. _____
32. _____
33. _____
34. _____
35. _____
36. _____

CHAPTER 9

TEST FORM C

NAME_____

CLASS_____ SCORE_____ GRADE_____

ANSWERS

1. Find $(f \circ g)(x)$ and $(g \circ f)(x)$ if $f(x) = x + 2x^2$ and $g(x) = 2x - 3$.

2. Determine whether $f(x) = x + 10$ is one to one.

1._____

2._____

Find a formula for the inverse of each function.

3. $f(x) = 2x - 6$ 4. $g(x) = (8x + 1)^3$

3._____

4._____

Graph.

5. $f(x) = 2^x - 4$ 6. $g(x) = \log_4 x$

   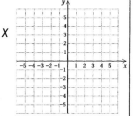

5. See graph.

6. See graph.

Simplify.

7. $\log_5 25$ 8. $\log_{36} 6$ 9. $4^{\log_4 12}$

7._____

8._____

9._____

Convert to logarithmic equations.

10. $3^{-5} = \dfrac{1}{243}$ 11. $144^{1/2} = 12$

10._____

11._____

Convert to exponential equations.

12. $m = \log_4 16$ 13. $\log_4 64 = 3$

12._____

13._____

14. Express in terms of logarithms of a, b, and c. $\log \dfrac{ab^2}{c^3}$

14._____

15. Express as a single logarithm: $\dfrac{1}{3} \log_a x - 3 \log_a z$

15._____

Simplify.

16. $\log_r r$ 17. $\log_t t^{0.052}$ 18. $\log_{137} 1$

16._____

17._____

18._____

If $\log_a 8 = 2.079$, $\log_a 5 = 1.609$, and $\log_a 3 = 1.099$, find the following.

19. $\log_a \dfrac{5}{3}$ 20. $\log_a 15$ 21. $\log_a 64$

19._____

20._____

21._____

CHAPTER 9

TEST FORM C

NAME_____

ANSWERS

State the domain and the range of each function.

22. $f(x) = e^x + 3$ 23. $g(x) = \ln(x-1)$

Solve. Where appropriate, include the approximation to the nearest ten-thousandth.

22._____

23._____

24. $5^x = \dfrac{1}{125}$ 25. $\log_x 125 = 3$ 26. $\log_{49} x = \dfrac{1}{2}$

24._____

27. $\log x = -2$ 28. $7^{4-9x} = 49$ 29. $8^x = 5.2$

25._____

30. $\ln x = \dfrac{5}{8}$ 31. $\log(x-4) + \log(x+4) = \log 4$

26._____

32. The average walking speed R of people living in a city of population P, in thousands, is given by $R = 0.37 \ln P + 0.05$, where R is in feet per second.

27._____

    a) The population of Boston, Massachusetts is 555,450. Find the average walking speed.

28._____

    b) A city's average walking speed is 2.50 ft / sec. Find the population.

29._____

33. On Monday, Jensen began monitoring the size of a bacteria colony. She noted that the mass of the colony grew exponentially from 0.5 grams at 8AM to 5 grams at 2 PM.

30._____

    a) Find the value for k and write an exponential function that estimates the mass M of the bacteria colony t hours after 8AM Monday.

31._____

    b) What will the mass be at 6PM Monday?

    c) To the nearest hour, at what time will the mass of the growing colony reach 15 grams?

32._____

33._____

34. An investment with interest compounded continuously doubled itself in 18 years. What is the interest rate?

34._____

35. Solve. $\log_5 |2x - 3| = 3$

35._____

36. If $\log_a x = 2$, $\log_a y = 3$, and $\log_a z = 4$, then find $\log_a \dfrac{\sqrt[3]{x^2 z^{-1}}}{\sqrt{yz}}$.

36._____

CHAPTER 9
TEST FORM D

NAME_____

CLASS_____ SCORE_____ GRADE_____

ANSWERS

1. Find $(f \circ g)(x)$ and $(g \circ f)(x)$ if $f(x) = x - 2x^2$ and $g(x) = 3x - 2$.

2. Determine whether $f(x) = |x - 7|$ is one to one.

Find a formula for the inverse of each function.

3. $f(x) = x + 7$

4. $g(x) = (8x - 3)^3$

Graph.

5. $f(x) = 3^x - 1$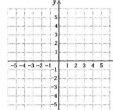

6. $g(x) = \log_5 x$

Simplify.

7. $\log_6 6$

8. $\log_{25} 5$

9. $3^{\log_3 19}$

Convert to logarithmic equations.

10. $2^{-6} = \dfrac{1}{64}$

11. $4^{1/2} = 2$

Convert to exponential equations.

12. $m = \log_5 25$

13. $\log_6 36 = 2$

14. Express in terms of logarithms of a, b, and c. $\log \dfrac{a^{1/2} b^{1/3}}{c}$

15. Express as a single logarithm: $\dfrac{1}{2} \log_a x - 2 \log_a z$

Simplify.

16. $\log_s s$

17. $\log_t t^{38}$

18. $\log_d 1$

If $\log_a 8 = 2.079$, $\log_a 5 = 1.609$, and $\log_a 3 = 1.099$, find the following.

19. $\log_a \dfrac{5}{8}$

20. $\log_a 120$

21. $\log_a 9$

1._____

2._____

3._____

4._____

5. See graph.

6. See graph.

7._____

8._____

9._____

10._____

11._____

12._____

13._____

14._____

15._____

16._____

17._____

18._____

19._____

20._____

21._____

CHAPTER 9

TEST FORM D

NAME_____

ANSWERS

State the domain and the range of each function.

22. $f(x) = e^x - 3$ 23. $g(x) = \ln(x+1)$

22._____

Solve. Where appropriate, include the approximation to the nearest ten-thousandth.

23._____

24. $2^x = \dfrac{1}{64}$ 25. $\log_x 128 = 7$ 26. $\log_{36} x = \dfrac{1}{2}$

24._____

27. $\log x = -1$ 28. $6^{5-2x} = \dfrac{1}{36}$ 29. $9^x = 6.2$

25._____

30. $\ln x = \dfrac{7}{8}$ 31. $\log(x-5) + \log(x+1) = \log 7$

26._____

32. The average walking speed R of people living in a city of population P, in thousands, is given by $R = 0.37 \ln P + 0.05$, where R is in feet per second.

27._____

   a) The population of Honolulu, Hawaii is 395,790. Find the average walking speed.

28._____

   b) A city's average walking speed is 2.70 ft / sec. Find the population.

29._____

33. On Monday, Speezak began monitoring the size of a bacteria colony. He noted that the mass of the colony grew exponentially from 2 grams at 8AM to 3.7 grams at 2 PM.

30._____

   a) Find the value for k and write an exponential function that estimates the mass M of the bacteria colony t hours after 8AM Monday.

31._____

   b) What will the mass be at 6PM Monday?

   c) To the nearest hour, at what time will the mass of the growing colony reach 15 grams?

32._____

33._____

34. An investment with interest compounded continuously doubled itself in 16 years. What is the interest rate?

34._____

35. Solve. $\log_6 |2x - 4| = 3$

35._____

36. If $\log_a x = 2$, $\log_a y = 3$, and $\log_a z = 4$, then find $\log_a \dfrac{\sqrt[3]{y^2 x}}{\sqrt{x^{-1} z^{-1}}}$.

36._____

192

CHAPTER 9
TEST FORM E

NAME_____

CLASS_____ SCORE_____ GRADE_____

1. Find $(f \circ g)(x)$ and $(g \circ f)(x)$ if $f(x) = 2x + x^2$ and $g(x) = 2 - 3x$.

2. Determine whether $f(x) = x^3 - 2$ is one to one.

Find a formula for the inverse of each function.

3. $f(x) = 5x - 2$

4. $g(x) = (8x + 4)^3$

Graph.

5. $f(x) = 3^x - 2$

6. $g(x) = \log_6 x$

Simplify.

7. $\log_5 625$

8. $\log_{81} 9$

9. $3^{\log_3 10}$

Convert to logarithmic equations.

10. $3^{-4} = \dfrac{1}{81}$

11. $169^{1/2} = 13$

Convert to exponential equations.

12. $m = \log_6 36$

13. $\log_5 125 = 3$

14. Express in terms of logarithms of a, b, and c. $\log \dfrac{a^{1/3} b^{1/2}}{c}$

15. Express as a single logarithm: $\dfrac{1}{3} \log_a x + 2 \log_a z$

Simplify.

16. $\log_t t$

17. $\log_t t^{1012}$

18. $\log_{1000} 1$

If $\log_a 4 = 1.386$, $\log_a 5 = 1.609$, and $\log_a 6 = 1.792$, find the following.

19. $\log_a \dfrac{2}{3}$

20. $\log_a 20$

21. $\log_a 16$

ANSWERS

1._____
2._____
3._____
4._____
5. See graph.
6. See graph.
7._____
8._____
9._____
10._____
11._____
12._____
13._____
14._____
15._____
16._____
17._____
18._____
19._____
20._____
21._____

CHAPTER 9

TEST FORM E

NAME_____

ANSWERS

22._____

23._____

24._____

25._____

26._____

27._____

28._____

29._____

30._____

31._____

32._____

33._____

34._____

35._____

36._____

State the domain and the range of each function.

22. $f(x) = 1 - e^x$

23. $g(x) = \ln(2x - 1)$

Solve. Where appropriate, include the approximation to the nearest ten-thousandth.

24. $10^x = \dfrac{1}{10,000}$

25. $\log_x 216 = 3$

26. $\log_{25} x = \dfrac{1}{2}$

27. $\log x = 0$

28. $2^{6-3x} = 128$

29. $2^x = 2.6$

30. $\ln x = \dfrac{9}{8}$

31. $\log(x-3) + \log(x+3) = \log 6$

32. The average walking speed R of people living in a city of population P, in thousands, is given by $R = 0.37 \ln P + 0.05$, where R is in feet per second.

   a) The population of White Plains, New York is 49,940. Find the average walking speed.

   b) A city's average walking speed is 2.90 ft / sec. Find the population.

33. On Monday, Feldspar began monitoring the size of a bacteria colony. He noted that the mass of the colony grew exponentially from 2.5 grams at 8AM to 4.8 grams at 2 PM.

   a) Find the value for k and write an exponential function that estimates the mass M of the bacteria colony t hours after 8AM Monday.

   b) What will the mass be at 6PM Monday?

   c) To the nearest hour, at what time will the mass of the growing colony reach 15 grams?

34. An investment with interest compounded continuously doubled itself in 14 years. What is the interest rate?

35. Solve. $\log_5 |2x + 9| = 4$

36. If $\log_a x = 2$, $\log_a y = 3$, and $\log_a z = 4$, then find $\log_a \dfrac{\sqrt{xyz}}{\sqrt[3]{x^2 y^2}}$.

CHAPTER 9
TEST FORM F

NAME_____

CLASS_____ SCORE_____ GRADE_____

1. Find $(f \circ g)(x)$ and $(g \circ f)(x)$ if $f(x) = -2x + x^2$ and $g(x) = 3 - 2x$.

2. Determine whether $f(x) = (x-2)^2$ is one to one.

Find a formula for the inverse of each function.

3. $f(x) = 6x + 7$

4. $g(x) = (64x - 3)^3$

Graph.

5. $f(x) = 3^x - 4$

6. $g(x) = \log_7 x$

Simplify.

7. $\log_4 64$

8. $\log_{225} 15$

9. $4^{\log_4 28}$

Convert to logarithmic equations.

10. $5^{-4} = \dfrac{1}{625}$

11. $81^{1/2} = 9$

Convert to exponential equations.

12. $m = \log_7 49$

13. $\log_4 256 = 4$

14. Express in terms of logarithms of a, b, and c. $\log \dfrac{a^3 b^2}{c}$

15. Express as a single logarithm: $\dfrac{1}{4} \log_a x + \dfrac{3}{2} \log_a z$

Simplify.

16. $\log_u u$

17. $\log_t t^{3.14}$

18. $\log_f 1$

If $\log_a 4 = 1.386$, $\log_a 5 = 1.609$, and $\log_a 6 = 1.792$, find the following.

19. $\log_a \dfrac{3}{2}$

20. $\log_a 120$

21. $\log_a 80$

ANSWERS

1._____
2._____
3._____
4._____
5. See graph.
6. See graph.
7._____
8._____
9._____
10._____
11._____
12._____
13._____
14._____
15._____
16._____
17._____
18._____
19._____
20._____
21._____

CHAPTER 9

TEST FORM F

NAME_____

ANSWERS

State the domain and the range of each function.

22. $f(x) = 2 - e^x$
23. $g(x) = \ln(2x+1)$

Solve. Where appropriate, include the approximation to the nearest ten-thousandth.

24. $2^x = \dfrac{1}{128}$
25. $\log_x 243 = 5$
26. $\log_{16} x = \dfrac{1}{2}$

27. $\log x = 1$
28. $8^{7-4x} = \dfrac{1}{64}$
29. $3^x = 2.5$

30. $\ln x = \dfrac{11}{8}$
31. $\log(x-4) + \log(x+2) = \log 7$

32. The average walking speed R of people living in a city of population P, in thousands, is given by $R = 0.37 \ln P + 0.05$, where R is in feet per second.

   a) The population of Athens, Ohio is 21,700. Find the average walking speed.

   b) A city's average walking speed is 1.90 ft / sec. Find the population.

33. On Monday, Ambrose began monitoring the size of a bacteria colony. She noted that the mass of the colony grew exponentially from 1.7 grams at 8AM to 5.8 grams at 2 PM.

   a) Find the value for k and write an exponential function that estimates the mass M of the bacteria colony t hours after 8AM Monday.

   b) What will the mass be at 6PM Monday?

   c) To the nearest hour, at what time will the mass of the growing colony reach 15 grams?

34. An investment with interest compounded continuously doubled itself in 12 years. What is the interest rate?

35. Solve. $\log_4 |2x - 6| = 4$

36. If $\log_a x = 2$, $\log_a y = 3$, and $\log_a z = 4$, then find $\log_a \dfrac{\sqrt[3]{xyz}}{\sqrt[3]{x^2 z^2}}$.

CHAPTER 9   NAME_____

TEST FORM G   CLASS_____ SCORE_____ GRADE_____

ANSWERS

1. Find $(f \circ g)(x)$ if $f(x) = x - x^2$ and $g(x) = 2x - 3$.

   a) $-4x^2 - 15x - 12$ b) $-9x^2 + 14x + 6$
   c) $-9x^2 - 15x + 6$ d) $-4x^2 + 14x - 12$

1._____

2. Find a formula for the inverse of $g(x) = \left(\dfrac{2x-1}{2}\right)^3$.

   a) $g^{-1}(x) = 2\sqrt[3]{x} + 5$ b) $g^{-1}(x) = 2\sqrt[3]{2x} + 1$

   c) $g^{-1}(x) = 3\sqrt[3]{2x} - 1$ d) $g^{-1}(x) = 3\sqrt[3]{x} - 5$

3. Which graph represents $f(x) = \left(\dfrac{3}{2}\right)^x - \dfrac{9}{2}$?

2._____

a)
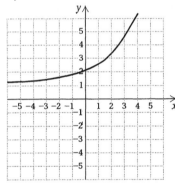

b)
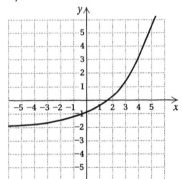

c)
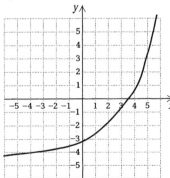

d)
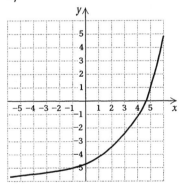

3._____

4. Simplify $\log_3 243$.

   a) 6   b) 3   c) 2   d) 5

4._____

197

CHAPTER 9

TEST FORM G

NAME_____

ANSWERS

5. _____

5. Convert to a logarithmic equation. $4^{-2} = \dfrac{1}{16}$

   a) $\log_6 \dfrac{1}{16} = -3$ b) $\log_6 \dfrac{1}{216} = -3$

   c) $\log_4 \dfrac{1}{16} = -2$ d) $\log_4 \dfrac{1}{216} = -2$

6. _____

6. Convert to a logarithmic equation. $196^{1/2} = 14$

   a) $\log_{196} 14 = \dfrac{1}{2}$ b) $\log_{289} 14 = \dfrac{1}{2}$

   c) $\log_{289} 17 = \dfrac{1}{2}$ d) $\log_{196} 17 = \dfrac{1}{2}$

7. _____

7. Convert to an exponential equation. $m = \log_8 64$

   a) $8^m = 64$ b) $9^m = 81$ c) $9^m = 64$ d) $8^m = 81$

8. Convert to an exponential equation. $\log_3 27 = 3$

   a) $2^4 = 16$ b) $27^4 = 16$ c) $3^3 = 27$ d) $16^2 = 4$

8. _____

9. Express in terms of logs of a, b, and c. $\log \dfrac{a^2 b}{c^3}$

   a) $2 \log a + \log b - 3 \log c$ b) $\log a + \dfrac{1}{2} \log b - \dfrac{1}{3} \log c$

9. _____

   c) $\log a + 2 \log b - \dfrac{1}{3} \log c$ d) $2 \log a + \log b - \dfrac{1}{3} \log c$

10. Express as a single logarithm. $\dfrac{3}{4} \log_a x - 3 \log_a z$

   a) $\log_a \dfrac{x^4}{z^3}$ b) $\log_a \dfrac{x^{3/4}}{z^3}$ c) $\log_a \dfrac{x^2}{z^3}$ d) $\log_a \dfrac{x^2}{z^{1/2}}$

10. _____

CHAPTER 9

TEST FORM G

NAME_____

11. Simplify. $\log_v v$

   a) w b) v c) 1   d) 0

12. Simplify. $\log_t t^{100}$

   a) G b) t c) 0.01   d) 100

13. Simplify. $\log_g 1$

   a) h b) 1   c) 0   d) g

14. Given $\log_a 4 = 1.386$, $\log_a 5 = 1.609$, and $\log_a 6 = 1.792$, find $\log_a 100$.

   a) 4.604   b) 3.584   c) 4.787   d) 3.218

15. What is the range of $f(x) = 3 - e^x$?

   a) $(3, \infty)$ b) $(-\infty, 3)$ c) $(1, \infty)$ d) $(-\infty, 1)$

16. What is the domain of $g(x) = \ln(x-2)$?

   a) $(-2, 2)$ b) $(-2, \infty)$ c) $(2, \infty)$ d) $(-\infty, 2)$

17. Solve. $3^x = \dfrac{1}{81}$

   a) 4   b) -3 c) -4 d) 3

18. Solve. $\log_x 64 = 3$

   a) 16   b) 4   c) 9   d) 3

19. Solve. $\log_9 x = \dfrac{1}{2}$

   a) 9   b) 10   c) 18   d) 3

ANSWERS

11._____

12._____

13._____

14._____

15._____

16._____

17._____

18._____

19._____

CHAPTER 9

TEST FORM G

NAME_____

ANSWERS

20._____

21._____

22._____

23._____

24._____

25._____

26._____

27._____

20. Solve. $4^x = 2.4$ Give exact answer in terms of common logarithms.

   a) $\dfrac{\log 2.3}{\log 5}$ b) $\dfrac{\log 2.4}{\log 4}$ c) $\dfrac{\log 4}{\log 2.4}$ d) $\dfrac{\log 5}{\log 2.3}$

21. Solve and give answer to the nearest ten-thousandth. $\ln x = \dfrac{13}{8}$

   a) 5.0789   b) 6.5208   c) 6.5272   d) 5.0784

22. Solve. $\log(x-6) + \log(x+3) = \log 10$

   a) 2   b) 4   c) 5   d) 7

23. The population in India in 2000 was 1.014 billion, and was growing exponentially by 1.58% per year. When will India's population reach 1.5 billion?

   a) 2017   b) 2025   c) 2031   d) 2022

24. How old is an animal bone that has lost 22% of its carbon-14?
   (use $P(t) = P_0 e^{-0.00012\,t}$.)

   a) 1860 yrs   b) 2070 yrs   c) 3080 yrs   d) 4290 yrs

25. An investment with interest compounded continuously doubled itself in 10 years. What is the interest rate?

   a) 6.93%   b) 4.32%   c) 8.66%   d) 7.25%

26. Solve. $\log_3 |2x+7| = 3$

   a) $-27, 37$ b) $-17, 10$ c) $-37, 27$ d) 10, 17

27. If $\log_a x = 2$, $\log_a y = 3$, and $\log_a z = 4$, then find $\log_a \dfrac{\sqrt[3]{x^2 y^2}}{\sqrt[3]{x^{-1} y z}}$.

   a) $\dfrac{13}{2}$ b) $\dfrac{4}{3}$ c) $\dfrac{5}{3}$ d) $\dfrac{5}{13}$

200

CHAPTER 9   NAME_____

TEST FORM H   CLASS_____ SCORE_____ GRADE_____

ANSWERS

1. Find $(f \circ g)(x)$ if $f(x) = -x + x^2$ and $g(x) = 3x - 2$.

   a) $-4x^2 - 15x - 12$ b) $-9x^2 + 14x + 6$
   c) $9x^2 - 15x + 6$ d) $-4x^2 + 14x - 12$

1._____

2. Find a formula for the inverse of $g(x) = \left(\dfrac{x+5}{3}\right)^3$

   a) $g^{-1}(x) = 2\sqrt[3]{x} + 5$ b) $g^{-1}(x) = 2\sqrt[3]{2x} + 1$

   c) $g^{-1}(x) = 3\sqrt[3]{2x} - 1$ d) $g^{-1}(x) = 3\sqrt[3]{x} - 5$

2._____

3. Which graph represents $f(x) = \left(\dfrac{3}{2}\right)^x - \dfrac{9}{4}$?

   a)
   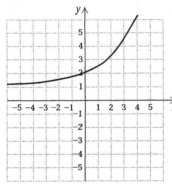

   b)
   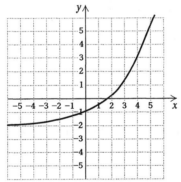

   c)
   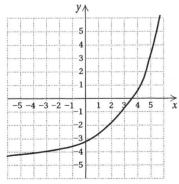

   d)
   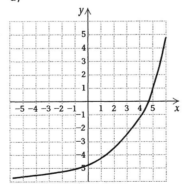

3._____

4._____

4. Simplify $\log_2 32$.

   a) 6   b) 3   c) 5   d) 2

CHAPTER 9

TEST FORM H

NAME_____

ANSWERS

5. _____

6. _____

7. _____

8. _____

9. _____

10. _____

5. Convert to a logarithmic equation. $6^{-3} = \dfrac{1}{216}$

a) $\log_6 \dfrac{1}{16} = -3$ b) $\log_6 \dfrac{1}{216} = -3$

c) $\log_4 \dfrac{1}{16} = -2$ d) $\log_4 \dfrac{1}{216} = -2$

6. Convert to a logarithmic equation. $289^{1/2} = 17$

a) $\log_{196} 14 = \dfrac{1}{2}$ b) $\log_{289} 14 = \dfrac{1}{2}$

c) $\log_{289} 17 = \dfrac{1}{2}$ d) $\log_{196} 17 = \dfrac{1}{2}$

7. Convert to an exponential equation. $m = \log_9 81$

a) $8^m = 64$ b) $9^m = 81$ c) $9^m = 64$ d) $8^m = 81$

8. Convert to an exponential equation. $\log_2 16 = 4$

a) $2^4 = 16$ b) $27^4 = 16$ c) $3^3 = 27$ d) $16^2 = 4$

9. Express in terms of logs of a, b, and c. $\log \dfrac{a\, b^{1/2}}{c^{1/3}}$

a) $2\log a + \log b - 3\log c$ b) $\log a + \dfrac{1}{2}\log b - \dfrac{1}{3}\log c$

c) $\log a + 2\log b - \dfrac{1}{3}\log c$ d) $2\log a + \log b - \dfrac{1}{3}\log c$

10. Express as a single logarithm. $2\log_a x = \dfrac{1}{2}\log_a z$

a) $\log_a \dfrac{x^4}{z^3}$ b) $\log_a \dfrac{x^{3/4}}{z^3}$ c) $\log_a \dfrac{x^2}{z^3}$ d) $\log_a \dfrac{x^2}{z^{1/2}}$

CHAPTER 9
TEST FORM H

NAME_____

ANSWERS

11. Simplify. $\log_w w$

   a) w b) 1 c) v d) 0

11._____

12. Simplify. $\log_t t^G$

   a) G b) t c) 0.01 d) 100

12._____

13. Simplify. $\log_g 1$

   a) 0 b) 1 c) h d) g

13._____

14. Given $\log_a 4 = 1.386$, $\log_a 5 = 1.609$, and $\log_a 6 = 1.792$, find $\log_a 36$.

   a) 4.604 b) 3.584 c) 4.787 d) 3.218

14._____

15. What is the range of $f(x) = e^x + 1$?

   a) $(3, \infty)$ b) $(-\infty, 3)$ c) $(1, \infty)$ d) $(-\infty, 1)$

15._____

16. What is the domain of $g(x) = \ln(x+2)$?

   a) $(-2, 2)$ b) $(-2, \infty)$ c) $(2, \infty)$ d) $(-\infty, 2)$

16._____

17. Solve. $4^x = \dfrac{1}{64}$

   a) 4 b) -3 c) -4 d) 3

17._____

18. Solve. $\log_x 81 = 2$

   a) 16 b) 4 c) 9 d) 3

18._____

19. Solve. $\log_{100} x = \dfrac{1}{2}$

   a) 9 b) 10 c) 18 d) 3

19._____

CHAPTER 9

TEST FORM H

NAME_____

ANSWERS

20._____

20. Solve. $5^x = 2.3$ Give exact answer in terms of common logarithms.

  a) $\dfrac{\log 2.3}{\log 5}$ b) $\dfrac{\log 2.4}{\log 4}$ c) $\dfrac{\log 4}{\log 2.4}$ d) $\dfrac{\log 5}{\log 2.3}$

21. Solve and give answer to the nearest ten-thousandth. $\ln x = \dfrac{15}{8}$

  a) 5.0789   b) 6.5208   c) 6.5272   d) 5.0784

21._____

22. Solve. $\log(x-1) + \log(x+1) = \log 2$

  a) 2   b) 4   c) 5   d) 7

22._____

23. The population in Malaysia in 2000 was 21.8 million, and was growing exponentially by 1.98% per year. When will Malaysia's population reach 30 million?

  a) 2017   b) 2025   c) 2031   d) 2022

23._____

24. How old is an animal bone that has lost 20% of its carbon-14?
(use $P(t) = P_0 e^{-0.00012t}$.)

  a) 1860 yr   b) 2070 yr   c) 3080 yr   d) 4290 yr

24._____

25. An investment with interest compounded continuously doubled itself in 8 years. What is the interest rate?

  a) 6.93%   b) 4.32%   c) 8.66%   d) 7.25%

25._____

26. Solve. $\log_4 |2x + 10| = 3$

  a) –27, 37   b) –17, 10   c) –37, 27   d) 10, 17

26._____

27. If $\log_a x = 2$, $\log_a y = 3$, and $\log_a z = 4$, then find $\log_a \dfrac{\sqrt[3]{x^2 y^2 z^2}}{\sqrt{y z^{-1}}}$.

  a) $\dfrac{13}{2}$ b) $\dfrac{4}{3}$ c) $\dfrac{5}{3}$ d) $\dfrac{5}{13}$

27._____

CHAPTER 10

TEST FORM A

NAME_____

CLASS_____ SCORE_____ GRADE_____

Find the distance between each pair of points. Where appropriate, find an approximation to three decimal places.

1. $(1, -3)$ and $(-8, 8)$
2. $(8, -a)$ and $(-8, a)$

Find the midpoint of the segment with the given endpoints.

3. $(1, -3)$ and $(-8, 8)$
4. $(8, -a)$ and $(-8, a)$

Find the center and the radius of each circle.

5. $(x + 2)^2 + (y - 9)^2 = 9$
6. $x^2 + y^2 + 2x - 8y + 1 = 0$

Classify the equation as a circle, an ellipse, a parabola, or a hyperbola. Then graph.

7. $y = x^2 - 8x + 12$

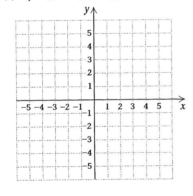

8. $x^2 + y^2 + 2x + 2y + 1 = 0$

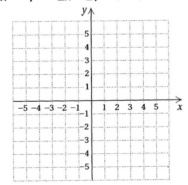

9. $\dfrac{x^2}{4} - \dfrac{y^2}{16} = 1$

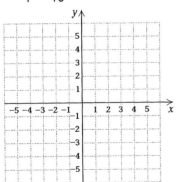

10. $x^2 + 36y^2 = 36$

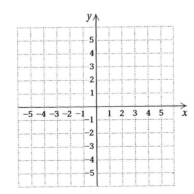

ANSWERS

1._____

2._____

3._____

4._____

5._____

6._____

7. See graph.

8. See graph.

9. See graph.

10. See graph.

CHAPTER 10

TEST FORM A

ANSWERS

11. See graph.

12. See graph.

13. _____

14. _____

15. _____

16. _____

17. _____

18. _____

19. _____

20. _____

21. _____

22. _____

23. _____

Classify the equation as a circle, an ellipse, a parabola, or a hyperbola. Then graph.

11. $xy = 3$

12. $x = -y^2 + 2y$

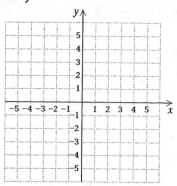
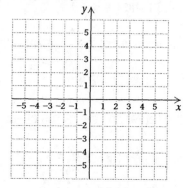

Solve.

13. $\dfrac{x^2}{9} + \dfrac{y^2}{4} = 1,$
    $2x + 3y = 6$

14. $x^2 + y^2 = 25,$
    $\dfrac{x^2}{25} - \dfrac{y^2}{16} = 1$

15. $x^2 - y^2 = 7,$
    $xy = 12$

16. $x^2 + y^2 = 7,$
    $x^2 = y^2 + 3$

17. A rectangle with diagonal of length $\sqrt{58}$ has an area of 21. Find the dimensions of the rectangle.

18. Two squares are such that the sum of their areas is 9 m^2 and the difference of their areas is 5 m^2. Find the length of a side of each square.

19. A rectangle has a diagonal of length 25 m and a perimeter of 70 m. Find the dimensions of the rectangle.

20. Shabaz invested a certain amount of money for one year and earned $50 in interest. Jasper invested $250 more at an interest rate that was 1% less than Shabaz, but earned the same amount of interest. Find Shabaz' principal and interest rate.

21. Find an equation of the ellipse passing through $(-1, 3)$ and $(5, 3)$ with vertices at $(2, -1)$ and $(2, 7)$.

22. Find the point on the y-axis that is equidistant from $(-9, 2)$ and $(4, -6)$.

23. The sum of two numbers is 20, and their product is 10. Find the sum of the reciprocals of the numbers.

CHAPTER 10

TEST FORM B

NAME_____

CLASS_____ SCORE_____ GRADE_____

ANSWERS

Find the distance between each pair of points. Where appropriate, find an approximation to three decimal places.

1. $(2, -5)$ and $(-6, 7)$
2. $(7, -a)$ and $(-7, a)$

Find the midpoint of the segment with the given endpoints.

3. $(2, -5)$ and $(-6, 7)$
4. $(7, -a)$ and $(-7, a)$

Find the center and the radius of each circle.

5. $(x-2)^2 + (y+8)^2 = 16$
6. $x^2 + y^2 - 2x + 6y + 2 = 0$

Classify the equation as a circle, an ellipse, a parabola, or a hyperbola. Then graph.

7. $y = x^2 - 6x + 5$
8. $x^2 + y^2 + 2x + 4y + 1 = 0$

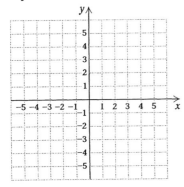
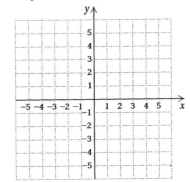

9. $\dfrac{x^2}{4} - \dfrac{y^2}{25} = 1$
10. $x^2 + 9y^2 = 36$

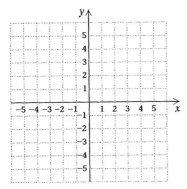
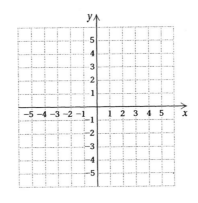

1._____

2._____

3._____

4._____

5._____

6._____

7. See graph._____

8. See graph._____

9. See graph._____

10. See graph._____

207

CHAPTER 10
TEST FORM B

NAME _____

ANSWERS

11. See graph.

12. See graph.

13. _____

14. _____

15. _____

16. _____

17. _____

18. _____

19. _____

20. _____

21. _____

22. _____

23. _____

Classify the equation as a circle, an ellipse, a parabola, or a hyperbola. Then graph.

11. $xy = 4$

12. $x = -y^2 - 2y$

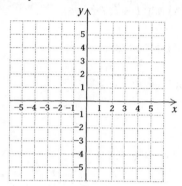
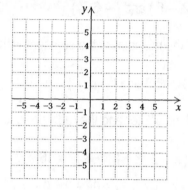

Solve.

13. $\dfrac{x^2}{16} + \dfrac{y^2}{25} = 1$,

$5x + 4y = 20$

14. $x^2 + y^2 = 16$,

$\dfrac{x^2}{16} - \dfrac{y^2}{9} = 1$

15. $x^2 - 4y^2 = 13$,

$xy = 21$

16. $x^2 + y^2 = 12$,

$x^2 = y^2 + 4$

17. A rectangle with diagonal of length $2\sqrt{29}$ has an area of 40. Find the dimensions of the rectangle.

18. Two squares are such that the sum of their areas is 16 m² and the difference of their areas is 6 m². Find the length of a side of each square.

19. A rectangle has a diagonal of length 13 m and a perimeter of 34 m. Find the dimensions of the rectangle.

20. Rufus invested a certain amount of money for one year and earned $60 in interest. Millie invested $200 more at an interest rate that was 1% less than Rufus, but earned the same amount of interest. Find Rufus' principal and interest rate.

21. Find an equation of the ellipse passing through (0, 2) and (6, 2) with vertices at (3, –2) and (3, 6).

22. Find the point on the y-axis that is equidistant from (–7, 4) and (5, –3).

23. The sum of two numbers is 24, and their product is 6. Find the sum of the reciprocals of the numbers.

CHAPTER 10
TEST FORM C

NAME_____

CLASS_____ SCORE_____ GRADE_____

ANSWERS

Find the distance between each pair of points. Where appropriate, find an approximation to three decimal places.

1. $(3, -7)$ and $(-4, 6)$
2. $(6, -a)$ and $(-6, a)$

Find the midpoint of the segment with the given endpoints.

3. $(3, -7)$ and $(-4, 6)$
4. $(6, -a)$ and $(-6, a)$

Find the center and the radius of each circle.

5. $(x + 2)^2 + (y - 7)^2 = 25$

6. $x^2 + y^2 + 4x - 4y + 3 = 0$

Classify the equation as a circle, an ellipse, a parabola, or a hyperbola. Then graph.

7. $y = x^2 - 4x - 1$

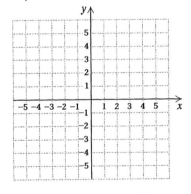

8. $x^2 + y^2 + 2x + 6y + 6 = 0$

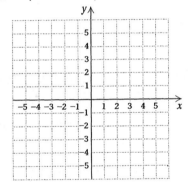

9. $\dfrac{x^2}{4} - \dfrac{y^2}{9} = 1$

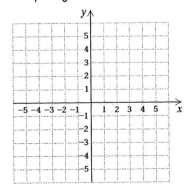

10. $x^2 + 4y^2 = 36$

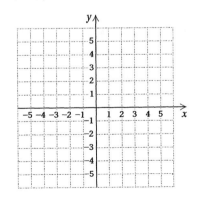

1._____

2._____

3._____

4._____

5._____

6._____

7. See graph.

8. See graph.

9. See graph.

10. See graph.

CHAPTER 10

TEST FORM C

NAME_____

ANSWERS

11. See graph.

12. See graph.

13._____

14._____

15._____

16._____

17._____

18._____

19._____

20._____

21._____

22._____

23._____

Classify the equation as a circle, an ellipse, a parabola, or a hyperbola. Then graph.

11. $xy = 5$

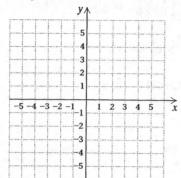

12. $x = -y^2 + 4y$

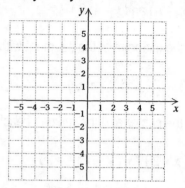

Solve.

13. $\dfrac{x^2}{4} + \dfrac{y^2}{16} = 1,$
    $4x + 2y = 8$

14. $x^2 + y^2 = 9,$
    $\dfrac{x^2}{9} - \dfrac{y^2}{4} = 1$

15. $x^2 - 3y^2 = 9,$
    $xy = 18$

16. $x^2 + y^2 = 5,$
    $x^2 = y^2 + 1$

17. A rectangle with diagonal of length $\sqrt{74}$ has an area of 45. Find the dimensions of the rectangle.

18. Two squares are such that the sum of their areas is 11 m² and the difference of their areas is 7 m². Find the length of a side of each square.

19. A rectangle has a diagonal of length 25 m and a perimeter of 62 m. Find the dimensions of the rectangle.

20. Nicole invested a certain amount of money for one year and earned $55 in interest. Jon invested $275 more at an interest rate that was 1% less than Nicole, but earned the same amount of interest. Find Nicole's principal and interest rate.

21. Find an equation of the ellipse passing through $(4, -1)$ and $(4, 3)$ with vertices at $(-1, 1)$ and $(9, 1)$.

22. Find the point on the y-axis that is equidistant from $(-5, 6)$ and $(4, -3)$.

23. The sum of two numbers is 35, and their product is 7. Find the sum of the reciprocals of the numbers.

CHAPTER 10 NAME_____

TEST FORM D CLASS_____ SCORE_____ GRADE_____

Find the distance between each pair of points. Where appropriate, find an approximation to three decimal places.

1. $(4, -9)$ and $(-2, 5)$

2. $(5, -a)$ and $(-5, a)$

Find the midpoint of the segment with the given endpoints.

3. $(4, -9)$ and $(-2, 5)$

4. $(5, -a)$ and $(-5, a)$

Find the center and the radius of each circle.

5. $(x-2)^2 + (y+6)^2 = 36$

6. $x^2 + y^2 - 4x + 2y + 4 = 0$

Classify the equation as a circle, an ellipse, a parabola, or a hyperbola. Then graph.

7. $y = x^2 - 2x + 3$

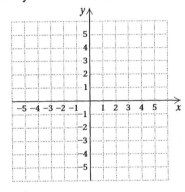

8. $x^2 + y^2 + 2x + 8y + 16 = 0$

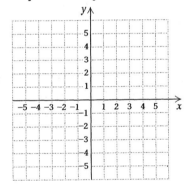

9. $\dfrac{x^2}{4} - y^2 = 1$

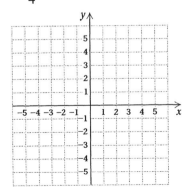

10. $4x^2 + 9y^2 = 144$

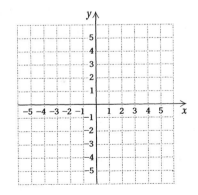

ANSWERS

1._____

2._____

3._____

4._____

5._____

6._____

7. See graph._____

8. See graph._____

9. See graph._____

10. See graph._____

CHAPTER 10

TEST FORM D

NAME_____

ANSWERS

11. See graph.

12. See graph.

13. _____

14. _____

15. _____

16. _____

17. _____

18. _____

19. _____

20. _____

21. _____

22. _____

23. _____

Classify the equation as a circle, an ellipse, a parabola, or a hyperbola. Then graph.

11. $xy = 6$

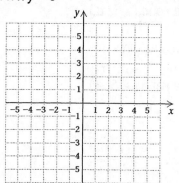

12. $x = -y^2 - 4y$

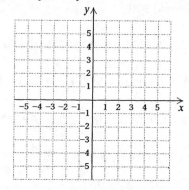

Solve.

13. $\dfrac{x^2}{25} + \dfrac{y^2}{9} = 1$,

$3x + 5y = 15$

14. $x^2 + y^2 = 4$,

$\dfrac{x^2}{4} - \dfrac{y^2}{2} = 1$

15. $x^2 - 7y^2 = 1$,

$xy = 24$

16. $x^2 + y^2 = 8$,

$x^2 = y^2 + 6$

17. A rectangle with diagonal of length $\sqrt{205}$ has an area of 78. Find the dimensions of the rectangle.

18. Two squares are such that the sum of their areas is 6 m² and the difference of their areas is 4 m². Find the length of a side of each square.

19. A rectangle has a diagonal of length 26 m and a perimeter of 68 m. Find the dimensions of the rectangle.

20. Jasmine invested a certain amount of money for one year and earned $66 in interest. Juan invested $220 more at an interest rate that was 1% less than Jasmine, but earned the same amount of interest. Find Jasmine's principal and interest rate.

21. Find an equation of the ellipse passing through (1, 2) and (1, 6) with vertices at (−4, 4) and (6, 4).

22. Find the point on the y-axis that is equidistant from (−3, 8) and (5, −6).

23. The sum of two numbers is 16, and their product is 2. Find the sum of the reciprocals of the numbers.

CHAPTER 10

TEST FORM E

NAME_____

CLASS_____ SCORE_____ GRADE_____

ANSWERS

Find the distance between each pair of points. Where appropriate, find an approximation to three decimal places.

1. $(5, -9)$ and $(-2, 4)$
2. $(4, -a)$ and $(-4, a)$

1._____

Find the midpoint of the segment with the given endpoints.

3. $(5, -9)$ and $(-2, 4)$
4. $(4, -a)$ and $(-4, a)$

2._____

Find the center and the radius of each circle.

5. $(x + 7)^2 + (y - 4)^2 = 36$
6. $x^2 + y^2 + 6x - 4y + 4 = 0$

3._____

4._____

Classify the equation as a circle, an ellipse, a parabola, or a hyperbola. Then graph.

5._____

7. $y = x^2 + 2x - 1$
8. $x^2 + y^2 + 6x + 8y + 21 = 0$

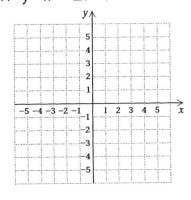 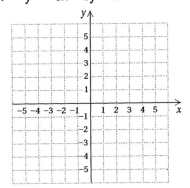

6._____

7. See graph.

8. See graph.

9. $\dfrac{y^2}{4} - x^2 = 1$
10. $25x^2 + 16y^2 = 400$

9. See graph.

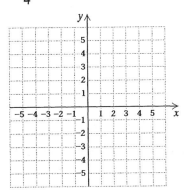 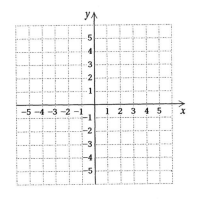

10. See graph.

CHAPTER 10
TEST FORM E

NAME

ANSWERS

11. See graph.

12. See graph.

13. _____

14. _____

15. _____

16. _____

17. _____

18. _____

19. _____

20. _____

21. _____

22. _____

23. _____

Classify the equation as a circle, an ellipse, a parabola, or a hyperbola. Then graph.

11. $xy = -6$

12. $x = -y^2 + 6y - 6$

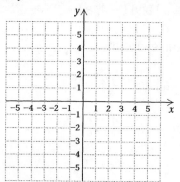
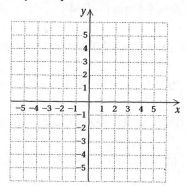

Solve.

13. $\dfrac{x^2}{4} + \dfrac{y^2}{25} = 1$,

$5x + 2y = 10$

14. $x^2 + y^2 = 25$,

$\dfrac{x^2}{25} - \dfrac{y^2}{9} = 1$

15. $x^2 - 2y^2 = -7$,

$xy = 20$

16. $x^2 + y^2 = 9$,

$x^2 = y^2 + 5$

17. A rectangle with diagonal of length $\sqrt{157}$ has an area of 66. Find the dimensions of the rectangle.

18. Two squares are such that the sum of their areas is 8 m^2 and the difference of their areas is 6 m^2. Find the length of a side of each square.

19. A rectangle has a diagonal of length 10 m and a perimeter of 28 m. Find the dimensions of the rectangle.

20. Trey invested a certain amount of money for one year and earned $65 in interest. Willie invested $325 more at an interest rate that was 1% less than Trey, but earned the same amount of interest. Find Trey's principal and interest rate.

21. Find an equation of the ellipse passing through $(-1, 4)$ and $(3, 4)$ with vertices at $(1, -1)$ and $(1, 9)$.

22. Find the point on the y-axis that is equidistant from $(-3, 2)$ and $(4, -7)$.

23. The sum of two numbers is 27, and their product is 9. Find the sum of the reciprocals of the numbers.

CHAPTER 10

TEST FORM F

NAME_____

CLASS_____ SCORE_____ GRADE_____

ANSWERS

Find the distance between each pair of points. Where appropriate, find an approximation to three decimal places.

1. $(6, -7)$ and $(-4, 3)$ 2. $(3, -a)$ and $(-3, a)$

Find the midpoint of the segment with the given endpoints.

3. $(6, -7)$ and $(-4, 3)$ 4. $(3, -a)$ and $(-3, a)$

Find the center and the radius of each circle.

5. $(x-7)^2 + (y+5)^2 = 25$

6. $x^2 + y^2 - 6x + 4y + 3 = 0$

Classify the equation as a circle, an ellipse, a parabola, or a hyperbola. Then graph.

7. $y = x^2 + 4x + 2$ 8. $x^2 + y^2 + 6x + 6y + 9 = 0$

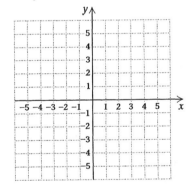
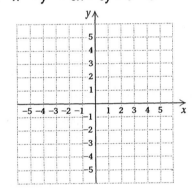

9. $\dfrac{y^2}{4} - \dfrac{x^2}{4} = 1$ 10. $25x^2 + 9y^2 = 225$

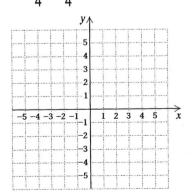
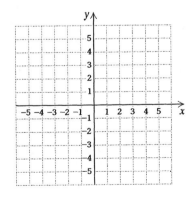

1._____

2._____

3._____

4._____

5._____

6._____

7. See graph.

8. See graph.

9. See graph.

10. See graph.

CHAPTER 10
TEST FORM F

ANSWERS

11. See graph.

12. See graph.

13. _____

14. _____

15. _____

16. _____

17. _____

18. _____

19. _____

20. _____

21. _____

22. _____

23. _____

Classify the equation as a circle, an ellipse, a parabola, or a hyperbola. Then graph.

11. $xy = -5$

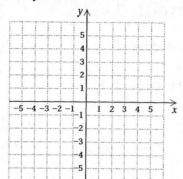

12. $x = -y^2 - 6y - 6$

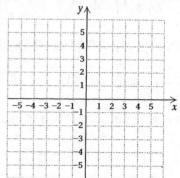

Solve.

13. $\dfrac{x^2}{9} + \dfrac{y^2}{16} = 1$,
    $4x + 3y = 12$

14. $x^2 + y^2 = 16$,
    $\dfrac{x^2}{16} - \dfrac{y^2}{4} = 1$

15. $x^2 - 3y^2 = -2$,
    $xy = 15$

16. $x^2 + y^2 = 16$,
    $x^2 = y^2 + 6$

17. A rectangle with diagonal of length 13 has an area of 60. Find the dimensions of the rectangle.

18. Two squares are such that the sum of their areas is 5 m² and the difference of their areas is 1 m². Find the length of a side of each square.

19. A rectangle has a diagonal of length 17 m and a perimeter of 46 m. Find the dimensions of the rectangle.

20. Wally invested a certain amount of money for one year and earned $78 in interest. Wanda invested $260 more at an interest rate that was 1% less than Wally, but earned the same amount of interest. Find Wally's principal and interest rate.

21. Find an equation of the ellipse passing through (2, 1) and (6, 1) with vertices at (4, −4) and (4, 6).

22. Find the point on the y-axis that is equidistant from (−5, 4) and (5, −2).

23. The sum of two numbers is 65, and their product is 13. Find the sum of the reciprocals of the numbers.

CHAPTER 10

TEST FORM G

NAME_____

CLASS_____ SCORE_____ GRADE_____

ANSWERS

Find the distance between each pair of points.

1. $(7, -5)$ and $(-6, 2)$

    a) $16\sqrt{17}$ b) $2\sqrt{109}$ c) $4\sqrt{17}$ d) $\sqrt{218}$

1._____

2. $(2, -a)$ and $(-2, a)$

    a) $2\sqrt{4+a^2}$ b) $2\sqrt{a}$ c) $\sqrt{2a}$ d) $2\sqrt{1+a^2}$

2._____

Find the midpoint of the segment with the given endpoints.

3._____

3. $(7, -5)$ and $(-6, 2)$

    a) $(0, -0.5)$ b) $(-1, 0)$ c) $(0.5, -1.5)$ d) $(0, -1)$

4. $(2, -a)$ and $(-2, a)$

    a) $(0, 0)$ b) $(2, 0)$ c) $(0, 2)$ d) $(1, 1)$

4._____

Find the center and the radius of each circle.

5. $(x+7)^2 + (y-6)^2 = 16$

    a) $(-7, 7)$; 3   b) $(7, -6)$; 4
    c) $(7, -7)$; 3   d) $(-7, 6)$; 4

5._____

6._____

6. $x^2 + y^2 + 8x - 4y + 2 = 0$

    a) $(4, 1)$; 4   b) $(-4, 2)$; $3\sqrt{2}$
    c) $(4, -2)$; $3\sqrt{2}$ d) $(4, -1)$; 4

7._____

Classify the equation as a circle, an ellipse, a parabola, or a hyperbola.

7. $y = x^2 + 6x + 6$

    a) parabola   b) ellipse   c) circle   d) hyperbola

8._____

8. $x^2 + y^2 + 6x + 4y + 4 = 0$

    a) ellipse   b) hyperbola   c) circle   d) parabola

CHAPTER 10
TEST FORM G

NAME_____

ANSWERS

Graph

9. $\dfrac{y^2}{4} - x^2 = 1$

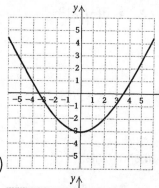
a)

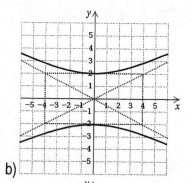
b)

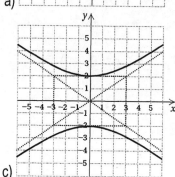
c)

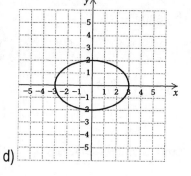
d)

9._____

10. $25x^2 + 4y^2 = 100$

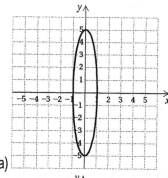
a)

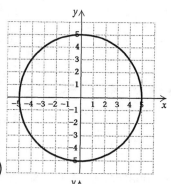
b)

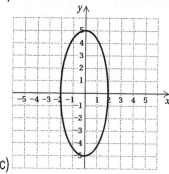
c)

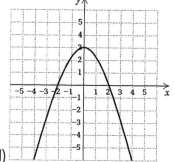
d)

10._____

CHAPTER 10

TEST FORM G

NAME_____

11. What is the equation corresponding to the following graph?

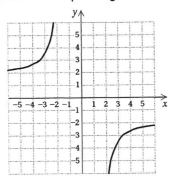

a) $\dfrac{y^2}{16} - \dfrac{x^2}{4} = 1$ b) $\dfrac{x^2}{4} - \dfrac{y^2}{16} = 1$

c) $xy = -13$ d) $xy = -4$

ANSWERS

11._____

12. What is the equation corresponding to the following graph?

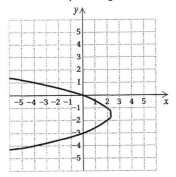

a) $y = x^2 - 3x$ b) $y = x^2 + 3x$

c) $x = -y^2 + 3y$ d) $x = -y^2 - 3y$

12._____

13. Find one solution to the system. $\dfrac{x^2}{16} + \dfrac{y^2}{4} = 1$,

$2x + 4y = 8$

a) $(2, 0)$ b) $(0, 4)$ c) $(0, 5)$ d) $(4, 0)$

13._____

14. Find one solution to the system. $x^2 + y^2 = 9$,

$\dfrac{x^2}{9} - \dfrac{y^2}{2} = 1$

a) $(2, 0)$ b) $(0, 3)$ c) $(0, -2)$ d) $(-3, 0)$

14._____

CHAPTER 10

TEST FORM G

NAME_____

ANSWERS

15._____

15. A rectangle with diagonal of length $\sqrt{65}$ has an area of 28. Find the dimensions of the rectangle.

a) 7 by 4   b) 24 by 3   c) 7 by 2   d) 12 by 3

16. Two squares are such that the sum of their areas is 12 m² and the difference of their areas is 4 m². Find the length of a side of each square.

a) $\sqrt{5}$ m, $\sqrt{2}$ m   b) $\sqrt{2}$ m, 5 m
c) $2\sqrt{2}$ m, 2 m   d) $2\sqrt{5}$ m, $2\sqrt{2}$ m

16._____

17. A rectangle has a diagonal of length 29 m and a perimeter of 82 m. Find the larger dimension of the rectangle.

a) 20 m   b) 12 m   c) 21 m   d) 9 m

17._____

18. Wendy invested a certain amount of money for one year and earned $70 in interest. Casper invested $350 more at an interest rate that was 1% less than Wendy, but earned the same amount of interest. Find Wendy's principal and interest rate.

18._____

a) $2000 ; 4%   b) $1400 ; 6%   c) $1250 ; 6%   d) $1400 ; 5%

19. Find an equation of the ellipse passing through (3, – 1) and (3, 5) with vertices at (– 1, 2) and (7, 2).

19._____

a) $\dfrac{(x-3)^2}{16}+\dfrac{(y-2)^2}{9}=1$ b) $\dfrac{(x+2)^2}{16}+\dfrac{(y+3)^2}{9}=1$

c) $\dfrac{(x-2)^2}{16}+\dfrac{(y-3)^2}{9}=1$ d) $\dfrac{(x-3)^2}{9}+\dfrac{(y-2)^2}{16}=1$

20. Find the point on the y-axis that is equidistant from (– 7, 6) and (4, – 2).

20._____

a) $\left(0,\dfrac{71}{30}\right)$ b) $\left(0,\dfrac{71}{16}\right)$ c) $\left(0,\dfrac{13}{6}\right)$ d) $\left(0,\dfrac{65}{16}\right)$

21. The sum of two numbers is 30, and their product is 5. Find the sum of the reciprocals of the numbers.

21._____

a) 7   b) 5   c) 6   d) 4

220

CHAPTER 10
TEST FORM H

NAME_____

CLASS_____ SCORE_____ GRADE_____

ANSWERS

Find the distance between each pair of points.

1. $(8, -3)$ and $(-8, 1)$

    a) $16\sqrt{17}$ b) $2\sqrt{109}$ c) $4\sqrt{17}$ d) $\sqrt{218}$

1._____

2. $(1, -a)$ and $(-1, a)$

    a) $2\sqrt{4 + a^2}$ b) $2\sqrt{a}$ c) $\sqrt{2a}$ d) $2\sqrt{1 + a^2}$

2._____

Find the midpoint of the segment with the given endpoints.

3. $(8, -3)$ and $(-8, 1)$

    a) $(0, -0.5)$ b) $(-1, 0)$ c) $(0.5, -1.5)$ d) $(0, -1)$

3._____

4. $(1, -a)$ and $(-1, a)$

    a) $(0, a)$ b) $(2, 0)$ c) $(0, 2)$ d) $(0, 0)$

4._____

Find the center and the radius of each circle.

5. $(x - 7)^2 + (y + 7)^2 = 9$

    a) $(-7, 7); 3$ b) $(7, -6); 4$
    c) $(7, -7); 3$ d) $(-7, 6); 4$

5._____

6. $x^2 + y^2 - 8x + 2y + 1 = 0$

    a) $(4, 1); 4$ b) $(-4, 2); 3\sqrt{2}$
    c) $(4, -2); 3\sqrt{2}$ d) $(4, -1); 4$

6._____

Classify the equation as a circle, an ellipse, a parabola, or a hyperbola.

7. $y = x^2 + 8x + 14$

    a) ellipse   b) parabola   c) circle   d) hyperbola

7._____

8. $x^2 + y^2 + 6x + 2y + 9 = 0$

    a) ellipse   b) circle   c) hyperbola   d) parabola

8._____

CHAPTER 10
TEST FORM H

NAME_____

ANSWERS

Graph

9. $\dfrac{y^2}{4} - \dfrac{x^2}{16} = 1$

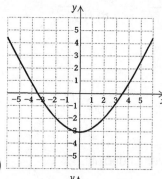
a)

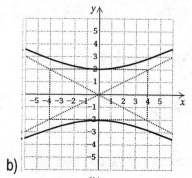
b)

9. _____

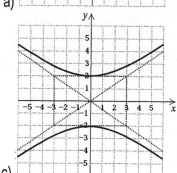
c)

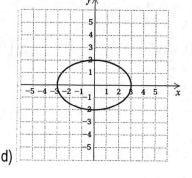
d)

10. $25x^2 + y^2 = 25$

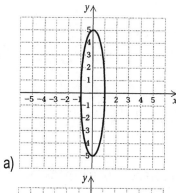
a)

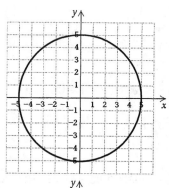
b)

10. _____

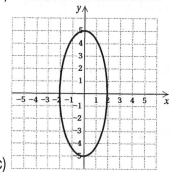
c)

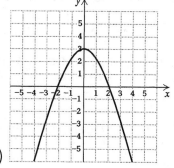
d)

CHAPTER 10

TEST FORM H

NAME_____

ANSWERS

11. What is the equation corresponding to the following graph?

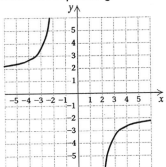

a) $\dfrac{y^2}{16} - \dfrac{x^2}{4} = 1$ b) $\dfrac{x^2}{4} - \dfrac{y^2}{16} = 1$

c) $xy = -13$ d) $xy = -4$

11._____

12. What is the equation corresponding to the following graph?

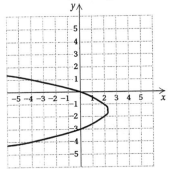

a) $y = x^2 - 3x$ b) $y = x^2 + 3x$

c) $x = -y^2 + 3y$ d) $x = -y^2 - 3y$

12._____

13. Find one solution to the system. $\dfrac{x^2}{25} + \dfrac{y^2}{16} = 1$,
$4x + 5y = 20$

a) $(2, 0)$ b) $(0, 4)$ c) $(0, 5)$ d) $(4, 0)$

13._____

14. Find one solution to the system. $x^2 + y^2 = 4$,
$\dfrac{x^2}{4} - \dfrac{y^2}{3} = 1$

a) $(2, 0)$ b) $(0, 3)$ c) $(0, -2)$ d) $(-3, 0)$

14._____

CHAPTER 10

TEST FORM H

NAME_____

ANSWERS

15._____

16._____

17._____

18._____

19._____

20._____

21._____

15. A rectangle with diagonal of length $3\sqrt{65}$ has an area of 72. Find the dimensions of the rectangle.

   a) 7 by 4   b) 24 by 3   c) 7 by 2   d) 12 by 3

16. Two squares are such that the sum of their areas is 7 m² and the difference of their areas is 3 m². Find the length of a side of each square.

   a) $\sqrt{5}$ m, $\sqrt{2}$ m   b) $\sqrt{2}$ m, 5 m
   c) $2\sqrt{2}$ m, 2 m   d) $2\sqrt{5}$ m, $2\sqrt{2}$ m

17. A rectangle has a diagonal of length 15 m and a perimeter of 42 m. Find the larger dimension of the rectangle.

   a) 20 m   b) 12 m   c) 21 m   d) 9 m

18. Wendell invested a certain amount of money for one year and earned $84 in interest. Berry invested $280 more at an interest rate that was 1% less than Wendell, but earned the same amount of interest. Find Wendell's principal and interest rate.

   a) $2000 ; 4%   b) $1400 ; 6%   c) $1250 ; 6%   d) $1400 ; 5%

19. Find an equation of the ellipse passing through (2, 0) and (2, 6) with vertices at (−2, 3) and (6, 3).

   a) $\dfrac{(x-3)^2}{16} + \dfrac{(y-2)^2}{9} = 1$ b) $\dfrac{(x+2)^2}{16} + \dfrac{(y+3)^2}{9} = 1$
   c) $\dfrac{(x-2)^2}{16} + \dfrac{(y-3)^2}{9} = 1$ d) $\dfrac{(x-3)^2}{9} + \dfrac{(y-2)^2}{16} = 1$

20. Find the point on the *y*-axis that is equidistant from (−9, 8) and (5, −7).

   a) $\left(0, \dfrac{71}{30}\right)$ b) $\left(0, \dfrac{71}{16}\right)$ c) $\left(0, \dfrac{13}{6}\right)$ d) $\left(0, \dfrac{65}{16}\right)$

21. The sum of two numbers is 21, and their product is 3. Find the sum of the reciprocals of the numbers.

   a) 7   b) 5   c) 6   d) 4

CHAPTER 11

TEST FORM A

NAME_____

CLASS_____ SCORE_____ GRADE_____

ANSWERS

1. Find the first five terms and the 16th term of a sequence with general term
$a_n = 5n - 2$.

1._____

2. Predict the general term of the sequence $\dfrac{1}{4}, \dfrac{1}{16}, \dfrac{1}{64}, \ldots$

2._____

3. Write out and evaluate: $\sum_{k=1}^{5}(2 - 3k)$

3._____

4. Rewrite using sigma notation: $(-1) + 4 + (-27) + 256 + (-3125)$

4._____

5. Find the 12th term, a_{12}, of the arithmetic sequence 10, 7, 4, ...

Assume arithmetic sequences for Questions 6 and 7.

5._____

6. Find the common difference d when $a_1 = 1$ and $a_7 = 4\dfrac{3}{4}$.

6._____

7. Find a_1 and d when $a_5 = 3$ and $a_{10} = -12$.

7._____

8. Find the sum of all multiples of 4 from 16 to 100 inclusive.

8._____

9. Find the 6th term of the geometric sequence 100, 25, $6\dfrac{1}{4}$, ...

9._____

10. Find the common ration of the geometric sequence 30, 20, $13\dfrac{1}{3}$, ...

10._____

11. Find the nth term of the geometric sequence $-4, 8, -16, \ldots$

11._____

12. Find the sum of the first nine terms of the geometric series
$(1 + x) + (3 + 3x) + (9 + 9x) + \ldots$

12._____

CHAPTER 11

TEST FORM A

NAME_____

ANSWERS

13._____

14._____

15._____

16._____

17._____

18._____

19._____

20._____

21._____

22._____

23._____

24._____

25._____

Determine whether each infinite geometric series has a limit. If a limit exists, find it.

13. $0.5 + 0.4 + 0.32 + \ldots$

14. $0.7 + 1.4 + 2.1 + \ldots$

15. $\$1900 + \$95 + \$4.75 + \ldots$

16. Find fractional notation for $0.15151515\ldots$

17. An auditorium has 16 seats in the first row, 20 seats in the second row, 24 seats in the third row, and so on for 20 rows. How many seats are in the 15th row?

18. In 1964, Francois began receiving cash gifts in the mail from an unknown person. A check would arrive without fail on August 13 every year. The first check was for $64, the check in 1965 was for $65, and so on until 2000 when Francois received an empty envelope on which the single word "*Sorry*" was printed in green crayon. How much in total had Francois received?

19. Each week the price of a $20,000 boat will be reduced 5% of the previous week's price. If we assume that it is not sold, what will be the price after 10 weeks?

20. Find the total rebound distance of a ball that is dropped from a height of 6 m, with each rebound two-fifths of the preceding one.

21. Simplify: $\binom{11}{4}$

22. Expand: $(x^2 - 2y)^5$

23. Find the 5th term in the expansion of $(a + x)^{10}$

24. Find a formula for the sum $S_n = \dfrac{1}{2} + 1 + \dfrac{3}{2} + \ldots + \dfrac{n}{2}$

25. Find the sum of the first n terms of $1 + \dfrac{2}{x} + \dfrac{4}{x^2} + \dfrac{8}{x^3} + \ldots$

CHAPTER 11　　　　　　　　　　　　　　　　　NAME_____

TEST FORM B　　　　　　　　　　　CLASS_____SCORE_____GRADE_____

1. Find the first five terms and the 16th term of a sequence with general term $a_n = 5n - 4$.

2. Predict the general term of the sequence $\frac{2}{5}, \frac{2}{25}, \frac{2}{125}, \ldots$

3. Write out and evaluate: $\sum_{k=1}^{5} (40 - 2^k)$

4. Rewrite using sigma notation: $\frac{1}{2} + 1 + \frac{3}{2} + 2 + \frac{5}{2}$

5. Find the 12th term, a_{12}, of the arithmetic sequence 16, 19, 22, ...

Assume arithmetic sequences for Questions 6 and 7.

6. Find the common difference d when $a_1 = \frac{3}{2}$ and $a_9 = 7\frac{1}{2}$.

7. Find a_1 and d when $a_5 = 4\frac{8}{9}$ and $a_{10} = -11\frac{5}{9}$.

8. Find the sum of all multiples of 5 from 10 to 200 inclusive.

9. Find the 6th term of the geometric sequence 96, 32, $10\frac{2}{3}$, ...

10. Find the common ration of the geometric sequence 60, 45, $33\frac{3}{4}$, ...

11. Find the nth term of the geometric sequence 2, -4, 8, ...

12. Find the sum of the first nine terms of the geometric series $(3 + 3x) + (6 + 6x) + (12 + 12x) + \ldots$

ANSWERS

1._____

2._____

3._____

4._____

5._____

6._____

7._____

8._____

9._____

10._____

11._____

12._____

CHAPTER 11

TEST FORM B

NAME_____

ANSWERS

Determine whether each infinite geometric series has a limit. If a limit exists, find it.

13. _____

13. $6 + 4.2 + 2.94 + \ldots$

14. $10 + 10.1 + 10.201 + \ldots$

14. _____

15. $\$4700 + \$282 + \$16.92 + \ldots$

16. Find fractional notation for $0.47474747\ldots$

15. _____

17. An auditorium has 25 seats in the first row, 27 seats in the second row, 29 seats in the third row, and so on for 20 rows. How many seats are in the 15th row?

16. _____

18. In 1965, Grady began receiving cash gifts in the mail from an unknown person. A check would arrive without fail on September 25 every year. The first check was for $65, the check in 1966 was for $66, and so on until 2000 when Grady received an empty envelope on which the single word "*Sorry*" was printed in green crayon. How much in total had Grady received?

17. _____

18. _____

19. _____

19. Each week the price of a $25,000 boat will be reduced 5% of the previous week's price. If we assume that it is not sold, what will be the price after 10 weeks?

20. _____

20. Find the total rebound distance of a ball that is dropped from a height of 7 m, with each rebound two-fifths of the preceding one.

21. _____

21. Simplify: $\binom{11}{6}$

22. _____

22. Expand: $(x^2 + 2y)^4$

23. _____

23. Find the 6th term in the expansion of $(a + x)^{10}$

24. _____

24. Find a formula for the sum $S_n = \frac{1}{3} + \frac{2}{3} + 1 + \ldots + \frac{n}{3}$

25. _____

25. Find the sum of the first *n* terms of $1 + \frac{3}{x^2} + \frac{9}{x^4} + \frac{27}{x^6} + \ldots$

CHAPTER 11 NAME_____

TEST FORM C   CLASS_____ SCORE_____ GRADE_____

ANSWERS

1. Find the first five terms and the 16th term of a sequence with general term
$a_n = 6n - 3$.

1._____

2. Predict the general term of the sequence $\dfrac{3}{2}, \dfrac{3}{4}, \dfrac{3}{8}, \ldots$

2._____

3. Write out and evaluate: $\displaystyle\sum_{k=1}^{5}(5 - 2^k)$

3._____

4. Rewrite using sigma notation: $(-2) + 4 + (-6) + 8 + (-10)$

4._____

5. Find the 12th term, a_{12}, of the arithmetic sequence $10, 5, 0, \ldots$

5._____

Assume arithmetic sequences for Questions 6 and 7.

6. Find the common difference d when $a_1 = 2$ and $a_7 = 7$.

6._____

7. Find a_1 and d when $a_5 = -20\dfrac{1}{2}$ and $a_{10} = -45\dfrac{1}{2}$.

7._____

8. Find the sum of all multiples of 6 from 30 to 180 inclusive.

8._____

9. Find the 6th term of the geometric sequence $10, 6\dfrac{2}{3}, 4\dfrac{4}{9}, \ldots$

9._____

10. Find the common ration of the geometric sequence $25, 16\dfrac{2}{3}, 11\dfrac{1}{9}, \ldots$

10._____

11. Find the nth term of the geometric sequence $\dfrac{10}{3}, \dfrac{20}{9}, \dfrac{40}{27}, \ldots$

11._____

12. Find the sum of the first nine terms of the geometric series
$(4 + 4x) + (12 + 12x) + (36 + 36x) + \ldots$

12._____

CHAPTER 11

TEST FORM C

NAME_____

ANSWERS

Determine whether each infinite geometric series has a limit. If a limit exists, find it.

13. _____

13. $0.4 + 0.24 + 0.144 + ...$

14. $1 + 1.5 + 2.25 + ...$

14. _____

15. $\$2300 + \$184 + \$14.72 + ...$

16. Find fractional notation for $0.63636363...$

15. _____

17. An auditorium has 40 seats in the first row, 46 seats in the second row, 52 seats in the third row, and so on for 20 rows. How many seats are in the 15th row?

16. _____

18. In 1970, Penelope began receiving cash gifts in the mail from an unknown person. A check would arrive without fail on October 10 every year. The first check was for $70, the check in 1971 was for $71, and so on until 2000 when Penelope received an empty envelope on which the single word "*Sorry*" was printed in green crayon. How much in total had Penelope received?

17. _____

18. _____

19. _____

19. Each week the price of a $30,000 boat will be reduced 5% of the previous week's price. If we assume that it is not sold, what will be the price after 8 weeks?

20. _____

20. Find the total rebound distance of a ball that is dropped from a height of 8 m, with each rebound two-fifths of the preceding one.

21. _____

21. Simplify: $\binom{11}{8}$

22. _____

22. Expand: $(x^2 - 2y)^5$

23. _____

23. Find the 7th term in the expansion of $(a + x)^{10}$

24. _____

24. Find a formula for the sum $S_n = 1 + 3 + 5 + ... + (2n - 1)$

25. _____

25. Find the sum of the first *n* terms of $1 + \dfrac{3}{x} + \dfrac{9}{x^2} + \dfrac{27}{x^3} + ...$

CHAPTER 11 NAME_____

TEST FORM D CLASS_____ SCORE_____ GRADE_____

1. Find the first five terms and the 16th term of a sequence with general term
   $a_n = 6n - 4$.

2. Predict the general term of the sequence $\frac{5}{3}, \frac{5}{9}, \frac{5}{27}, \ldots$

3. Write out and evaluate: $\sum_{k=1}^{5} (6-k)^k$

4. Rewrite using sigma notation: $1 + \frac{1}{4} + \frac{1}{9} + \frac{1}{16} + \frac{1}{25}$

5. Find the 12th term, a_{12}, of the arithmetic sequence 16, 21, 26, ...

Assume arithmetic sequences for Questions 6 and 7.

6. Find the common difference d when $a_1 = 1\frac{3}{4}$ and $a_9 = 25\frac{3}{4}$.

7. Find a_1 and d when $a_5 = -11\frac{1}{2}$ and $a_{10} = -15\frac{7}{8}$.

8. Find the sum of all multiples of 7 from 70 to 700 inclusive.

9. Find the 6th term of the geometric sequence 16, 8, 4, ...

10. Find the common ration of the geometric sequence 40, 30, $22\frac{1}{2}$, ...

11. Find the nth term of the geometric sequence $-1, \frac{3}{2}, -\frac{9}{4}, \ldots$

12. Find the sum of the first nine terms of the geometric series
    $(2+x) + (4+2x) + (8+4x) + \ldots$

ANSWERS

1._____

2._____

3._____

4._____

5._____

6._____

7._____

8._____

9._____

10._____

11._____

12._____

CHAPTER 11

TEST FORM D

ANSWERS

Determine whether each infinite geometric series has a limit. If a limit exists, find it.

13. $15 + 7.5 + 3.75 + ...$

14. $0.01 + 0.012 + 0.014 + ...$

15. $\$7200 + \$288 + \$11.52 + ...$

16. Find fractional notation for $2.28282828...$

17. An auditorium has 32 seats in the first row, 35 seats in the second row, 38 seats in the third row, and so on for 20 rows. How many seats are in the 15th row?

18. In 1971, Melinda began receiving cash gifts in the mail from an unknown person. A check would arrive without fail on November 9 every year. The first check was for $71, the check in 1972 was for $72, and so on until 2000 when Melinda received an empty envelope on which the single word "*Sorry*" was printed in green crayon. How much in total had Melinda received?

19. Each week the price of a $35,000 boat will be reduced 5% of the previous week's price. If we assume that it is not sold, what will be the price after 8 weeks?

20. Find the total rebound distance of a ball that is dropped from a height of 9 m, with each rebound two-fifths of the preceding one.

21. Simplify: $\binom{11}{10}$

22. Expand: $(x + 2y^2)^4$

23. Find the 8th term in the expansion of $(a + x)^{10}$

24. Find a formula for the sum $S_n = 6 + 12 + 18 + ... + 6n$

25. Find the sum of the first *n* terms of $1 + \dfrac{1}{2x} + \dfrac{1}{4x^2} + \dfrac{1}{8x^3} + ...$

CHAPTER 11

TEST FORM E

NAME_____

CLASS_____ SCORE_____ GRADE_____

ANSWERS

1. Find the first five terms and the 16th term of a sequence with general term
   $a_n = 4n - 1$.

   1._____

2. Predict the general term of the sequence $\dfrac{5}{2}, \dfrac{25}{4}, \dfrac{125}{8}, \ldots$

   2._____

3. Write out and evaluate: $\sum_{k=1}^{5} (2-k)^k$

   3._____

4. Rewrite using sigma notation: $1 + (-4) + 9 + (-16) + 25$

   4._____

5. Find the 12th term, a_{12}, of the arithmetic sequence $10, 3, -4, \ldots$

   5._____

Assume arithmetic sequences for Questions 6 and 7.

6. Find the common difference d when $a_1 = 3$ and $a_7 = 4\dfrac{1}{2}$.

   6._____

7. Find a_1 and d when $a_5 = -26\dfrac{1}{4}$ and $a_{10} = -61\dfrac{1}{4}$.

   7._____

8. Find the sum of all multiples of 8 from 32 to 144 inclusive.

   8._____

9. Find the 6th term of the geometric sequence $5, 1, \dfrac{1}{5}, \ldots$

   9._____

10. Find the common ration of the geometric sequence $1\dfrac{1}{2}, 2\dfrac{1}{4}, 3\dfrac{1}{8}, \ldots$

    10._____

11. Find the nth term of the geometric sequence $\dfrac{3}{8}, \dfrac{3}{16}, -\dfrac{3}{32}, \ldots$

    11._____

12. Find the sum of the first nine terms of the geometric series
    $(1 + 5x) + (2 + 10x) + (4 + 20x) + \ldots$

    12._____

CHAPTER 11

TEST FORM E

NAME_____

ANSWERS

Determine whether each infinite geometric series has a limit. If a limit exists, find it.

13._____

13. $2 + 1.6 + 1.28 + ...$

14. $15 - 30 + 60 - ...$

14._____

15. $\$3800 + \$190 + \$9.50 + ...$

16. Find fractional notation for $1.19191919...$

15._____

17. An auditorium has 15 seats in the first row, 18 seats in the second row, 21 seats in the third row, and so on for 15 rows. How many seats are in the 13th row?

16._____

18. In 1979, Sophie began receiving cash gifts in the mail from an unknown person. A check would arrive without fail on December 14 every year. The first check was for $79, the check in 1980 was for $80, and so on until 2000 when Sophie received an empty envelope on which the single word "*Sorry*" was printed in green crayon. How much in total had Sophie received?

17._____

18._____

19._____

19. Each week the price of a $40,000 boat will be reduced 5% of the previous week's price. If we assume that it is not sold, what will be the price after 10 weeks?

20._____

20. Find the total rebound distance of a ball that is dropped from a height of 6 m, with each rebound three-fifths of the preceding one.

21._____

21. Simplify: $\binom{14}{12}$

22._____

22. Expand: $(2x^2 - 2y)^5$

23._____

23. Find the 5th term in the expansion of $(a + x)^{13}$

24._____

24. Find a formula for the sum $S_n = 1 + 4 + 7 + ... + (3n - 2)$

25._____

25. Find the sum of the first n terms of $1 + \dfrac{2}{x^3} + \dfrac{4}{x^6} + \dfrac{8}{x^9} + ...$

CHAPTER 11

TEST FORM F

NAME_____

CLASS_____ SCORE_____ GRADE_____

ANSWERS

1. Find the first five terms and the 16th term of a sequence with general term
   $a_n = 4n - 3$.

   1._____

2. Predict the general term of the sequence $\frac{3}{4}, \frac{9}{16}, \frac{27}{64}, \ldots$

   2._____

3. Write out and evaluate: $\sum_{k=1}^{5}(1-k)^{k-2}$

   3._____

4. Rewrite using sigma notation: $2 + (-4) + 8 + (-16) + 32$

   4._____

5. Find the 12th term, a_{12}, of the arithmetic sequence 16, 13, 10, ...

   5._____

Assume arithmetic sequences for Questions 6 and 7.

6. Find the common difference d when $a_1 = 3\frac{1}{3}$ and $a_9 = 14$.

   6._____

7. Find a_1 and d when $a_5 = 95$ and $a_{10} = 88\frac{3}{4}$.

   7._____

8. Find the sum of all multiples of 9 from 27 to 180 inclusive.

   8._____

9. Find the 6th term of the geometric sequence 625, 125, 25, ...

10. Find the common ration of the geometric sequence $10, 13\frac{1}{3}, 17\frac{7}{9}, \ldots$

    9._____

11. Find the nth term of the geometric sequence $\frac{5}{2}, 1, \frac{2}{5}, \ldots$

    10._____

    11._____

12. Find the sum of the first nine terms of the geometric series
    $(3 + x) + (9 + 2x) + (27 + 4x) + \ldots$

    12._____

235

CHAPTER 11   NAME_____

TEST FORM F

ANSWERS

Determine whether each infinite geometric series has a limit. If a limit exists, find it.

13. _____

13. $0.9 + 0.63 + 0.441 + \ldots$

14. $-0.5 + 1 - 2 + \ldots$

14. _____

15. $\$2350 + \$141 + \$8.46 + \ldots$

16. Find fractional notation for $0.432432432\ldots$

15. _____

17. An auditorium has 20 seats in the first row, 23 seats in the second row, 26 seats in the third row, and so on for 25 rows. How many seats are in the 22nd row?

16. _____

18. In 1985, Chet began receiving cash gifts in the mail from an unknown person. A check would arrive without fail on January 21 every year. The first check was for $85, the check in 1986 was for $86, and so on until 2000 when Chet received an empty envelope on which the single word "*Sorry*" was printed in green crayon. How much in total had Chet received?

17. _____

18. _____

19. Each week the price of a $35,000 boat will be reduced 7% of the previous week's price. If we assume that it is not sold, what will be the price after 10 weeks?

19. _____

20. _____

20. Find the total rebound distance of a ball that is dropped from a height of 7 m, with each rebound three-fifths of the preceding one.

21. _____

21. Simplify: $\binom{14}{10}$

22. Expand: $(2x^2 + y)^4$

22. _____

23. _____

23. Find the 6th term in the expansion of $(a + x)^{13}$

24. _____

24. Find a formula for the sum $S_n = 1 + 6 + 11 + \ldots + (5n - 4)$

25. _____

25. Find the sum of the first n terms of $1 - \dfrac{x}{3} + \dfrac{x^2}{9} - \dfrac{x^3}{27} + \ldots$

236

CHAPTER 11

TEST FORM G

NAME_____

CLASS_____ SCORE_____ GRADE_____

ANSWERS

1. Find the 16th term of a sequence with general term $a_n = 3n + 2$.
   a) 50 b) 54 c) 48 d) 52

   1._____

2. Predict the general term of the sequence $\frac{18}{7}, \frac{18}{49}, \frac{18}{343}, \ldots$

   a) $18\left(\frac{1}{7}\right)^n$ b) $\left(\frac{4}{7}\right)^n$ c) $\left(\frac{18}{7}\right)^n$ d) $\left(\frac{4}{3}\right)^n$

   2._____

3. Evaluate $\sum_{k=1}^{5}(4 - 2^k)$

   a) $-\frac{11}{32}$ b) -36 c) $\frac{15}{32}$ d) -42

   3._____

4. Express using sigma notation: $8 + 16 + 32 + 64 + 128$

   a) $\sum_{k=1}^{5}\frac{2^k}{3^k}$ b) $\sum_{k=1}^{5} 2^{k+2}$

   c) $\sum_{k=1}^{5} 2^k$ d) $\sum_{k=1}^{5}\frac{2^k}{3^{k-1}}$

   4._____

5. Find the 12th term of the arithmetic sequence $10, 1, -8, \ldots$
   a) -152 b) -77 c) -160 d) -89

   5._____

6. Find the common difference of an arithmetic sequence when $a_1 = 4$ and $a_7 = 6\frac{2}{5}$.

   a) $2\frac{1}{4}$ b) $\frac{2}{5}$ c) $\frac{4}{5}$ d) $5\frac{1}{4}$

   6._____

7. Find a_1 of an arithmetic sequence when $a_5 = -46$ and $a_{10} = -91$.

   a) $-\frac{6}{5}$ b) -10 c) $-\frac{5}{6}$ d) 16

   7._____

CHAPTER 11   NAME_____

TEST FORM G

ANSWERS

8. Find the sum of all the multiples of 10 from 100 to 1000 inclusive.

   a) 5050   b) 14,850   c) 1485   d) 50,050

8._____

9. Find the 6th term of the geometric sequence $82, 41, 20\frac{1}{2}, \ldots$

   a) $\frac{16}{9}$ b) $4\frac{9}{16}$ c) $\frac{4}{9}$ d) $2\frac{9}{16}$

9._____

10. Find the common ratio of the geometric sequence $5, 7\frac{1}{2}, 11\frac{1}{4}, \ldots$

    a) $\frac{4}{3}$ b) $\frac{3}{4}$ c) $\frac{3}{2}$ d) $\frac{2}{3}$

10._____

11. Find the nth term of the geometric sequence $5, 25, 125, \ldots$

    a) $\left(-\frac{1}{4}\right)^n$ b) 4^n c) $\left(-\frac{5}{4}\right)^n$ d) 5^n

11._____

12. Find the sum of the first nine terms of the geometric series

    $(6 + 6x) + (18 + 12x) + (54 + 24x) + \ldots$

12._____

   a) $59046 + 3066x$ b) $87381 + 3066x$

   c) $87381 + 9841x$ d) $59046 + 9841x$

13._____

Determine whether each infinite geometric series has a limit. If a limit exists, find it.

13. $10 + 6 + 3.6 + \ldots$

    a) no limit   b) 0.2   c) 15   d) 25

14._____

14. $0.01 - 0.015 + 0.0225 - \ldots$

    a) 0.5   b) 0   c) no limit   d) 1

15._____

15. $\$4600 + \$368 + \$29.44 + \ldots$

    a) $2500   b) no limit   c) $5000   d) $5500

238

CHAPTER 11

TEST FORM G

NAME_____

ANSWERS

16. Find fractional notation for 0.518518518...

a) $\dfrac{76}{555}$ b) $\dfrac{45}{111}$ c) $\dfrac{518}{999}$ d) $\dfrac{89}{333}$

16._____

17. An auditorium has 15 seats in the first row, 17 seats in the second row, 19 seats in the third row, and so on for 35 rows. How many seats are in the 31st row?

a) 79   b) 75   c) 98   d) 87

17._____

18. In 1990, Lucas began receiving cash gifts in the mail from an unknown person. A check would arrive without fail on February 18 every year. The first check was for $90, the check in 1991 was for $91, and so on until 2000 when Lucas received an empty envelope on which the single word " Sorry " was printed in green crayon. How much in total had Lucas received?

a) $895   b) $945   c) $1890   d) $2175

18._____

19. Each week the price of a $30,000 boat will be reduced 7% of the previous week's price. If we assume that it is not sold, what will be the price after 8 weeks?

a) $18,051.03   b) $20,005.26
c) $17,876.12   d) $15,042.52

19._____

20. Find the total rebound distance of a ball that is dropped from a height of 8 m, with each rebound three-fifths of the preceding one.

20._____

a) 11 m   b) $10\dfrac{1}{2}$ m   c) $13\dfrac{1}{2}$ m   d) 12 m

CHAPTER 11

TEST FORM G

NAME_____

ANSWERS

21._____

21. Simplify: $\binom{14}{9}$

    a) 3003   b) 710   c) 2002   d) 1001

22. Expand $(2x - y^2)^5$

    a) $32x^5 - 32x^4y^2 + 24x^3y^4 - 16x^2y^6 + 8xy^8 - y^{10}$
    b) $16x^4 + 32x^3y^2 + 24x^2y^4 + 8xy^6 + y^8$
    c) $16x^4 + 32x^3y^2 + 80x^2y^4 + 24xy^6 + y^8$
    d) $32x^5 - 80x^4y^2 + 80x^3y^4 - 40x^2y^6 + 10xy^8 - y^{10}$

22._____

23._____

23. Find the 7th term in the expansion of $(a + x)^{13}$

    a) $1287a^7x^6$ b) $1716a^6x^7$ c) $1287a^7x^6$ d) $1716a^7x^6$

24. Find the formula for the sum $S_n = 1 + 7 + 13 + ... + 6n - 5$

24._____

    a) $3n^2 - 5n$ b) $3n^2 - 2n$ c) $5n^2$ d) $3n^2$

25. Find the sum of the first n terms of $2 - \dfrac{4}{x} + \dfrac{8}{x^2} - \dfrac{16}{x^3} + ...$

    a) $\dfrac{x^n + 4^n}{x^{n-1}(x+1)}$ b) $2\left[\dfrac{x^n + 2^n}{x^{n-1}(x+2)}\right]$

    c) $3\left[\dfrac{x^{2n} + 2^n}{x^{n-2}(x+2)}\right]$ d) $\dfrac{x^{3n} + 4^n}{x^{3n-3}(x^3+4)}$

25._____

CHAPTER 11   NAME_____

TEST FORM H   CLASS_____ SCORE_____ GRADE_____

ANSWERS

1. Find the 16th term of a sequence with general term $a_n = 3n + 4$.
   a) 50   b) 54   c) 48   d) 52

   1._____

2. Predict the general term of the sequence $\frac{4}{3}, \frac{16}{9}, \frac{64}{27}, \ldots$

   a) $18\left(\frac{1}{7}\right)^n$ b) $\left(\frac{4}{7}\right)^n$ c) $\left(\frac{18}{7}\right)^n$ d) $\left(\frac{4}{3}\right)^n$

   2._____

3. Evaluate $\sum_{k=1}^{5}\left(-\frac{1}{2}\right)^k$

   a) $-\frac{11}{32}$ b) -36 c) $\frac{15}{32}$ d) -42

   3._____

4. Express using sigma notation: $2 + \frac{4}{3} + \frac{8}{9} + \frac{16}{27} + \frac{32}{81}$

   a) $\sum_{k=1}^{5} \frac{2^k}{3^k}$ b) $\sum_{k=1}^{5} 2^{k+2}$

   c) $\sum_{k=1}^{5} 2^k$ d) $\sum_{k=1}^{5} \frac{2^k}{3^{k-1}}$

   4._____

   5._____

5. Find the 12th term of the arithmetic sequence 16, 0, −16, …
   a) −152   b) −77   c) −160   d) −89

6. Find the common difference of an arithmetic sequence when $a_1 = 2.25$ and $a_9 = 44.25$.

   6._____

   a) $2\frac{1}{4}$ b) $\frac{2}{5}$ c) $\frac{4}{5}$ d) $5\frac{1}{4}$

7. Find a_1 of an arithmetic sequence when $a_5 = -1\frac{41}{42}$ and $a_{10} = -3\frac{17}{42}$.

   7._____

   a) $-\frac{6}{5}$ b) -10 c) $-\frac{5}{6}$ d) 16

CHAPTER 11   NAME_____

TEST FORM H

ANSWERS

8. Find the sum of all the multiples of 11 from 99 to 198 inclusive.

   a) 5050   b) 14,850   c) 1485   d) 50,050

8._____

9. Find the 6th term of the geometric sequence 108, 36, 12, ...

   a) $\dfrac{16}{9}$ b) $4\dfrac{9}{16}$ c) $\dfrac{4}{9}$ d) $2\dfrac{9}{16}$

9._____

10. Find the common ratio of the geometric sequence $4\dfrac{1}{4}, 5\dfrac{2}{3}, 7\dfrac{5}{9}, ...$

    a) $\dfrac{4}{3}$ b) $\dfrac{3}{4}$ c) $\dfrac{3}{2}$ d) $\dfrac{2}{3}$

10._____

11. Find the nth term of the geometric sequence $-\dfrac{1}{4}, \dfrac{1}{16}, -\dfrac{1}{64}, ...$

    a) $\left(-\dfrac{1}{4}\right)^n$ b) 4^n c) $\left(-\dfrac{5}{4}\right)^n$ d) 5^n

11._____

12. Find the sum of the first nine terms of the geometric series

    $(1 + x) + (4 + 3x) + (16 + 9x) + ...$

12._____

   a) $59046 + 3066x$ b) $87381 + 3066x$

   c) $87381 + 9841x$ d) $59046 + 9841x$

13._____

Determine whether each infinite geometric series has a limit. If a limit exists, find it.

13. $0.1 + 0.05 + 0.025 + ...$

    a) no limit   b) 0.2   c) 15   d) 25

14._____

14. $-10 - 15 - 22.5 - ...$

    a) 0.5   b) no limit   c) -100 d) 1

15._____

15. $\$2400 + \$96 + \$3.84 + ...$

    a) $2500   b) no limit   c) $5000   d) $5500

242

CHAPTER 11   NAME_____

TEST FORM H

16. Find fractional notation for 0.267267267...

   a) $\dfrac{76}{555}$ b) $\dfrac{45}{111}$ c) $\dfrac{518}{999}$ d) $\dfrac{89}{333}$

16._____

17. An auditorium has 20 seats in the first row, 22 seats in the second row, 24 seats in the third row, and so on for 45 rows. How many seats are in the 40th row?

   a) 79   b) 75   c) 98   d) 87

17._____

18. In 1975, Maggie began receiving cash gifts in the mail from an unknown person. A check would arrive without fail on March 29 every year. The first check was for $75, the check in 1976 was for $76, and so on until 2000 when Maggie received an empty envelope on which the single word "Sorry" was printed in green crayon. How much in total had Maggie received?

   a) $895   b) $945   c) $1890   d) $2175

18._____

19. Each week the price of a $25,000 boat will be reduced 7% of the previous week's price. If we assume that it is not sold, what will be the price after 8 weeks?

   a) $18,051.03   b) $20,005.26
   c) $17,876.12   d) $15,042.52

19._____

20. Find the total rebound distance of a ball that is dropped from a height of 9 m, with each rebound three-fifths of the preceding one.

   a) 11 m   b) $10\dfrac{1}{2}$ m   c) $13\dfrac{1}{2}$ m   d) 12 m

20._____

CHAPTER 11

TEST FORM H

ANSWERS

21. _____

22. _____

23. _____

24. _____

25. _____

21. Simplify: $\binom{14}{6}$

    a) 3003   b) 710   c) 2002   d) 1001

22. Expand $(2x + y^2)^4$

    a) $32x^5 - 32x^4y^2 + 24x^3y^4 - 16x^2y^6 + 8xy^8 - y^{10}$
    b) $16x^4 + 32x^3y^2 + 24x^2y^4 + 8xy^6 + y^8$
    c) $16x^4 + 32x^3y^2 + 80x^2y^4 + 24xy^6 + y^8$
    d) $32x^5 - 80x^4y^2 + 80x^3y^4 - 40x^2y^6 + 10xy^8 - y^{10}$

23. Find the 8th term in the expansion of $(a + x)^{13}$

    a) $1287a^7x^6$ b) $1716a^6x^7$ c) $1287a^7x^6$ d) $1716a^7x^6$

24. Find the formula for the sum $S_n = 5 + 15 + 25 + \ldots + 10n - 5$

    a) $3n^2 - 5n$ b) $3n^2 - 2n$ c) $5n^2$ d) $3n^2$

25. Find the sum of the first n terms of $1 - \dfrac{4}{x^3} + \dfrac{16}{x^6} - \dfrac{64}{x^9} + \ldots$

    a) $\dfrac{x^n + 4^n}{x^{n-1}(x+1)}$ b) $2\left[\dfrac{x^n + 2^n}{x^{n-1}(x+2)}\right]$

    c) $3\left[\dfrac{x^{2n} + 2^n}{x^{n-2}(x+2)}\right]$ d) $\dfrac{x^{3n} + 4^n}{x^{3n-3}(x^3+4)}$

FINAL EXAMINATION   NAME_____

TEST FORM A   CLASS_____ SCORE_____ GRADE_____

ANSWERS

Chapter 1

Perform the indicated operation.

1. $8.10 + (-3.48)$ 2. $-\dfrac{3}{2} - \left(-\dfrac{2}{5}\right)$

1._____

3. Combine like terms. $8a^2b + 9ab^2 - 6ab^2 - 5a^2b + 1$

2._____

4. Solve. $15x + 17 = 28x - 22$

5. Simplify and write the answer in scientific notation. Use the correct number of significant digits. $(5.67 \times 10^{-5})(5.32 \times 10^{-2})$

3._____

Chapter 2

4._____

6. Graph. $y = x - 3$

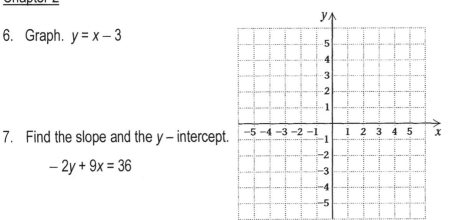

5._____

7. Find the slope and the y–intercept.
$-2y + 9x = 36$

6. See graph.

8. Find the slope of the line containing the following points. If the slope is undefined, state so. $(6, -5)$ and $(7, -12)$

7._____

8._____

9. Without graphing, determine whether the pair of lines is parallel, perpendicular, or neither.
$$4y - 5 = 2x$$
$$-2x + 4y = -2$$

9._____

10. Find an equation of the line containing $(-3, -4)$ and perpendicular to the line $6x - 5y = 30$.

10._____

245

FINAL EXAMINATION   NAME_____

TEST FORM A

ANSWERS	

Chapter 3

Solve, if possible, using the elimination method.

11._____

11. $3x - 2y = 35$
    $2x - 4y = 34$

12. $4y - 3x = 4$
    $6x + 5y = \dfrac{33}{4}$

12._____

13. Between his home mortgage, car loan, and credit card bill, Joel is $115,000 in debt. Each month, Joel's credit card accumulates 1.8% interest, his car loan 1.2% interest, and his mortgage 1% interest. After one month, his total accumulated interest is $1210. The interest on Joel's car loan was $30 more than the interest on his credit card. Find the amount of each of these three debts.

13._____

Solve. If the system's equations are dependent or if there is no solution, then state this.

14._____

14. $3x + 5y = 30$,
    $2y + 3z = 13$,
    $2x + 3y + 5z = 26$

15. $5x - 4z = 2$,
    $4x - 5y = -28$,
    $4y + 5z = 1$

15._____

Chapter 4

16._____
  See graph.

16. Graph the inequality and write the solution set in both set-builder and interval notation.

$8a - 4 \geq -8a + 6$ ⟵——|——⟶
                              0

17._____

17. Jan can rent a van for either $75 per day with unlimited mileage or $50 per day with 75 free miles and an extra charge of 20¢ for each mile over 75. For what numbers of miles traveled would the unlimited mileage plan save Jan money?

18._____

18. Find the intersection: $\{0, 1, 2, 3, 5\} \cap \{0, 2, 4, 6\}$

Solve and graph each solution set.

19. See graph.

19. $6x - 4 < 10$ or $x - 2 > 2$ ⟵——|——⟶
                                         0

20. See graph.

20. $|8 - 14x| = -4$ ⟵——|——⟶
                             0

246

FINAL EXAMINATION   NAME_____

TEST FORM A

Chapter 5

21. Given $P(x) = x^2 + 7x$, find and simplify $P(a-h) + P(-a)$.

21._____

22. Subtract. $(9y^2 - 8y - 6y^3) - (4y^2 - 4y - 3y^3)$

22._____

23. Multiply. $(-20x^2y^3)(-4x^2y)$

23._____

24. Factor. $50x - 10x^3$

24._____

25. Factor. $64y^2 - 25$

25._____

26. Find the domain of the function f given by $f(x) = \dfrac{11-x}{x^2 - 10x + 25}$.

26._____

Chapter 6

27. Simplify. $\dfrac{t-5}{t+4} \cdot \dfrac{4t+16}{5t^2 - 125}$

27._____

28. Subtract and simplify when possible. $\dfrac{3a^2}{a-b} - \dfrac{3b^2 + 6ab}{b-a}$

28._____

29. Solve. $\dfrac{t+16}{t^2 - t - 6} + \dfrac{2}{t-3} = \dfrac{4}{t+2}$

29._____

30. Divide. $(3x^4 + 7x^2 + x + 4) \div (x^2 + 1)$

30._____

31. If $f(x) = 5x^4 - 7x^3 + 3x - 9$, then use synthetic division to find $f(2)$.

31._____

32. Jeanie bicycles 34 km/h with no wind. Against the wind she bikes 27 km in the same time it takes to bike 41 km with the wind. What is the speed of the wind?

32._____

FINAL EXAMINATION NAME_____

TEST FORM A

ANSWERS

Chapter 7

Simplify. Assume that variables can represent any real number.

33. $\sqrt{x^2 - 20x + 100}$

34. $(5 + \sqrt{x})(5 - 3\sqrt{x})$

35. Rationalize the denominator: $\dfrac{\sqrt{5}}{6 + \sqrt{3}}$

36. Subtract: $(8 - 2i) - (4 - 7i)$

37. Divide and simplify to the form $a + bi$. $\dfrac{-4 + 2i}{7 - i}$

38. Simplify: i^7

Chapter 8

39. Solve. $x^2 + 3x + 1 = 0$

40. Solve $x^2 + 7x = 1$. Use a calculator to approximate the solutions with rational numbers.

41. Nisha and T'Mar can proofread a manuscript in 12 hours. Working alone, it takes T'Mar 10 hours longer than Nisha to proofread the same manuscript. How long would it take for Nisha to do this job by herself?

42. Determine the type of number that the solutions of $x^2 + 9x + 4 = 0$.

43. For the function $f(x) = 2x^2 + 2x + 1$
   a) find the vertex and the axis of symmetry;
   b) graph the function

44. Solve. $x + \dfrac{5}{x} > 0$

33._____

34._____

35._____

36._____

37._____

38._____

39._____

40._____

41._____

42._____

43.a)_____
   b) See graph.

44._____

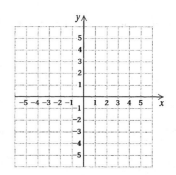

248

FINAL EXAMINATION NAME_____

TEST FORM A

Chapter 9

45. Find $(f \circ g)(x)$ and $(g \circ f)(x)$ if $f(x) = x - 2x^2$ and $g(x) = 3x - 2$.

46. Find a formula for the inverse of the function. $g(x) = (8x - 3)^3$

47. Convert to a logarithmic equation. $2^{-6} = \dfrac{1}{64}$

48. Express in terms of logarithms of a, b, and c. $\log \dfrac{a^{1/2} b^{1/3}}{c}$

If $\log_a 8 = 2.079$, $\log_a 5 = 1.609$, and $\log_a 3 = 1.099$, find the following.

49. $\log_a \dfrac{5}{8}$ 50. $\log_a 120$

51. Solve. $\log_x 128 = 7$

Chapter 10

52. Find the midpoint of the segment with the given endpoints.

$(4, -9)$ and $(-2, 5)$

53. Classify the equation as a circle, an ellipse, a parabola, or a hyperbola. Then graph.

$x^2 + y^2 + 2x + 8y + 16 = 0$

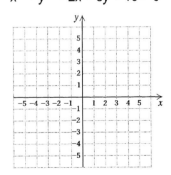

ANSWERS

45. _____

46. _____

47. _____

48. _____

49. _____

50. _____

51. _____

52. _____

53. See graph._____

FINAL EXAMINATION   NAME_____

TEST FORM A

ANSWERS

Chapter 10 (Continued)

54. Solve. $x^2 - 7y^2 = 1,$
             $xy = 24$

54._____

55. A rectangle with diagonal of length $\sqrt{205}$ has an area of 78. Find the dimensions of the rectangle.

55._____

Chapter 11

56._____

56. Write out and evaluate: $\sum_{k=1}^{5} (6-k)^k$

57._____

57. Find the common ration of the geometric sequence $40, 30, 22\frac{1}{2}, ...$

58. Determine whether the infinite geometric series has a limit. If a limit exists, find it.

$$0.01 + 0.012 + 0.0144 + ...$$

58._____

59. Find fractional notation for 2.28282828...

59._____

60. Simplify: $\begin{pmatrix} 11 \\ 10 \end{pmatrix}$

60._____

FINAL EXAMINATION   NAME_____

TEST FORM B   CLASS____ SCORE____ GRADE____

ANSWERS

Chapter 1

Perform the indicated operation.

1. $9.21 + (-4.59)$ 2. $-\dfrac{3}{5} - \left(-\dfrac{5}{2}\right)$

3. Combine like terms. $3a^2b - 11ab^2 + 4ab^2 + 4a^2b - 2$

4. Solve. $16x - 18 = 27x - 7$

5. Simplify and write the answer in scientific notation. Use the correct number of significant digits. $(4.78 \times 10^{-5})(6.41 \times 10^{-2})$

Chapter 2

6. Graph. $y = -\dfrac{1}{2}x + 2$

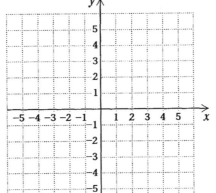

7. Find the slope and the y–intercept.
   $4y - 11x = 22$

8. Find the slope of the line containing the following points. If the slope is undefined, state so. $(1, 1)$ and $(1, 3)$

9. Without graphing, determine whether the pair of lines is parallel, perpendicular, or neither.
   $$5y + 10 = x$$
   $$-5x + 25y = -12$$

10. Find an equation of the line containing $(5, 2)$ and perpendicular to the line $5x - 6y = 30$.

1._____

2._____

3._____

4._____

5._____

6. See graph._____

7._____

8._____

9._____

10._____

FINAL EXAMINATION   NAME_____

TEST FORM B

ANSWERS   Chapter 3

                         Solve, if possible, using the elimination method.

11._____

                         11. $2x - 4y = 32$ 12. $4y + 5x = -1$
                              $4x + 5y = 25$ $2x + 2y = -\dfrac{3}{10}$

12._____

                         13. Between her home mortgage, car loan, and credit card bill, Helda is
                              $28,000 in debt. Each month, Helda's credit card accumulates 1.8%
                              interest, her car loan 1.2% interest, and her mortgage 1% interest. After
                              one month, her total accumulated interest is $352. The interest on Helda's
13._____   credit card was $8 more than the interest on her mortgage. Find the
                              amount of each of these three debts.

                         Solve. If the system's equations are dependent or if there is no solution, then
14._____   state this.

                         14. $x + 6z = -3$,   15. $-4x - 5y = -4$,
                              $4x + 2z = 10$, $5y + 4z = -4$,
15._____ $6x + 4y + 2z = 16$ $-5x + 4z = -9$

                         Chapter 4

16._____   16. Graph the inequality and write the solution set in both set-builder and
   See graph.   interval notation.
                                    $2a - 5 \geq -7a + 7$ <------|------>
                                                                     0

17._____
                         17. Dan can rent a van for either $75 per day with unlimited mileage or $55 per
                              day with 100 free miles and an extra charge of 20¢ for each mile over 100.
                              For what numbers of miles traveled would the unlimited mileage plan save
18._____   Dan money?

                         18. Find the intersection: $\{0, 5, 10, 15, 20\} \cap \{10, 20, 30, 40\}$

19. _See graph._
                         Solve and graph each solution set.

                         19. $5x - 3 < 7$ or $x - 1 > 2$ <------|------>
                                                                    0
20. _See graph._
                         20. $|9 - 16x| = -3$ <------|------>
                                                                   0

252

FINAL EXAMINATION   NAME_____

TEST FORM B

Chapter 5

21. Given $P(x) = x^2 + 8x$, find and simplify $P(a-h) + P(-a)$.

21._____

22. Subtract. $(7y^2 - 2y - 5y^3) - (2y^2 - 5y - 5y^3)$

22._____

23. Multiply. $(-16x^2 y^3)(-6x^2 y)$

23._____

24. Factor. $44x^2 - 11x^4$

24._____

25. Factor. $144y^2 - 49$

25._____

26. Find the domain of the function f given by $f(x) = \dfrac{12 - x}{x^2 - 12x + 36}$.

26._____

Chapter 6

27. Simplify. $\dfrac{t-5}{t+1} \cdot \dfrac{6t+6}{7t^2 - 175}$

27._____

28. Subtract and simplify when possible. $\dfrac{7a^2}{a-b} - \dfrac{7b^2 - 14ab}{b-a}$

28._____

29. Solve. $\dfrac{t+18}{t^2-t-2} + \dfrac{1}{t-2} = \dfrac{4}{t+1}$

29._____

30. Divide. $(4x^4 + 8x^2 + x + 5) \div (x^2 + 2)$

30._____

31. If $f(x) = 2x^4 - 9x^3 + 6x - 5$, then use synthetic division to find $f(3)$.

31._____

32. Simon bicycles 37 km/h with no wind. Against the wind he bikes 75 km in the same time it takes to bike 110 km with the wind. What is the speed of the wind?

32._____

253

FINAL EXAMINATION   NAME_____

TEST FORM B

ANSWERS

33._____

34._____

35._____

36._____

37._____

38._____

39._____

40._____

41._____

42._____

43.a)_____
   b) See graph.

44._____

Chapter 7

Simplify. Assume that variables can represent any real number.

33. $\sqrt{x^2 - 12x + 36}$

34. $(6 + \sqrt{x})(4 - 2\sqrt{x})$

35. Rationalize the denominator: $\dfrac{\sqrt{3}}{7 - \sqrt{2}}$

36. Subtract: $(9 - i) - (6 - 8i)$

37. Divide and simplify to the form $a + bi$. $\dfrac{9 + 5i}{-2 + i}$

38. Simplify: i^{15}

Chapter 8

39. Solve. $x^2 + 3x + 2 = 0$

40. Solve $x^2 + x = 7$. Use a calculator to approximate the solutions with rational numbers.

41. Lauren and Humphrey can scrub a floor in 3 hours. Working alone, it takes Humphrey 3.25 hours longer than Lauren to scrub the same floor. How long would it take for Lauren to do this job by herself?

42. Determine the type of number that the solutions of $x^2 + 8x - 9 = 0$.

43. For the function $f(x) = 2x^2 + 2x + 2$
   a) find the vertex and the axis of symmetry;
   b) graph the function

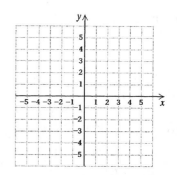

44. Solve. $x - \dfrac{5}{x} < 0$

254

FINAL EXAMINATION   NAME_____

TEST FORM B

Chapter 9

45. Find $(f \circ g)(x)$ and $(g \circ f)(x)$ if $f(x) = 2x + x^2$ and $g(x) = 2 - 3x$.

45. _____

46. Find a formula for the inverse of the function. $g(x) = (8x + 4)^3$

46. _____

47. Convert to a logarithmic equation. $3^{-4} = \dfrac{1}{81}$

47. _____

48. Express in terms of logarithms of a, b, and c. $\log \dfrac{a^{1/3} b^{1/2}}{c}$

48. _____

If $\log_a 4 = 1.386$, $\log_a 5 = 1.609$, and $\log_a 6 = 1.792$, find the following..

49. $\log_a \dfrac{2}{3}$ 50. $\log_a 20$

49. _____

51. Solve. $\log_x 216 = 3$

50. _____

Chapter 10

52. Find the midpoint of the segment with the given endpoints.

$(5, -9)$ and $(-2, 4)$

51. _____

53. Classify the equation as a circle, an ellipse, a parabola, or a hyperbola. Then graph.

$x^2 + y^2 + 6x + 8y + 21 = 0$

52. _____

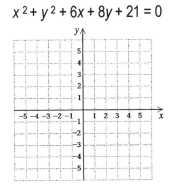

53. ____See graph.____

FINAL EXAMINATION   NAME_____

TEST FORM B

ANSWERS | Chapter 10 (Continued)

54. Solve. $x^2 - 2y^2 = -7$,
              $xy = 20$

54._____

55. A rectangle with diagonal of length $\sqrt{157}$ has an area of 66. Find the dimensions of the rectangle.

55._____

Chapter 11

56._____

56. Write out and evaluate: $\sum_{k=1}^{5}(2-k)^k$

57._____

57. Find the common ration of the geometric sequence $1\frac{1}{2}, 2\frac{1}{4}, 3\frac{1}{8}, \ldots$

58. Determine whether the infinite geometric series has a limit. If a limit exists, find it.

$$15 - 30 + 60 - \ldots$$

58._____

59._____

59. Find fractional notation for 1.19191919...

60._____

60. Simplify: $\binom{14}{12}$

FINAL EXAMINATION   NAME_____

TEST FORM C   CLASS_____ SCORE_____ GRADE_____

ANSWERS

Chapter 1

Perform the indicated operation.

1. $0.32 + (-5.60)$ 2. $-\dfrac{3}{7} - \left(-\dfrac{7}{2}\right)$ 1._____

3. Combine like terms. $7a^2b - 13ab^2 + 2ab^2 - 3a^2b - 3$

4. Solve. $17x - 10 = 26x - 46$ 2._____

5. Simplify and write the answer in scientific notation. Use the correct number of significant digits. $(3.89 \times 10^{-4})(7.50 \times 10^{-3})$ 3._____

Chapter 2

6. Graph. $y = \dfrac{1}{2}x - 4$

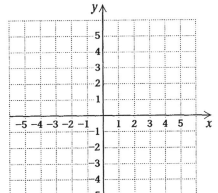

4._____

5._____

7. Find the slope and the y-intercept.

   $6y + 13x = -78$

6. See graph.

8. Find the slope of the line containing the following points. If the slope is undefined, state so. $(-3, 5)$ and $(-4, 5)$

7._____

8._____

9. Without graphing, determine whether the pair of lines is parallel, perpendicular, or neither.

   $6y - 4 = -x$
   $2x + 10y = 5$

9._____

10. Find an equation of the line containing $(5, -2)$ and perpendicular to the line $4x - 7y = 28$.

10._____

FINAL EXAMINATION   NAME_____

TEST FORM C

ANSWERS   Chapter 3

Solve, if possible, using the elimination method.

11._____

11. $4x + 5y = 34$ 12. $5y + 4x = 0$
    $3x - 10y = 53$ $4x + 3y = -\dfrac{6}{5}$

12._____

13. Between his home mortgage, car loan, and credit card bill, Leon is $212,000 in debt. Each month, Leon's credit card accumulates 1.8% interest, his car loan 1.2% interest, and his mortgage 1% interest. After one month, his total accumulated interest is $2174. The interest on Leon's credit card was $6 more than the interest on his car loan. Find the amount of each of these three debts.

13._____

Solve. If the system's equations are dependent or if there is no solution, then state this.

14._____

14. $-4y + z = 19$,   15. $4x - 3y = 11$,
    $-3x + 2y = -14$, $-3x + 4z = 2$,
    $2x + 6y + 3z = -11$ $4y + 3z = 2$

15._____

Chapter 4

16._____
   See graph.

16. Graph the inequality and write the solution set in both set-builder and interval notation.

$9a - 6 \geq -5a + 8$

17._____

17. Fran can rent a van for either $75 per day with unlimited mileage or $60 per day with 150 free miles and an extra charge of 15¢ for each mile over 150. For what numbers of miles traveled would the unlimited mileage plan save Fran money?

18._____

18. Find the intersection: $\{3, 6, 9, 12, 15\} \cap \{9, 18, 27, 36\}$

19. See graph.

Solve and graph each solution set.

19. $4x - 1 < 7$ or $x - 6 > -4$

20. See graph.

20. $|1 - 7x| = -2$

258

FINAL EXAMINATION   NAME_____

TEST FORM C

Chapter 5

21. Given $P(x) = x^2 + 37x$, find and simplify $P(a - h) + P(-a)$.

22. Subtract. $(5y^2 - 4y - 4y^3) - (8y^2 - 6y - 7y^3)$

23. Multiply. $(-12x^2 y^3)(-8x^2 y)$

24. Factor. $36x^3 - 12x^5$

25. Factor. $36y^2 - 25$

26. Find the domain of the function f given by $f(x) = \dfrac{13 - x}{x^2 - 14x + 49}$.

Chapter 6

27. Simplify. $\dfrac{t-4}{t+2} \cdot \dfrac{2t+4}{4t^2 - 64}$

28. Subtract and simplify when possible. $\dfrac{2a^2}{a-b} - \dfrac{2b^2 + 2ab}{b-a}$

29. Solve. $\dfrac{t+20}{t^2 + 8t - 9} + \dfrac{9}{t-1} = \dfrac{5}{t+9}$

30. Divide. $(5x^4 + 6x^2 + 4x + 6) \div (x^2 + 3)$

31. If $f(x) = 2x^4 - 9x^3 + 6x - 5$, then use synthetic division to find $f(4)$.

32. Leontine bicycles 33 km/h with no wind. Against the wind she bikes 75 km in the same time it takes to bike 123 km with the wind. What is the speed of the wind?

ANSWERS

21._____

22._____

23._____

24._____

25._____

26._____

27._____

28._____

29._____

30._____

31._____

32._____

FINAL EXAMINATION

NAME_____

TEST FORM C

ANSWERS

Chapter 7

Simplify. Assume that variables can represent any real number.

33. $\sqrt{x^2 - 10x + 25}$

34. $(5 + \sqrt{x})(3 - \sqrt{x})$

35. Rationalize the denominator: $\dfrac{\sqrt{2}}{10 + \sqrt{3}}$

36. Subtract: $(10 - 3i) - (2 - 9i)$

37. Divide and simplify to the form $a + bi$. $\dfrac{-10 - 4i}{1 + 3i}$

38. Simplify: i^{28}

Chapter 8

39. Solve. $x^2 + 3x + 3 = 0$

40. Solve $x^2 + 3x = 3$. Use a calculator to approximate the solutions with rational numbers.

41. Katherine and Spencer can weed the garden in 35 minutes. Working alone, it takes Spencer 24 minutes longer than Katherine to weed the garden. How long would it take for Katherine to do this job by herself?

42. Determine the type of number that the solutions of $x^2 + 7x + 10 = 0$.

43. For the function $f(x) = 2x^2 + 2x + 3$
  a) find the vertex and the axis of symmetry;
  b) graph the function

43.a)_____
b) 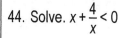 See graph.

44. Solve. $x + \dfrac{4}{x} < 0$

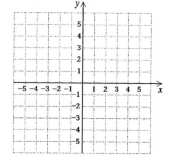

260

FINAL EXAMINATION   NAME_____

TEST FORM C

Chapter 9

45. Find $(f \circ g)(x)$ and $(g \circ f)(x)$ if $f(x) = -2x + x^2$ and $g(x) = 3 - 2x$.

46. Find a formula for the inverse of the function. $g(x) = (64x - 3)^3$

47. Convert to a logarithmic equation. $5^{-4} = \dfrac{1}{625}$

48. Express in terms of logarithms of a, b, and c. $\log \dfrac{a^3 b^2}{c}$

If $\log_a 4 = 1.386$, $\log_a 5 = 1.609$, and $\log_a 6 = 1.792$, find the following.

49. $\log_a \dfrac{3}{2}$ 50. $\log_a 120$

51. Solve. $\log_x 243 = 5$

Chapter 10

52. Find the midpoint of the segment with the given endpoints.

$(6, -7)$ and $(-4, 3)$

53. Classify the equation as a circle, an ellipse, a parabola, or a hyperbola. Then graph.

$x^2 + y^2 + 6x + 6y + 9 = 0$

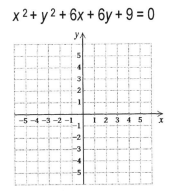

ANSWERS

45. _____

46. _____

47. _____

48. _____

49. _____

50. _____

51. _____

52. _____

53. See graph.

FINAL EXAMINATION   NAME_____

TEST FORM C

ANSWERS   Chapter 10 (Continued)

54. Solve. $x^2 - 3y^2 = -2,$
              $xy = 15$

54._____

55. A rectangle with diagonal of length 13 has an area of 60. Find the dimensions of the rectangle.

55._____

Chapter 11

56._____

56. Write out and evaluate: $\sum_{k=1}^{5} (1-k)^{k-2}$

57._____

57. Find the common ration of the geometric sequence $10, 13\frac{1}{3}, 17\frac{7}{9}, \ldots$

58. Determine whether the infinite geometric series has a limit. If a limit exists, find it.

58._____

$$-0.5 + 1 - 2 + \ldots$$

59._____

59. Find fractional notation for 0.432432432...

60._____

60. Simplify: $\binom{14}{10}$

FINAL EXAMINATION
TEST FORM D

NAME_____

CLASS_____ SCORE_____ GRADE_____

ANSWERS

1. For an arithmetic sequence, find the common difference d when $a_1 = 1$ and $a_7 = 4\frac{3}{4}$.

1._____

2. Find the distance between the points $(1, -3)$ and $(-8, 8)$.

2._____

3. Find a formula for the inverse of the function $f(x) = 3x - 4$

3._____

4. Solve. $\quad x^2 + x + 2 = 0$

4._____

Simplify. Assume that variables can represent any real number.

5. $\dfrac{\sqrt[5]{a^4}}{\sqrt[4]{a^3}}$

6. $\dfrac{t-2}{t+1} \cdot \dfrac{5t+5}{6t^2 - 24}$

5._____

7. Given $P(x) = x^2 - 3x$, find and simplify $P(a+h) - P(a)$.

6._____

8. Graph the inequality and write the solution set in both set-builder and interval notation.

$9(2-x) > 5x + 2$

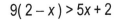

7._____

8. See graph._____

9. Solve, if possible, using the substitution method.

$7x - 5y = 8,$
$y = x - 2$

9._____

10. Find the common ration of the geometric sequence $\quad 30, 20, 13\frac{1}{3}, \ldots$

10._____

11. Add. $\quad 5.87 + (-0.15)$

11._____

263

FINAL EXAMINATION   NAME_____

TEST FORM D

ANSWERS

12. Graph. $y = -2x + 1$

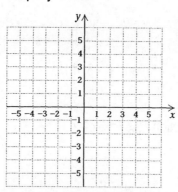

13. Graph. $g(x) = \log_2 x$

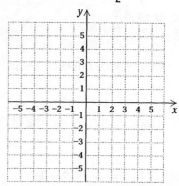

12. __See graph.__

13. __See graph.__

14. Classify the equation as a circle, an ellipse, a parabola, or a hyperbola. Then graph. $x = -y^2 + 2y$

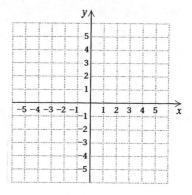

14. __See graph.__

15. _____

15. Solve $x^2 + x = 3$. Use a calculator to approximate the solutions with rational numbers.

16. _____

16. If $f(x) = \sqrt{10 - 5x}$, determine the domain of f.

17. _____

17. Find the LCD: $\dfrac{2x}{x^2 + x - 6}$, $\dfrac{x + 5}{x^2 + 2x - 8}$

18. _____

18. Subtract. $(5y^2 - 2y - 9y^3) - (2y^2 - y - 7y^3)$

FINAL EXAMINATION   NAME_____

TEST FORM D

	ANSWERS
19. Greg can rent a van for either $50 per day with unlimited mileage or $35 per day with 75 free miles and an extra charge of 25¢ for each mile over 75. For what numbers of miles traveled would the unlimited mileage plan save Greg money?	19._____
20. The perimeter of a rectangle is 24. The length of the rectangle is eight less than three times the width. Find the dimensions of the rectangle.	20._____
21. Find the slope and the y-intercept. $-8y - 3x = 18$	21._____
22. Subtract. $-\dfrac{1}{3} - \left(-\dfrac{3}{5}\right)$	22._____
23. Determine whether the infinite geometric series has a limit. If a limit exists, find it. $0.5 + 0.4 + 0.32 + ...$	23._____
24. Solve $x^2 + y^2 = 25,$ $\dfrac{x^2}{25} - \dfrac{y^2}{16} = 1$	24._____
25. Convert to an exponential equation. $\log_2 128 = 7$	25._____
26. Elsworth and Priscilla can eat an entire chocolate cream pie in 42 minutes. Eating alone, it takes Priscilla 80 minutes longer than Elsworth to eat the same type of pie. How long would it take for Elsworth to eat the pie by himself?	26._____
27. Solve. $x = \sqrt{2x - 1} + 2$	27._____
28. Simplify. $\dfrac{\dfrac{x^2 + 3x - 10}{x^2 - 100}}{\dfrac{x^2 - 3x - 40}{x^2 - 20x + 100}}$	28._____
29. Multiply. $(5a - 3b)(2a + 4b)$	29._____

FINAL EXAMINATION

TEST FORM D

NAME_____

ANSWERS

30._____

31._____

32._____

33._____

34._____

35._____

36._____

37._____

38._____

39._____

40._____

41._____

30. Find the union: $\{1, 2, 3, 4, 5\} \cup \{0, 1, 2, 3\}$

31. Solve. If the system's equations are dependent or if there is no solution, then state this.
$$x + 5y + 2z = -1,$$
$$5x + 2y + z = 2,$$
$$2x + y + 5z = 3$$

32. Find the slope of the line containing the following points. If the slope is undefined, state so.
$$(-2, 0) \text{ and } (2, 6)$$

33. Combine like terms. $\quad a^2b - 3ab^2 + 12ab^2 + 8a^2b + 1$

34. Find the total rebound distance of a ball that is dropped from a height of 6 m, with each rebound two-fifths of the preceding one.

35. A rectangle with diagonal of length $\sqrt{58}$ has an area of 21. Find the dimensions of the rectangle.

36. Solve. Where appropriate, include the approximation to the nearest ten-thousandth.
$$3^x = \frac{1}{27}$$

37. Determine the type of number that the solutions of $\quad x^2 + 12x + 27 = 0$.

38. Multiply. Write the answer in the form $a + bi$. $\quad (2 - i)^2$

39. David can paint a bedroom in 0.75 hr. Jay can paint the same bedroom in 1.25 hr. How long will it take them, working together, to paint the room?

Factor.

40. $14x - 7x^3$

41. $81y^2 - 1$

FINAL EXAMINATION   NAME_____

TEST FORM D

	ANSWERS

42. Solve using matrices. $5x - 4y = -3,$
    $8x + 6y = 14$

42._____

43. Using function notation, write a slope-intercept equation for the line containing $(1, 3)$ and $(2, 5)$.

43._____

44. Solve. $12x - 37 = 31x + 20$

44._____

45. Find the 5th term in the expansion of $(a + x)^{10}$.

45._____

46. Find the center and the radius of the circle $(x + 2)^2 + (y - 9)^2 = 9$.

46._____

47. Solve. $6^x = 3.2$

47._____

48. For the function $f(x) = 2x^2 + x + 1$
    a) find the vertex and the axis of symmetry;
    b) graph the function.

48.a)_____
b) See graph._____

49. Simplify: i^{19}

49._____

50. Divide. $(3x^4 + 8x^2 + 4x + 1) \div (x^2 + 1)$

50._____

51. Solve. $2y^2 = 8$

51._____

52. Solve and graph the solution set.

   $|2x - 1| < 2.5$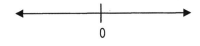

 See graph._____

267

FINAL EXAMINATION

TEST FORM D

NAME_____

ANSWERS

53._____

54._____

55._____

56._____

57._____

58._____

59. See graph._____

60.a)_____
   b)_____
   c)_____
   d)_____
   e)_____

53. Find the equilibrium point for the demand and supply functions

$$D(p) = 76 - 7p \quad \text{and} \quad S(p) = 54 + 4p$$

54. Without graphing, determine whether the pair of lines is parallel, perpendicular, or neither.
$$y = -5x + 2,$$
$$5y - x = 3$$

55. Simplify. Do not use negative exponents in the answer. $(-9xy^{-4})^{-2}$

56. Find fractional notation for 0.15151515...

57. An investment with interest compounded continuously doubled itself in 22 years. What is the interest rate?

58. Rationalize the denominator: $\dfrac{\sqrt{10}}{2 + \sqrt{7}}$

59. Solve and graph the solution set.

$-2x > 8$ or $3x > -9$

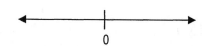

60. Unethicon Inc. produces cloned hens. For the first year, the fixed set-up costs are $15,000. The variable costs for cloning each hen are $10. The revenue from each hen is $35. Find the following.

a) The total cost $C(x)$ of producing x hens.
b) The total revenue $R(x)$ from the sale of x hens.
c) The total profit $P(x)$ from the production and sale of x hens.
d) The profit or loss from production and sale of 300 hens; of 900 hens.
e) The break-even point.

FINAL EXAMINATION   NAME_____

TEST FORM E   CLASS_____ SCORE_____ GRADE_____

ANSWERS

1. For an arithmetic sequence, find the common difference d when $a_1 = \frac{3}{2}$ and $a_9 = 7\frac{1}{2}$.

1._____

2. Find the distance between the points $(2, -5)$ and $(-6, 7)$.

2._____

3. Find a formula for the inverse of the function $f(x) = 3x + 5$

3._____

4. Solve. $x^2 + 2x + 1 = 0$

Simplify. Assume that variables can represent any real number.

4._____

5. $\dfrac{\sqrt[5]{a^3}}{\sqrt[4]{a^2}}$

6. $\dfrac{t-3}{t+2} \cdot \dfrac{2t+4}{3t^2 - 27}$

5._____

7. Given $P(x) = x^2 - 4x$, find and simplify $P(a+h) - P(a)$.

6._____

8. Graph the inequality and write the solution set in both set-builder and interval notation.

$8(3 - x) > 6x + 3$

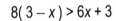

7._____

8. See graph._____

9. Solve, if possible, using the substitution method.

$2x + 3y = -6,$
$y = 4x + 3$

9._____

10. Find the common ration of the geometric sequence $60, 45, 33\frac{3}{4}, \ldots$

10._____

11. Add. $6.98 + (-1.26)$

11._____

269

FINAL EXAMINATION

TEST FORM E

NAME_____

ANSWERS

12. ___See graph.___

13. ___See graph.___

14. ___See graph.___

15. _____

16. _____

17. _____

18. _____

12. Graph. $y = 2x - 1$

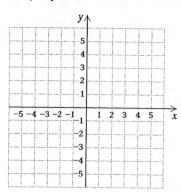

13. Graph. $g(x) = \log_3 x$

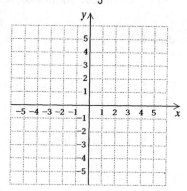

14. Classify the equation as a circle, an ellipse, a parabola, or a hyperbola. Then graph. $x = -y^2 - 2y$

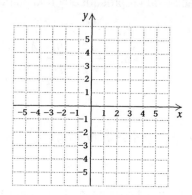

15. Solve $x^2 + 3x = 7$. Use a calculator to approximate the solutions with rational numbers.

16. If $f(x) = \sqrt{9 - 3x}$, determine the domain of f.

17. Find the LCD: $\dfrac{3x}{x^2 - 2x - 8}$, $\dfrac{x + 4}{x^2 - x - 12}$

18. Subtract. $(3y^2 - 4y - 8y^3) - (8y^2 - 2y - 9y^3)$

270

FINAL EXAMINATION   NAME_____

TEST FORM E

	ANSWERS
19. Dodie can rent a van for either $50 per day with unlimited mileage or $40 per day with 100 free miles and an extra charge of 25¢ for each mile over 100. For what numbers of miles traveled would the unlimited mileage plan save Dodie money?	19._____
20. The perimeter of a rectangle is 64. The length of the rectangle is eight more than twice the width. Find the dimensions of the rectangle.	20._____
21. Find the slope and the y-intercept. $\quad -6y + 5x = -3$	21._____
22. Subtract. $\quad -\dfrac{2}{3} - \left(-\dfrac{3}{7}\right)$	22._____
23. Determine whether the infinite geometric series has a limit. If a limit exists, find it. $\quad 6 + 4.2 + 2.94 + \ldots$	23._____
24. Solve $\quad x^2 + y^2 = 16,$ $\quad \dfrac{x^2}{16} - \dfrac{y^2}{9} = 1$	24._____
25. Convert to an exponential equation. $\log_3 9 = 2$	25._____
26. Jean Paul and Maurice can bake a dozen loaves of bread in 3 hours. Working alone, it takes Maurice 2.5 hours longer than Jean Paul to bake this much bread. How long would it take for Jean Paul to bake the bread by himself?	26._____
27. Solve. $\quad x = \sqrt{15x - 9} - 3$	27._____
28. Simplify. $\quad \dfrac{\dfrac{x^2 + 2x - 15}{x^2 - 81}}{\dfrac{x^2 - 2x - 35}{x^2 - 18x + 81}}$	28._____
29. Multiply. $\quad (a - 4b)(3a + 5b)$	29._____

FINAL EXAMINATION

TEST FORM E

NAME_____

ANSWERS

30. Find the union: $\{2, 4, 6, 8, 10\} \cup \{1, 2, 3, 4\}$

30._____

31. Solve. If the system's equations are dependent or if there is no solution, then state this.
$$6x + 5y + 4z = 43,$$
$$3x + 2y + z = 16,$$
$$x - 2y + 3z = 8$$

31._____

32. Find the slope of the line containing the following points. If the slope is undefined, state so.
$$(5, -1) \text{ and } (8, 5)$$

32._____

33. Combine like terms. $\quad 9a^2b - 5ab^2 + 10ab^2 - 7a^2b + 2$

33._____

34. Find the total rebound distance of a ball that is dropped from a height of 7 m, with each rebound two-fifths of the preceding one.

34._____

35. A rectangle with diagonal of length $2\sqrt{29}$ has an area of 40. Find the dimensions of the rectangle.

35._____

36. Solve. Where appropriate, include the approximation to the nearest ten-thousandth.
$$4^x = \frac{1}{16}$$

36._____

37. Determine the type of number that the solutions of $\quad x^2 + 11x - 10 = 0$.

37._____

38. Multiply. Write the answer in the form $a + bi$. $\quad (3 + i)^2$

38._____

39. Joe can paint a bedroom in 1.25 hr. Sara can paint the same bedroom in 1.75 hr. How long will it take them, working together, to paint the room?

39._____

Factor.

40._____

40. $24x^2 - 8x^4$ $\qquad$ 41. $100y^2 - 121$

41._____

FINAL EXAMINATION   NAME_____

TEST FORM E

	ANSWERS
42. Solve using matrices. $6x + 5y = 17,$ $9x - 7y = 11$	42._____
43. Using function notation, write a slope-intercept equation for the line containing $(2, -2)$ and $(3, -4)$.	43._____
44. Solve. $13x + 56 = 30x + 22$	44._____
45. Find the 6th term in the expansion of $(a + x)^{10}$.	45._____
46. Find the center and the radius of the circle $(x - 2)^2 + (y + 8)^2 = 16$.	46._____
47. Solve. $7^x = 4.2$	47._____
48. For the function $f(x) = 2x^2 + x + 2$ a) find the vertex and the axis of symmetry; b) graph the function.	48.a)_____ b) See graph._____
49. Simplify: i^{25}	49._____
50. Divide. $(4x^4 + 7x^2 + 3x + 2) \div (x^2 + 2)$	50._____
51. Solve. $3y^2 = 27$	51._____
52. Solve and graph the solution set. $\|2x - 2\| < 3$ 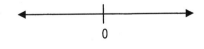	52. See graph._____

273

FINAL EXAMINATION

TEST FORM E

NAME_____

ANSWERS

53. Find the equilibrium point for the demand and supply functions

$$D(p) = 87 - 8p \quad \text{and} \quad S(p) = 48 + 5p$$

53._____

54. Without graphing, determine whether the pair of lines is parallel, perpendicular, or neither.

$$3y = -6x + 4,$$
$$6y + 3x = 5$$

54._____

55. Simplify. Do not use negative exponents in the answer. $(-3x^2 y^{-3})^{-3}$

55._____

56. Find fractional notation for 0.47474747...

56._____

57. An investment with interest compounded continuously doubled itself in 20 years. What is the interest rate?

57._____

58. Rationalize the denominator: $\dfrac{\sqrt{7}}{3 - \sqrt{6}}$

58._____

59. Solve and graph the solution set.

$$-2x > 6 \text{ or } 3x > -6$$

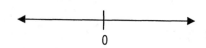

59. See graph._____

60. Unethicon Inc. produces cloned sheep. For the first year, the fixed set-up costs are $24,000. The variable costs for cloning each hen are $15. The revenue from each sheep is $45. Find the following.
 a) The total cost $C(x)$ of producing x sheep.
 b) The total revenue $R(x)$ from the sale of x sheep.
 c) The total profit $P(x)$ from the production and sale of x sheep.
 d) The profit or loss from production and sale of 250 sheep; of 1000 sheep.
 e) The break-even point.

60.a)_____
   b)_____
   c)_____
   d)_____
   e)_____

FINAL EXAMINATION NAME_____

TEST FORM F   CLASS_____ SCORE_____ GRADE_____

ANSWERS

1. For an arithmetic sequence, find the common difference d when $a_1 = 2$ and $a_7 = 7$.

2. Find the distance between the points $(3, -7)$ and $(-4, 6)$.

1._____

3. Find a formula for the inverse of the function $f(x) = 2x - 6$.

2._____

4. Solve. $x^2 + 2x + 2 = 0$

3._____

Simplify. Assume that variables can represent any real number.

4._____

5. $\dfrac{\sqrt[5]{a^6}}{\sqrt[4]{a}}$

6. $\dfrac{t-4}{t+3} \cdot \dfrac{3t+9}{4t^2 - 64}$

5._____

7. Given $P(x) = x^2 - 41x$, find and simplify $P(a+h) - P(a)$.

6._____

8. Graph the inequality and write the solution set in both set-builder and interval notation.

   $7(4-x) > 7x + 4$

7._____

9. Solve, if possible, using the substitution method.

8. See graph.

   $3x - 4y = 10,$
   $y = -2x + 5$

9._____

10. Find the common ration of the geometric sequence $25, 16\dfrac{2}{3}, 11\dfrac{1}{9}, ...$

10._____

11. Add. $7.09 + (-2.37)$

11._____

275

FINAL EXAMINATION

TEST FORM F

NAME_____

ANSWERS

12. Graph. $y = -x + 3$

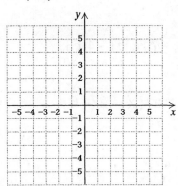

13. Graph. $g(x) = \log_4 x$

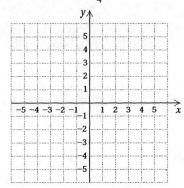

12. __See graph.__

13. __See graph.__

14. Classify the equation as a circle, an ellipse, a parabola, or a hyperbola. Then graph. $x = -y^2 + 4y$

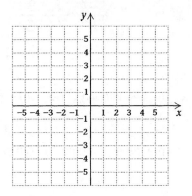

14. __See graph.__

15. _____

15. Solve $x^2 + 5x = 5$. Use a calculator to approximate the solutions with rational numbers.

16. _____

16. If $f(x) = \sqrt{15 - 3x}$, determine the domain of f.

17. _____

17. Find the LCD: $\dfrac{4x}{x^2 - x - 6}$, $\dfrac{x + 3}{x^2 + x - 12}$

18. _____

18. Subtract. $(y^2 - 6y - 7y^3) - (6y^2 - 3y - y^3)$

276

FINAL EXAMINATION NAME_____

TEST FORM F

19. Stan can rent a van for either $50 per day with unlimited mileage or $45 per day with 150 free miles and an extra charge of 25¢ for each mile over 150. For what numbers of miles traveled would the unlimited mileage plan save Stan money?

19._____

20. The perimeter of a rectangle is 84. The length of the rectangle is two less than three times the width. Find the dimensions of the rectangle.

20._____

21. Find the slope and the y – intercept. $-4y - 7x = -5$

21._____

22. Subtract. $-\dfrac{3}{4} - \left(-\dfrac{4}{5}\right)$

22._____

23. Determine whether the infinite geometric series has a limit. If a limit exists, find it.
$$0.4 + 0.24 + 0.144 + \ldots$$

23._____

24. Solve $x^2 + y^2 = 9,$
$\dfrac{x^2}{9} - \dfrac{y^2}{4} = 1$

24._____

25. Convert to an exponential equation. $\log_4 64 = 3$

25._____

26. Gunther and Gerta can peel a bushel of potatoes in 48 minutes. Working alone, it takes Gerta 28 minutes longer than Gunther to peel the potatoes. How long would it take for Gunther to do this job by himself?

26._____

27. Solve. $x = \sqrt{3x - 12} + 4$

27._____

28. Simplify. $\dfrac{\dfrac{x^2 - 16}{x^2 - 64}}{\dfrac{x^2 - 2x - 24}{x^2 - 16x + 64}}$

28._____

29. Multiply. $(2a - 5b)(4a + b)$

29._____

FINAL EXAMINATION

TEST FORM F

NAME_____

ANSWERS

30._____

31._____

32._____

33._____

34._____

35._____

36._____

37._____

38._____

39._____

40._____

41._____

30. Find the union: $\{1, 3, 5, 7, 9\} \cup \{1, 2, 3, 4\}$

31. Solve. If the system's equations are dependent or if there is no solution, then state this.
$$-2x + 7y - 2z = 3,$$
$$5x - 2y + 7z = 11,$$
$$-2x + 5y - 2z = 1$$

32. Find the slope of the line containing the following points. If the slope is undefined, state so.
$$(0, 4) \text{ and } (-2, -2)$$

33. Combine like terms. $\quad 2a^2b + 7ab^2 - 8ab^2 + 6a^2b + 3$

34. Find the total rebound distance of a ball that is dropped from a height of 8 m, with each rebound two-fifths of the preceding one.

35. A rectangle with diagonal of length $\sqrt{74}$ has an area of 45. Find the dimensions of the rectangle.

36. Solve. Where appropriate, include the approximation to the nearest ten-thousandth.
$$5^x = \frac{1}{125}$$

37. Determine the type of number that the solutions of $\quad x^2 + 10x + 26 = 0$.

38. Multiply. Write the answer in the form $a + bi$. $\quad (3 + 2i)^2$

39. Jodi can paint a bedroom in 1.5 hr. Sara can paint the same bedroom in 2.75 hr. How long will it take them, working together, to paint the room?

Factor.

40. $36x^3 - 9x^5$

41. $121y^2 - 64$

FINAL EXAMINATION   NAME_____

TEST FORM F

ANSWERS

42. Solve using matrices. $7x + 6y = 32$,
$-x + 8y = 22$

42._____

43. Using function notation, write a slope-intercept equation for the line containing $(-3, 4)$ and $(2, -6)$.

43._____

44. Solve. $14x + 5 = 29x + 35$

44._____

45. Find the 7th term in the expansion of $(a + x)^{10}$.

45._____

46. Find the center and the radius of the circle $(x + 2)^2 + (y - 7)^2 = 25$.

46._____

47. Solve. $8^x = 5.2$

47._____

48. For the function $f(x) = 2x^2 + x + 3$
   a) find the vertex and the axis of symmetry;
   b) graph the function.

48.a)_____
   b) See graph._____

49. Simplify: i^{32}

49._____

50. Divide. $(5x^4 + 6x^2 + 2x + 3) \div (x^2 + 3)$

50._____

51. Solve. $4y^2 = 64$

51._____

52. Solve and graph the solution set.

$|3x - 1| < 3.5$

52. See graph._____

FINAL EXAMINATION

TEST FORM F

NAME_____

ANSWERS

53._____

54._____

55._____

56._____

57._____

58._____

59. See graph.

60. a)_____
    b)_____
    c)_____
    d)_____
    e)_____

53. Find the equilibrium point for the demand and supply functions

$$D(p) = 98 - 9p \quad \text{and} \quad S(p) = 53 + 6p$$

54. Without graphing, determine whether the pair of lines is parallel, perpendicular, or neither.
$$5y = -7x + 6,$$
$$7y - 5x = 7$$

55. Simplify. Do not use negative exponents in the answer. $(-8x^3 y^{-2})^{-2}$

56. Find fractional notation for 0.63636363...

57. An investment with interest compounded continuously doubled itself in 18 years. What is the interest rate?

58. Rationalize the denominator: $\dfrac{\sqrt{6}}{5 - \sqrt{5}}$

59. Solve and graph the solution set.

$-2x > 4$ or $3x > -9$

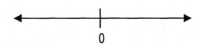

60. Unethicon Inc. produces cloned cows. For the first year, the fixed set-up costs are $33,250. The variable costs for cloning each cow are $22. The revenue from each cow is $57. Find the following.

a) The total cost $C(x)$ of producing x cows.
b) The total revenue $R(x)$ from the sale of x cows.
c) The total profit $P(x)$ from the production and sale of x cows.
d) The profit or loss from production and sale of 575 cows; of 1500 cows.
e) The break-even point.

280

FINAL EXAMINATION   NAME_____

TEST FORM G   CLASS_____ SCORE_____ GRADE_____

ANSWERS

1. Find the common ratio of the geometric sequence $4\frac{1}{4}, 5\frac{2}{3}, 7\frac{5}{9}, \ldots$

   a) $\frac{4}{3}$ b) $\frac{3}{4}$ c) $\frac{3}{2}$ d) $\frac{2}{3}$

   1._____

2. Find the midpoint of the segment with the endpoints $(8, -3)$ and $(-8, 1)$.

   a) $(0, -0.5)$ b) $(-1, 0)$ c) $(0.5, -1.5)$ d) $(0, -1)$

   2._____

3. Find $(f \circ g)(x)$ if $f(x) = -x + x^2$ and $g(x) = 3x - 2$.

   a) $-4x^2 - 15x - 12$ b) $-9x^2 + 14x + 6$
   c) $9x^2 - 15x + 6$ d) $-4x^2 + 14x - 12$

   3._____

4. Solve. $x^2 + 4x + 2$

   a) $2 \pm \sqrt{3}$ b) $2 \pm \sqrt{2}$ c) $-2 \pm \sqrt{2}$ d) $-2 \pm \sqrt{3}$

   4._____

5. Simplify. $\sqrt{x^2 - 4x + 4}$

   a) $x - 2$ b) $x + 3$ c) $x + 2$ d) $x - 3$

   5._____

6. Simplify. $\frac{t-2}{t+4} \cdot \frac{4t+16}{2t^2-8}$

   a) $\frac{1}{t+2}$ b) $3t$ c) $\frac{1}{t+3}$ d) $\frac{2}{t+2}$

   6._____

7. Find and simplify $P(a+h) - P(a)$ given $P(x) = x^2 - 2x$.

   a) $h^2 - 6h + a^2$ b) $h^2 - 2h + 2ah$
   c) $h^2 - 2h - 6ah + 2a^2$ d) $h^2 + 2ah + 2a^2$

   7._____

8. Solve. $6a - 8 \leq -a + 1$

   a) $\{a \mid a \leq \frac{9}{7}\}$ b) $\{a \mid a \leq \frac{7}{9}\}$

   c) $\{a \mid a \geq \frac{9}{7}\}$ d) $\{a \mid a \geq \frac{7}{9}\}$

   8._____

281

FINAL EXAMINATION

TEST FORM G

NAME_____

ANSWERS

9. Solve, if possible, using the elimination method.

$$x + 2y = 15$$
$$3x + 5y = 44$$

What is the y - coordinate?

9._____

a) 1 b) – 1 c) 0 d) not possible

10. Which graph represents $\frac{2}{3}x + 3$?

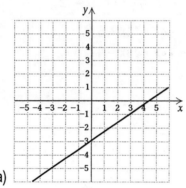
a)

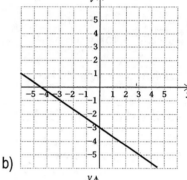
b)

10._____

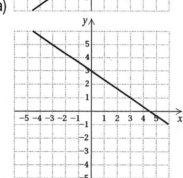
c)

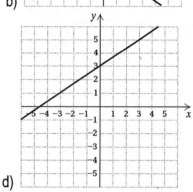
d)

11._____

11. Subtract: 87.6 – 71.9 .

a) 14.7 b) 16.3 c) 6.3 d) 15.7

12. Determine whether each infinite geometric series has a limit. If a limit exists, find it.

$$-10 - 15 - 22.5 - \ldots$$

12._____

a) 0.5 b) no limit c) – 100 d) 1

FINAL EXAMINATION   NAME_____

TEST FORM G

13. What is the equation corresponding to the following graph?

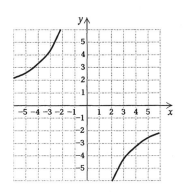

   a) $\dfrac{y^2}{16} - \dfrac{x^2}{4} = 1$ b) $\dfrac{x^2}{4} - \dfrac{y^2}{16} = 1$

   c) $xy = -13$ d) $xy = -4$

13._____

14. Simplify $\log_2 32$.

   a) 6   b) 3   c) 5   d) 2

14._____

15. Solve. $x^2 + 7x = 5$ Use a calculator to approximate the solutions with rational numbers.

   a) 0.65331193,   b) -5.1925824,
      7.65331193   0.1925824

   c) -0.1925824,   d) -7.65331193,
      5.65331193   0.65331193

15._____

16. Simplify. $\dfrac{\sqrt[5]{a^6}}{\sqrt[4]{a^3}}$

   a) $\sqrt[20]{a^5}$ b) $\sqrt[20]{a^9}$ c) $\sqrt[9]{a^7}$ d) $\sqrt{a^5}$

16._____

FINAL EXAMINATION

TEST FORM G

ANSWERS

17. Subtract and simplify. $\dfrac{4a^2}{a-b} - \dfrac{4b^2 - 8ab}{b-a}$

17._____

    a) $4(a+b)$ b) $\dfrac{4(a+b)^2}{a-b}$ c) $4(a-b)$ d) $\dfrac{4(a-b)^2}{a-b}$

18. Multiply. $(2a - 5b)(3a + 5b)$

    a) $6a^2 - 5ab + 25b^2$ b) $6a^2 - 15ab - 25b^2$

18._____

    c) $6a^2 + 15ab - 25b^2$ d) $6a^2 - 5ab - 25b^2$

19. Hans can rent a van for either $75 per day with unlimited mileage or $45 per day with 100 free miles and an extra charge of 15¢ for each mile over 100. For what numbers of miles traveled would the unlimited mileage plan save Hans money?

19._____

    a) at least 350   b) at most 300   c) at least 300   d) at most 350

20. The perimeter of a rectangle is 70. The length of the rectangle is five more than three times the width. Find the width of the rectangle.

20._____

    a) 5   b) 13   c) 7   d) 10

21. Find the slope of the line containing the points $(-2, 7)$ and $(1, -2)$.

21._____

    a) -3 b) -4 c) 3   d) 4

22. Divide: $\dfrac{-5.8}{-2.9}$.

22._____

    a) 2   b) -7.6 c) -2 d) 3

23. Find the total rebound distance of a ball that is dropped from a height of 9 m, with each rebound three-fifths of the preceding one.

23._____

    a) 11 m   b) $10\tfrac{1}{2}$ m   c) $13\tfrac{1}{2}$ m   d) 12 m

24. Two squares are such that the sum of their areas is 7 m^2 and the difference of their areas is 3 m^2. Find the length of a side of each square.

24._____

    a) $\sqrt{5}$ m, $\sqrt{2}$ m   b) $\sqrt{2}$ m, 5 m
    c) $2\sqrt{2}$ m, 2 m   d) $2\sqrt{5}$ m, $2\sqrt{2}$ m

FINAL EXAMINATION   NAME_____

TEST FORM G

ANSWERS

25. Given $\log_a 4 = 1.386$, $\log_a 5 = 1.609$, and $\log_a 6 = 1.792$, find $\log_a 36$.

   a) 4.604   b) 3.584   c) 4.787   d) 3.218

25._____

26. Determine the type of number that the solutions of $x^2 + 5x + 7 = 0$ will be.

   a) no solution   b) complex   c) rational   d) irrational

26._____

27. Rationalize the denominator: $\dfrac{\sqrt{5}}{5 + \sqrt{6}}$

   a) $\dfrac{7\sqrt{5} + \sqrt{15}}{19}$ b) $\dfrac{7\sqrt{3} + \sqrt{15}}{42}$

   c) $\dfrac{5\sqrt{5} - \sqrt{30}}{19}$ d) $\dfrac{5\sqrt{3} + \sqrt{30}}{42}$

27._____

28. Edie can paint a bedroom in 3 hr. Zeb can paint the same room in 3.75 hr. How long will it take them, working together, to paint the bedroom?

   a) $\dfrac{2}{13}$ hr   b) $\dfrac{20}{13}$ hr   c) $\dfrac{20}{3}$ hr   d) $\dfrac{5}{3}$ hr

28._____

29. Solve. $x^2 - 63 = 2x$

   a) $-7, 9$ b) $-9, -7$ c) $7, 9$ d) $-9, 7$

29._____

30. Find the intersection. $\{4, 6, 8, 9, 10\} \cap \{0, 2, 4, 6\}$

   a) $\{2, 4\}$ b) $\{0, 2, 4, 6, 8, 9, 10\}$

   c) $\{4, 6\}$ d) $\{0, 2, 4, 6\}$

30._____

FINAL EXAMINATION

TEST FORM G

NAME_____

ANSWERS

31._____

31. Solve.
$$-x + 5y - z = 45,$$
$$-5x - 5y + 2z = -50,$$
$$2x + y - 2z = -25$$

What is the y – coordinate?

a) 10   b) 15   c) 5   d) 25

32._____

32. Find an equation in point-slope form of the line with slope -1 and containing $(-8, -10)$.

a) $y - 10 = x - 8$ b) $y + 10 = -(x + 8)$

c) $y + 8 = -(x + 10)$ d) $y + 10 = 8 - x$

33._____

33. Solve. $19x + 6 = 24x - 19$

a) -5 b) 0   c) 10   d) 5

34._____

34. Find the 8th term in the expansion of $(a + x)^{13}$

a) $1287a^7x^6$ b) $1716a^6x^7$ c) $1287a^7x^6$ d) $1716a^7x^6$

35._____

35. A rectangle has a diagonal of length 15 m and a perimeter of 42 m. Find the larger dimension of the rectangle.

a) 20 m   b) 12 m   c) 21 m   d) 9 m

36. An investment with interest compounded continuously doubled itself in 8 years. What is the interest rate?

a) 6.93%   b) 4.32%   c) 8.66%   d) 7.25%

36._____

FINAL EXAMINATION   NAME_____

TEST FORM G

37. Find the equation corresponding to the following graph.

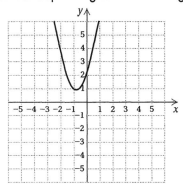

a) $-2x^2 + 3x + 2$ b) $2x^2 - 3x + 2$
c) $2x^2 + 3x + 2$ d) $2x^2 + 3x - 2$

38. Solve: $\sqrt{7x-5} + \sqrt{15x+4} = \sqrt{30x+31}$

a) 3   b) -4 c) 5   d) 4

39. If $f(x) = 2x^4 - 9x^3 + 6x - 5$, then use synthetic division to find $f(6)$.

a) 679   b) 150   c) -467 d) 650

40. Solve. $3x^3 + 8x = -25x^2$

a) $-8, -\dfrac{1}{3}, 0$ b) $-8, 0$ c) $-8, -\dfrac{1}{3}$ d) $0, \dfrac{1}{3}, 8$

41. Identify the graph of the solution set for $-3 \leq -5t - 8 < 2$

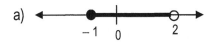

42. Find the equilibrium point for the demand and supply functions
$D(p) = 165 - 14p$ and $S(p) = 90 + 11p$

a) ($5, 106)   b) ($2, 116)   c) ($3, 123)   d) ($2, 106)

ANSWERS

37._____

38._____

39._____

40._____

41._____

42._____

FINAL EXAMINATION

TEST FORM G

NAME_____

ANSWERS

43._____

44._____

45._____

46._____

47._____

48._____

49._____

50._____

43. Which line is parallel to $8y - 1 = -3x$?

   a) $3x + 8y = 3$ b) $3x + 2y = 8$
   c) $8x - 3y = 3$ d) $2x + 8y = -3$

44. Simplify using scientific notation and correct number of significant figures:
$$\frac{1.8 \times 10^3}{8.5 \times 10^{-8}}.$$

   a) 2.1×10^{10} b) 7.6×10^{-5} c) 2.1×10^9 d) 1.6×10^{-9}

45. Find the distance between the points $(8, -3)$ and $(-8, 1)$.

   a) $16\sqrt{17}$ b) $2\sqrt{109}$ c) $4\sqrt{17}$ d) $\sqrt{218}$

46. Rosie's Palm Pirates, a manufacturer of electronic organizers, estimates that when x hundred Palm Pirates are made, the average cost per unit is given by $C(x) = 0.3x^2 - 2.94x + 8.183$, where C is in hundreds of dollars. What is the minimum cost per unit?

   a) $106 b) $98 c) $10.60 d) $0.98

47. Alyson bicycles 32 km/h with no wind. Against the wind she bikes 33km in the same time it takes to bike 63 km with the wind. What is the speed of the wind?

   a) 10 km/h b) 9.5 km/h c) 8.5 km/h d) 8 km/h

48. Solve. $|-4t - 6| \geq 10$

   a) $\{t \mid t \leq -1 \text{ or } t \geq 4\}$ b) $\{t \mid t \leq 1 \text{ and } t \geq 4\}$
   c) $\{t \mid t \leq -4 \text{ or } t \geq 1\}$ d) $\{t \mid t \leq -6 \text{ or } t \geq -4\}$

49. Given that $g(x) = -2x - 3$ and $h(x) = x^2 + 1$, find $(g \cdot h)(3)$.

   a) 72 b) -72 c) -90 d) 90

50. Solve. $5^x = 2.3$ Give exact answer in terms of common logarithms.

   a) $\dfrac{\log 2.3}{\log 5}$ b) $\dfrac{\log 2.4}{\log 4}$ c) $\dfrac{\log 4}{\log 2.4}$ d) $\dfrac{\log 5}{\log 2.3}$

FINAL EXAMINATION NAME_____

TEST FORM H   CLASS_____ SCORE_____ GRADE_____

ANSWERS

1. Find the common ratio of the geometric sequence $5, 7\frac{1}{2}, 11\frac{1}{4}, \ldots$

   a) $\frac{4}{3}$ b) $\frac{3}{4}$ c) $\frac{3}{2}$ d) $\frac{2}{3}$

   1._____

2. Find the midpoint of the segment with the endpoints $(7, -5)$ and $(-6, 2)$.

   a) $(0, -0.5)$ b) $(-1, 0)$ c) $(0.5, -1.5)$ d) $(0, -1)$

   2._____

3. Find $(f \circ g)(x)$ if $f(x) = x - x^2$ and $g(x) = 2x - 3$.

   a) $-4x^2 - 15x - 12$ b) $-9x^2 + 14x + 6$
   c) $9x^2 - 15x + 6$ d) $-4x^2 + 14x - 12$

   3._____

4. Solve. $x^2 + 4x + 1$

   a) $2 \pm \sqrt{3}$ b) $2 \pm \sqrt{2}$ c) $-2 \pm \sqrt{2}$ d) $-2 \pm \sqrt{3}$

   4._____

5. Simplify. $\sqrt{x^2 - 6x + 9}$

   a) $x - 2$ b) $x + 3$ c) $x + 2$ d) $x - 3$

   5._____

6. Simplify. $\dfrac{t-3}{t+3} \cdot \dfrac{3t+9}{3t^2 - 27}$

   a) $\dfrac{1}{t+5}$ b) $\dfrac{1}{t+3}$ c) $\dfrac{2}{t+3}$ d) $\dfrac{1}{t-3}$

   6._____

7. Find and simplify $P(a-h) - P(-a)$ given $P(x) = x^2 + 6x$.

   a) $h^2 - 6h + a^2$ b) $h^2 - 6h - 2ah + 2a^2$
   c) $h^2 - 2h - 6ah + 2a^2$ d) $h^2 + 2ah + 2a^2$

   7._____

8. Solve. $4a - 7 \geq -3a + 9$

   a) $\{a \mid a \leq \frac{7}{16}\}$ b) $\{a \mid a \leq \frac{16}{7}\}$

   c) $\{a \mid a \geq \frac{16}{7}\}$ d) $\{a \mid a \geq \frac{7}{16}\}$

   8._____

FINAL EXAMINATION   NAME_____

TEST FORM H

ANSWERS

9. Solve, if possible, using the elimination method.

$$3x - 10y = 46$$
$$x + 2y = 10$$

What is the *y* - coordinate?

9._____

a) –1   b) 1   c) 0   d) not possible

10. Which graph represents $-\frac{2}{3}x - 2$?

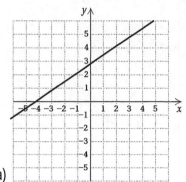
a)

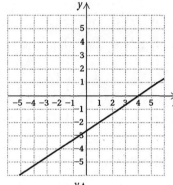
b)

10._____

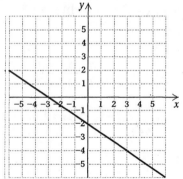
c)

d)

11._____

11. Subtract: 78.5 – 82.8 .

a) –4.3   b) –161.3   c) –3.4   d) 4.3

12. Determine whether each infinite geometric series has a limit. If a limit exists, find it.

$$0.01 - 0.015 + 0.0225 - \ldots$$

12._____

a) 0.5   b) 0   c) no limit   d) 1

FINAL EXAMINATION

TEST FORM H

NAME_____

ANSWERS

13. What is the equation corresponding to the following graph?

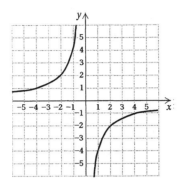

a) $\dfrac{y^2}{16} - \dfrac{x^2}{4} = 1$

b) $\dfrac{x^2}{4} - \dfrac{y^2}{16} = 1$

c) $xy = -13$

d) $xy = -4$

13._____

14. Simplify $\log_3 243$.

a) 6 b) 3 c) 2 d) 5

14._____

15. Solve. $x^2 + 5x = 1$ Use a calculator to approximate the solutions with rational numbers.

a) 0.65331193, 7.65331193

b) −5.1925824, 0.1925824

c) −0.1925824, 5.65331193

d) −7.65331193, 0.65331193

15._____

16. Simplify. $\dfrac{\sqrt[5]{a^3}}{\sqrt[4]{a}}$

a) $\sqrt[20]{a^5}$ b) $\sqrt[20]{a^7}$ c) $\sqrt[9]{a^7}$ d) $\sqrt{a^5}$

16._____

FINAL EXAMINATION

TEST FORM H

NAME_____

ANSWERS

17._____

17. Subtract and simplify. $\dfrac{5a^2}{a-b} - \dfrac{5b^2+10ab}{b-a}$

   a) $\dfrac{5(a-b)^2}{a-b}$ b) $\dfrac{5(a+b)^2}{a-b}$ c) $\dfrac{4(a+b)^2}{a-b}$ d) $\dfrac{5(a+2b)^2}{a-b}$

18. Multiply. $(a-4b)(4a+b)$

   a) $4a^2 - 15ab + 4b^2$ b) $4a^2 - 15ab - 4b^2$
   c) $4a^2 + 15ab - 4b^2$ d) $4a^2 + 17ab + 4b^2$

18._____

19. Nan can rent a van for either $50 per day with unlimited mileage or $35 per day with 75 free miles and an extra charge of 20¢ for each mile over 75. For what numbers of miles traveled would the unlimited mileage plan save Nan money?

19._____

   a) at least 150   b) at most 150   c) at least 250   d) at most 350

20. The perimeter of a rectangle is 12. The length of the rectangle is five less than four times the width. Find the width of the rectangle.

20._____

   a) 4   b) 3   c) 6   d) 2

21. Find the slope of the line containing the points $(4, -3)$ and $(6, 5)$.

21._____

   a) 4   b) -4 c) 0.25   d) -0.25

22. Divide. $\dfrac{-11.4}{-3.8}$

22._____

   a) -15.2 b) -7.6 c) -3 d) 3

23. Find the total rebound distance of a ball that is dropped from a height of 8 m, with each rebound three-fifths of the preceding one.

23._____

   a) 11 m   b) $10\dfrac{1}{2}$ m   c) $13\dfrac{1}{2}$ m   d) 12 m

24. Two squares are such that the sum of their areas is 12 m² and the difference of their areas is 4 m². Find the length of a side of each square.

24._____

   a) $\sqrt{5}$ m, $\sqrt{2}$ m   b) $\sqrt{2}$ m, 5 m
   c) $2\sqrt{2}$ m, 2 m   d) $2\sqrt{5}$ m, $2\sqrt{2}$ m

FINAL EXAMINATION   NAME_____

TEST FORM H

25. Given $\log_a 4 = 1.386$, $\log_a 5 = 1.609$, and $\log_a 6 = 1.792$, find $\log_a 100$.

    a) 4.604   b) 3.584   c) 4.787   d) 3.218

26. Determine the type of number that the solutions of $x^2 + 6x + 7 = 0$ will be.

    a) no solution   b) complex   c) rational   d) irrational

27. Rationalize the denominator: $\dfrac{\sqrt{3}}{7 + \sqrt{5}}$

    a) $\dfrac{7\sqrt{5} + \sqrt{15}}{19}$ b) $\dfrac{7\sqrt{3} + \sqrt{15}}{42}$

    c) $\dfrac{5\sqrt{5} - \sqrt{30}}{19}$ d) $\dfrac{5\sqrt{3} + \sqrt{30}}{42}$

28. Edward can paint a bedroom in 2.5 hr. Jeb can paint the same room in 4 hr. How long will it take them, working together, to paint the bedroom?

    a) $\dfrac{20}{13}$ hr   b) 2 hr   c) $\dfrac{7}{3}$ hr   d) 1.5 hr

29. Solve. $x^2 - 10 = 3x$

    a) –1, 7   b) –2, –5   c) 2, 5   d) –2, 5

30. Find the intersection. $\{1, 2, 3, 5, 7\} \cap \{0, 2, 4, 6\}$

    a) { 2 }   b) { 0, 1, 2, 3, 4, 5, 6, 7 }
    c) { 0, 2, 4, 6 }   d) { 1, 2, 3, 5, 7 }

ANSWERS

25._____

26._____

27._____

28._____

29._____

30._____

FINAL EXAMINATION

TEST FORM H

NAME_____

ANSWERS

31._____

31. Solve.
$$2x + 6y + 4z = -30,$$
$$6x + 4y + 2z = -32,$$
$$4x + 2y + 6z = -34$$

What is the y-coordinate?

a) 10   b) -10 c) -2 d) 6

32._____

32. Find an equation in point-slope form of the line with slope 3 and containing $(-6, -7)$.

a) $y - 7 = 3(x - 6)$
b) $y + 7 = 3(x + 6)$
c) $y + 7 = -3(x + 6)$
d) $y + 6 = 3(x + 7)$

33._____

33. Solve. $18x + 5 = 25x + 5$

a) 5   b) 0   c) 10   d) -5

34. Find the 7th term in the expansion of $(a + x)^{13}$

a) $1287a^7x^6$ b) $1716a^6x^7$ c) $1287a^7x^6$ d) $1716a^7x^6$

34._____

35. A rectangle has a diagonal of length 29 m and a perimeter of 82 m. Find the larger dimension of the rectangle.

a) 20 m   b) 12 m   c) 21 m   d) 9 m

35._____

36. An investment with interest compounded continuously doubled itself in 10 years. What is the interest rate?

a) 6.93%   b) 4.32%   c) 8.66%   d) 7.25%

36._____

FINAL EXAMINATION   NAME_____

TEST FORM H

37. Find the equation corresponding to the following graph.

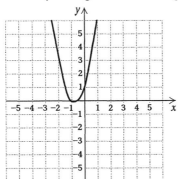

  a) $-2x^2 + 3x + 1$ b) $2x^2 + 3x + 1$
  c) $2x^2 + 3x + 2$ d) $2x^2 + 3x - 2$

38. Solve: $\sqrt{4x-7} + \sqrt{2x+8} = \sqrt{9x+13}$

  a) 3   b) -4 c) 5   d) 4

39. If $f(x) = 2x^4 - 9x^3 + 6x - 5$, then use synthetic division to find $f(5)$.

  a) 150   b) 159   c) -304 d) 10

40. Solve. $2x^3 + 21x = -17x^2$

  a) $7, \frac{2}{3}, 0$ b) $\frac{3}{2}, 0$ c) $0, 7$ d) $-7, -\frac{3}{2}, 0$

41. Identify the graph of the solution set for $1 \leq -4t - 7 < 3$

  a) 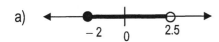   b)

  c) 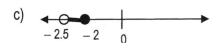   d)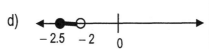

42. Find the equilibrium point for the demand and supply functions
  $D(p) = 132 - 13p$ and $S(p) = 86 + 10p$

  a) ($5, 106)   b) ($2, 116)   c) ($3, 126)   d) ($2, 106)

ANSWERS

37._____

38._____

39._____

40._____

41._____

42._____

FINAL EXAMINATION

TEST FORM H

NAME _____

ANSWERS

43. _____

44. _____

45. _____

46. _____

47. _____

48. _____

49. _____

50. _____

43. Which line is parallel to $7y - 13 = -2x$?

a) $2x + 7y = 7$ b) $7x + 2y = 2$
c) $2x - 7y = 7$ d) $2x + 7y = -7$

44. Simplify using scientific notation and correct number of significant figures:
$$\frac{7.4 \times 10^2}{9.7 \times 10^{-8}}.$$

a) 1.4×10^9 b) 7.6×10^9 ...... c) 6.7×10^9 d) 1.6×10^{-6}

45. Find the distance between the points $(7, -5)$ and $(-6, 2)$.

a) $16\sqrt{17}$ b) $2\sqrt{109}$ c) $4\sqrt{17}$ d) $\sqrt{218}$

46. Rosie's Palm Pirates, a manufacturer of electronic organizers, estimates that when x hundred Palm Pirates are made, the average cost per unit is given by $C(x) = 0.4x^2 - 3.68x + 9.524$, where C is in hundreds of dollars. What is the minimum cost per unit?

a) $106   b) $98   c) $10.60   d) $0.98

47. Erik bicycles 36 km/h with no wind. Against the wind he bikes 27 km in the same time it takes to bike 45 km with the wind. What is the speed of the wind?

a) 11 km/h   b) 8 km/h   c) 9 km/h   d) 10 km/h

48. Solve. $|-3t - 5| \geq 9$

a) $\{t \mid t \leq -\frac{14}{3} \text{ or } t \geq \frac{4}{3}\}$ b) $\{t \mid t \leq -\frac{14}{3} \text{ and } t \geq \frac{4}{3}\}$
c) $\{t \mid t \geq -14 \text{ or } t \leq 3\}$ d) $\{t \mid t \leq 3 \text{ or } t \geq -14\}$

49. Given that $g(x) = -3x - 2$ and $h(x) = x^2 + 2$, find $(g \cdot h)(3)$.

a) -55 b) -121 c) 121 d) 0

50. Solve. $4^x = 2.4$ Give exact answer in terms of common logarithms.

a) $\frac{\log 2.3}{\log 5}$ b) $\frac{\log 2.4}{\log 4}$ c) $\frac{\log 4}{\log 2.4}$ d) $\frac{\log 5}{\log 2.3}$

Answers for Chapter 1 Tests: FORM A

1. $5xy$
2. 188
3. 10.6 cm^2
4. -45
5. -15.2
6. 5.72
7. -42.4
8. -46.5
9. -11.75
10. $\dfrac{4}{15}$
11. $\dfrac{2}{3}$
12. 3
13. $-\dfrac{4}{3}$
14. 3
15. $y + 8x$ or $x8 + y$
16. $7y$
17. $9a^2b + 9ab^2 + 1$
18. $5x + 32$
19. -3
20. $\{t \mid t \in \mathbb{R}\}$ identity
21. $m = \dfrac{T}{g-a}$
22. 65
23. $22, 24, 26$
24. $-17x - 6$
25. $16b - 31$
26. $-\dfrac{14y^2}{x^8}$
27. $-\dfrac{1}{25}$
28. $\dfrac{y^4}{81x^2}$
29. $\dfrac{x^6 y^2}{4}$
30. 1
31. 1.71×10^{-6}
32. 5.8×10^9
33. 2.8×10^{-3}
34. 3.6×10^8 km
35. $25^c x^{4ac} y^{(2bc+2c)}$
36. $-2a^{16}$
37. $\dfrac{10y}{3x^4}$

Answers for Chapter 1 Tests: FORM B

1. $5x + 2y$
2. ~~242~~ 98
3. 15.75 cm^2
4. -43
5. -13.4
6. 5.72
7. -22.4
8. -48.5
9. -20.72
10. $-\dfrac{5}{21}$
11. $\dfrac{9}{14}$
12. 4
13. $-\dfrac{3}{2}$
14. 4
15. $5y + x$ or $x + y5$
16. $9y$
17. $2a^2b + 5ab^2 + 2$
18. $6x + 27$
19. 2
20. $\varnothing$ contradiction
21. $G = \dfrac{gR^2}{M}$
22. 70
23. $9, 11, 13$
24. $-10x - 19$
25. $24b - 32$
26. $\dfrac{24y^2}{x^8}$
27. $-\dfrac{1}{1000}$
28. $-\dfrac{y^9}{27x^6}$
29. $\dfrac{x^8}{9y^4}$
30. 1
31. 2.34×10^{-6}
32. 1.4×10^{10}
33. 5.7×10^2
34. $6.9 \times 10^8 \text{ km}$
35. $64^c x^{9ac} y^{(3bc-3c)}$
36. $-3a^{14}$
37. $8y$

Answers for Chapter 1 Tests: FORM C

1. $5(x+y)$
2. 50
3. 10.5 cm^2
4. -41
5. -11.6
6. 4.72
7. -3.3
8. -50.5
9. -31.85
10. $\dfrac{1}{20}$
11. $\dfrac{1}{6}$
12. 5
13. $-\dfrac{12}{5}$
14. $\dfrac{43}{2}$
15. $y + 7x$ or $x7 + y$
16. $11y$
17. $8a^2b - ab^2 + 3$
18. $9x + 16$
19. -2
20. $\{\,t \mid t \in \mathbb{R}\,\}$ identity
21. $P = 2 - \dfrac{W}{d^2 k}$
22. 75
23. 18, 20, 22
24. $-13x - 2$
25. $32b - 37$
26. $-\dfrac{36y^2}{x^4}$
27. $-\dfrac{1}{49}$
28. $\dfrac{y^4}{64x^6}$
29. $\dfrac{16x^2}{y^2}$
30. 1
31. 2.77×10^{-6}
32. 5.7×10^9
33. 1.1×10^{-3}
34. 9.4×10^8 km
35. $81^c x^{16ac} y^{(4bc + 8c)}$
36. $-4a^{12}$
37. $\dfrac{x^4}{y}$

Answers for Chapter 1 Tests: FORM D

1. $2xy - 5$
2. -67
3. 8.75 cm^2
4. -39
5. -9.8
6. 4.62
7. 16.7
8. -27.6
9. -37.74
10. $-\dfrac{11}{10}$
11. $\dfrac{5}{8}$
12. 6
13. $-\dfrac{10}{3}$
14. -13
15. $4y + x$ or $x + y4$
16. $13y$
17. $3a^2b + 3ab^2 + 1$
18. $4x + 25$
19. 3
20. $\varnothing$ contradiction
21. $v = c - \dfrac{r}{t_B - t_A}$
22. 80
23. $13, 15, 17$
24. $-10x - 5$
25. $29b - 42$
26. $\dfrac{35}{x^6 y^9}$
27. $-\dfrac{1}{125}$
28. $-\dfrac{y^3}{64x^{12}}$
29. $\dfrac{x^4}{25y^6}$
30. 1
31. 3.02×10^{-6}
32. 1.6×10^{10}
33. 1.3×10^3
34. 1.4×10^9 km
35. $32^c x^{25ac} y^{(5bc-10c)}$
36. $-5a^{10}$
37. $\dfrac{5x^6}{y}$

Answers for Chapter 1 Tests: FORM E

1. $2xy$
2. 257
3. 29.6 cm^2
4. -37
5. -8.0
6. 4.62
7. 36.7
8. -59.8
9. -46.62
10. $\dfrac{19}{10}$
11. $\dfrac{5}{21}$
12. 5
13. $-\dfrac{10}{3}$
14. $\dfrac{110}{3}$
15. $y + 6x$ or $x6 + y$
16. $15y$
17. $7a^2b - 7ab^2 - 2$
18. 35
19. -1
20. $\varnothing$ contradiction
21. $v = \dfrac{c}{k}\left(1 - \dfrac{a}{bB}\right)$
22. 85
23. $8, 10, 12$
24. $-10x - 8$
25. $22b - 37$
26. $-\dfrac{40}{x^{11}y^{11}}$
27. $-\dfrac{1}{81}$
28. $\dfrac{1}{49x^{10}y^2}$
29. $\dfrac{25x^3}{y^6}$
30. 1
31. 3.06×10^{-6}
32. 6.9×10^9
33. 4.2×10^{-4}
34. $4.9 \times 10^9 \text{ km}$
35. $64^{c}x^{36ac}y^{(6bc+18c)}$
36. $-6a^8$
37. $\dfrac{4x^6}{y^2}$

Answers for Chapter 1 Tests: FORM F

1. $2x + 5y$
2. 228
3. 37.8 cm^2
4. -35
5. -6.2
6. -5.28
7. -24.3
8. -59.0
9. -48.75
10. $\dfrac{43}{14}$
11. $\dfrac{1}{4}$
12. 4
13. $-\dfrac{8}{5}$
14. $-\dfrac{404}{9}$
15. $3y + x$ or $x + y3$
16. $17y$
17. $4a^2b - 11ab^2 - 3$
18. $8x - 9$
19. 4
20. $\{t \mid t \in \mathbb{R}\}$ identity
21. $k = \dfrac{RT}{6\pi NPD}$
22. 90
23. $31, 33, 35$
24. $-7x - 4$
25. $34b - 43$
26. $\dfrac{36}{x^{11}y^{13}}$
27. $-\dfrac{1}{64}$
28. $-\dfrac{x^9}{8y^6}$
29. $\dfrac{x^9}{64y^{15}}$
30. 1
31. 2.92×10^{-6}
32. 1.4×10^{10}
33. 7.1×10^{-3}
34. 8.8×10^9 km
35. $81^c x^{28ac} y^{(4bc-12c)}$
36. $-7a^6$
37. $\dfrac{6x^4}{y}$

Answers for Chapter 1 Tests: FORM G

1. c
2. d
3. d
4. a
5. b
6. a
7. b
8. b
9. c
10. d
11. c
12. c
13. a
14. d
15. d
16. c
17. b
18. a
19. a
20. a
21. c
22. b
23. b
24. b
25. d
26. a
27. b
28. c
29. d
30. a

Answers for Chapter 1 Tests: FORM H

1. b
2. a
3. a
4. d
5. c
6. d
7. c
8. c
9. b
10. a
11. b
12. b
13. c
14. b
15. b
16. a
17. d
18. b
19. c
20. b
21. a
22. d
23. c
24. d
25. c
26. c
27. a
28. a
29. b
30. d

Answers for Chapter 2 Tests: FORM A

1. yes
2. no
3.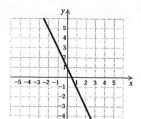
4.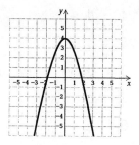
5.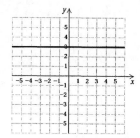
6.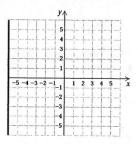

7. (a) $f(3) = 4$
   (b) $\{x \mid 0 \leq x \leq 5\}$
   (c) $x = 1, 4$
   (d) $\{f \mid 0 \leq f \leq 6\}$

8. $27.2 million
9. 101 fights
10. $\dfrac{2}{5}$; $(0, -13)$
11. $-\dfrac{3}{8}$; $\left(0, -\dfrac{9}{4}\right)$
12. $\dfrac{3}{2}$
13. 2
14. 7.5 calories/minute

15. $f(x) = -3x - 4$
16.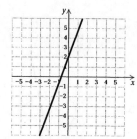
17. $x = -2$
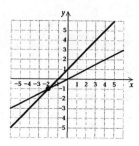

18. a, b
19. $y - 4 = 15(x - 3)$
20. $f(x) = 2x + 1$
21. parallel
22. perpendicular
23. $y = \dfrac{9}{2}x - \dfrac{19}{2}$
24. $y = -\dfrac{2}{9}x + \dfrac{14}{3}$
25. (a) 12
    (b) -544
    (c) $\{x \mid x \in \mathbb{R} \text{ and } x \neq -\dfrac{5}{9}\}$
26. (a) $C(m) = 0.1m + 40$
    (b) $65
27. $s = -8r + 42$
28. $h(x) = 8x - 5$

Answers for Chapter 2 Tests: FORM B

1. yes
2. no
3.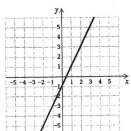
4.

5.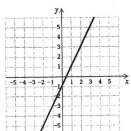
6.

7. (a) $f(2) = 3$
   (b) $\{x \mid -5 \leq x \leq 4\}$
   (c) $x = -4$
   (d) $\{f \mid -2 \leq f \leq 5\}$
8. $20.9 million
9. 12.4 hours
10. $-\dfrac{3}{7}$; $(0, 12)$
11. $\dfrac{5}{6}$; $\left(0, \dfrac{1}{2}\right)$
12. 2
13. undefined
14. 3.33 calories/minute

15. $f(x) = 8x - 5$
16.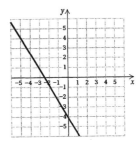
17. $x = -\dfrac{4}{3}$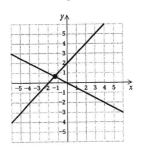
18. a, c
19. $y - 7 = -13(x - 5)$
20. $f(x) = -2x + 2$
21. parallel
22. neither
23. $y = \dfrac{8}{3}x - 12$
24. $y = -\dfrac{3}{8}x - \dfrac{23}{8}$
25. (a) 11
    (b) -480
    (c) $\{x \mid x \in \mathbb{R} \text{ and } x \neq -\dfrac{3}{4}\}$
26. (a) $C(m) = 0.2m + 35$
    (b) $115
27. $s = -\dfrac{7}{2}r + 24$
28. $h(x) = 7x - 6$

Answers for Chapter 2 Tests: FORM C

1. yes
2. no
3.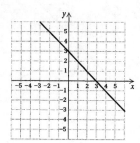
4.
5.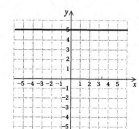
6.

7. (a) $f(2) = 2$
   (b) $\{x \mid -\infty < x < \infty\}$
   (c) $x = 0$
   (d) $\{f \mid -\infty < f \leq 3\}$
8. $255.4 thousand
9. $62
10. $\dfrac{4}{5}$; $(0, 11)$
11. $-\dfrac{7}{4}$; $\left(0, \dfrac{5}{4}\right)$
12. 3
13. 0
14. 5 dollars/day
15. $f(x) = -x + 6$
16.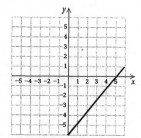
17. $x = -2$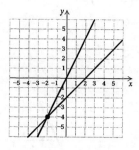
18. a, b
19. $y + 5 = 11(x - 4)$
20. $f(x) = -2x - 2$
21. neither
22. perpendicular
23. $y = \dfrac{7}{4}x + \dfrac{37}{4}$
24. $y = -\dfrac{4}{7}x + \dfrac{16}{7}$
25. (a) 10
    (b) -420
    (c) $\{x \mid x \in \mathbb{R} \text{ and } x \neq -1\}$
26. (a) $C(m) = 0.3m + 30$
    (b) $93
27. $s = \dfrac{1}{2}r + \dfrac{1}{2}$
28. $h(x) = 6x - 7$

Answers for Chapter 2 Tests: FORM D

1. yes
2. no
3.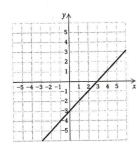
4.
5.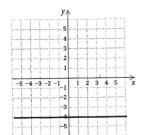
6.
7. (a) $f(2) = 3$
   (b) $\{x \mid -4 \leq x < \infty\}$
   (c) $x = -2$
   (d) $\{f \mid 0 \leq f < \infty\}$
8. $0.5 thousand
9. $75
10. $-\dfrac{5}{7}; (0, -10)$
11. $\dfrac{9}{2}; (0, -18)$
12. -7
13. 3
14. -10 dollars/day
    -100 " "

15. $f(x) = 4x + 7$
16.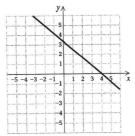
17. $x = \dfrac{1}{3}$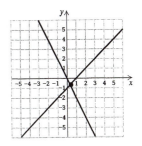
18. a, c
19. $y + 8 = -9(x - 6)$
20. $f(x) = 2x - 1$
21. parallel
22. perpendicular
23. $y = \dfrac{6}{5}x - \dfrac{2}{5}$
24. $y = -\dfrac{5}{6}x - \dfrac{13}{2}$
25. (a) 9
    (b) -364
    (c) $\{x \mid x \in \mathbb{R} \text{ and } x \neq -\dfrac{4}{3}\}$
26. (a) $C(m) = 0.3m + 27$
    (b) $57
27. $s = \dfrac{4}{5}r - \dfrac{7}{5}$
28. $h(x) = 4x - 9$

Answers for Chapter 2 Tests: FORM E

1. yes
2. no
3.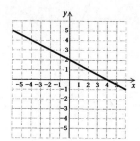
4.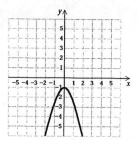
5.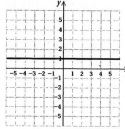
6.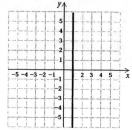
7. (a) $f(2) = -3$
   (b) $\{x \mid -6 \leq x \leq 3\}$
   (c) $x = -6, -1$
   (d) $\{f \mid -5 \leq f \leq \frac{3}{2}\}$
8. $21.1 million
9. 146,250 people
10. $-\frac{5}{2}$; $(0, 9)$
11. $\frac{11}{4}$; $\left(0, \frac{11}{2}\right)$
12. undefined
13. -4
14. 8 winks/hour
15. $f(x) = -10x - 8$
16.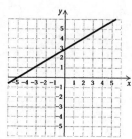
17. $x = 2$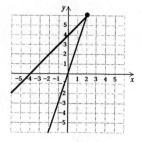
18. a, b
19. $y - 6 = 7(x + 5)$
20. $f(x) = x + 3$
21. parallel
22. neither
23. $y = \frac{5}{6}x - \frac{13}{6}$
24. $y = -\frac{6}{5}x + 8$
25. (a) 8
    (b) -312
    (c) $\{x \mid x \in \mathbb{R} \text{ and } x \neq -\frac{9}{5}\}$
26. (a) $C(m) = 0.2m + 30$
    (b) $80
27. $s = -\frac{4}{5}r + \frac{66}{5}$
28. $h(x) = 3x - 1$

Answers for Chapter 2 Tests: FORM F

1. yes
2. no
3.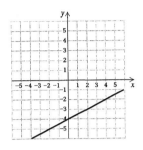
4.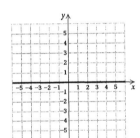
5.
6.

7. (a) $f(2) = -1$
   (b) $\{x \mid -5 \leq x \leq 6\}$
   (c) $x = -2, 0, 6$
   (d) $\{f \mid -2 \leq f \leq 2\}$
8. $17.5 thousand
9. $180
10. $\dfrac{7}{3}$; $(0, 8)$
11. $-\dfrac{13}{6}$; $(0, -13)$
12. 0
13. -1
14. 2 minutes/month
15. $f(x) = -3x - 4$
16.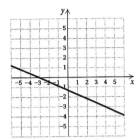
17. $x = 1$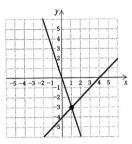
18. a, c
19. $y - 8 = -5(x + 7)$
20. $f(x) = -x + 4$
21. neither
22. perpendicular
23. $y = \dfrac{4}{7}x - \dfrac{34}{7}$
24. $y = -\dfrac{7}{4}x + \dfrac{27}{4}$
25. (a) 7
    (b) -156
    (c) $\{x \mid x \in \mathbb{R} \text{ and } x \neq -\dfrac{1}{4}\}$
26. (a) $C(m) = 0.1m + 35$
    (b) $110
27. $s = -\dfrac{1}{2}r + \dfrac{15}{2}$
28. $h(x) = 2x - 2$

309

Answers for Chapter 2 Tests: FORM G

1. c
2. d
3. a
4. b
5. b
6. a
7. a
8. d
9. c
10. c
11. b
12. a
13. b
14. b
15. d
16. a
17. a
18. c

Answers for Chapter 2 Tests: FORM H

1. b
2. d
3. c
4. c
5. a
6. b
7. a
8. d
9. d
10. a
11. b
12. a
13. c
14. c
15. b
16. c
17. a
18. d

Answers for Chapter 3 Tests: FORM A

1. $\left(\dfrac{15}{4}, \dfrac{25}{12}\right)$
2. $(-1, -3)$
3. $(6, -7)$
4. $\left(\dfrac{5}{3}, -\dfrac{3}{2}\right)$
5. width = 5; length = 7
6. $100,000 mortgage; $10,000 car loan; $5000 credit card
7. $\left(\dfrac{13}{21}, \dfrac{25}{21}, \dfrac{2}{3}\right)$
8. $\left(\dfrac{1}{2}, -\dfrac{1}{2}, \dfrac{1}{2}\right)$
9. $(4, 5, 6)$
10. $(3, -5, 4)$
11. $(1, 2)$
12. $(3, 4, 2)$
13. 26
14. 18
15. $(-5, 5)$
16. 5 hours
17. ($2, 62)
18. a. $C(x) = 15{,}000 + 10x$
    b. $R(x) = 35x$
    c. $P(x) = 25x - 15{,}000$
    d. $P(300) = -7500;\ P(900) = 7500$
    e. $x = 600$
19. $m = \dfrac{2}{3};\ b = 5$
20. 320 adults', 100 seniors', 150 children's

Answers for Chapter 3 Tests: FORM B

1. $\left(\dfrac{19}{5}, -\dfrac{8}{25}\right)$
2. $\left(-\dfrac{15}{14}, -\dfrac{9}{7}\right)$
3. $(7, -6)$
4. $\left(\dfrac{1}{2}, \dfrac{8}{3}\right)$
5. width = 8; length = 24
6. $150,000 mortgage; $12,000 car loan; $6000 credit card
7. $(5, -3, 2)$
8. $(2, 3, 4)$
9. $(6, 5, 4)$
10. $(-1, 3, -2)$
11. $(2, 1)$
12. $(-1, -3, -5)$
13. -26
14. 18
15. $(4, -4)$
16. 5.25 hours
17. ($3, 63)
18. a. $C(x) = 24{,}000 + 15x$
    b. $R(x) = 45x$
    c. $P(x) = 30x - 24{,}000$
    d. $P(250) = -16{,}500;\ P(1000) = 6000$
    e. $x = 800$
19. $m = 3;\ b = -8$
20. 480 adults', 150 seniors', 300 children's

Answers for Chapter 3 Tests: FORM C

1. $\left(\dfrac{1}{2}, \dfrac{11}{4}\right)$
2. $\left(\dfrac{30}{11}, -\dfrac{5}{11}\right)$
3. $(8, -5)$
4. $\left(\dfrac{3}{4}, \dfrac{5}{3}\right)$
5. width = 11; length = 31
6. $20,000 mortgage; $7000 car loan; $3000 credit card
7. $(0, 3, 8)$
8. $\left(\dfrac{1}{2}, 1, \dfrac{3}{2}\right)$
9. $(-1, 2, -3)$
10. $(4, -6, 5)$
11. $(2, 3)$
12. $(2, 4, 6)$
13. 57
14. -240
15. $(-3, 3)$
16. 6 hours
17. $(\$3, 71)$
18. a. $C(x) = 32{,}250 + 22x$
    b. $R(x) = 57x$
    c. $P(x) = 35x - 32{,}2500$
    d. $P(575) = -13{,}125$; $P(1500) = 19{,}250$
    e. $x = 950$
19. $m = \dfrac{4}{5}$; $b = -3$
20. 200 adults', 55 seniors', 100 children's

Answers for Chapter 3 Tests: FORM D

1. $\left(\dfrac{11}{5}, \dfrac{47}{5}\right)$
2. $\left(\dfrac{1}{5}, -\dfrac{13}{5}\right)$
3. $(9, -4)$
4. $\left(\dfrac{1}{3}, \dfrac{5}{4}\right)$
5. width = 14; length = 30
6. $100,000 mortgage; $10,000 car loan; $5000 credit card
7. $(-6, -4, -2)$
8. $(-5, 0, 5)$
9. $(3, 5, 1)$
10. $(-2, 4, -3)$
11. $(3, 2)$
12. $(1, 0, -1)$
13. -57
14. -18
15. $(2, -2)$
16. 7 hours
17. $(\$4, 93)$
18. a. $C(x) = 51{,}300 + 51x$
    b. $R(x) = 108x$
    c. $P(x) = 57x - 51{,}300$
    d. $P(250) = -8550$; $P(1000) = 5700$
    e. $x = 900$
19. $m = -8$; $b = 10$
20. 350 adults', 110 seniors', 150 children's

Answers for Chapter 3 Tests: FORM E

1. $\left(-\frac{32}{9}, \frac{22}{9}\right)$
2. $\left(-\frac{11}{2}, -\frac{1}{2}\right)$
3. $(10, -3)$
4. $\left(-\frac{2}{5}, \frac{1}{4}\right)$
5. width = 10; length = 37
6. $10,000 mortgage; $12,000 car loan; $6000 credit card
7. $(1, 2, 1)$
8. $(3, 2, 1)$
9. $(3, 0, -1)$
10. $(1, 0, -1)$
11. $(3, 4)$
12. $(3, 2, 1)$
13. 0
14. -47
15. $(-1, 1)$
16. 7.75 hours
17. $(\$2, 86)$
18. a. $C(x) = 52{,}290 + 47x$
    b. $R(x) = 130x$
    c. $P(x) = 83x - 52{,}290$
    d. $P(500) = -10{,}790$; $P(700) = 5810$
    e. $x = 630$
19. $m = -\frac{1}{6}$; $b = 6$
20. 480 adults', 130 seniors', 200 children's

Answers for Chapter 3 Tests: FORM F

1. $\left(-\frac{33}{13}, \frac{41}{13}\right)$
2. $\left(\frac{5}{12}, -\frac{8}{3}\right)$
3. $(11, -2)$
4. $\left(-\frac{3}{4}, \frac{3}{5}\right)$
5. width = 6; length = 21
6. $200,000 mortgage; $7000 car loan; $5000 credit card
7. $(10, 5, 0)$
8. $(5, 3, 7)$
9. $(2, -4, 3)$
10. $(2, -1, 2)$
11. $(4, 3)$
12. $(-3, -2, -1)$
13. -24
14. -6
15. $(2, -2)$
16. 6 hours
17. $(\$4, 116)$
18. a. $C(x) = 36{,}225 + 16x$
    b. $R(x) = 79x$
    c. $P(x) = 63x - 36{,}225$
    d. $P(400) = -11{,}025$; $P(800) = 14{,}175$
    e. $x = 575$
19. $m = 5$; $b = -7$
20. 200 adults', 60 seniors', 100 children's

Answers for Chapter 3 Tests: FORM G

21. c
22. d
23. a
24. a
25. d
26. d
27. d
28. c
29. b
30. b
31. a
32. c
33. c
34. b
35. a
36. d
37. d
38. a
39. c

Answers for Chapter 3 Tests: FORM H

1. d
2. a
3. a
4. b
5. c
6. d
7. b
8. b
9. a
10. c
11. c
12. d
13. b
14. a
15. c
16. c
17. d
18. b
19. c

Answers for Chapter 4 Tests: FORM A

1. $\{x \mid x < 6\}$; $(-\infty, 6)$
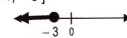

2. $\{y \mid y > -90\}$; $(-90, \infty)$

3. $\{y \mid y \leq -3\}$; $(-\infty, -3]$

4. $\{a \mid a \leq \frac{4}{9}\}$; $(-\infty, \frac{4}{9}]$

5. $\{x \mid x < \frac{8}{7}\}$; $(-\infty, \frac{8}{7})$

6. $\{x \mid x \geq -\frac{4}{13}\}$; $[-\frac{4}{13}, \infty)$

7. $\{x \mid x > -1\}$

8. ~~at least~~ more than 135 miles

9. at most ~~5.5~~ $5\frac{5}{6}$ hours (5 hr 50 min. = 5:50 hrs.)

10. $\{1, 2, 3\}$

11. $\{0, 1, 2, 3, 4, 5\}$

12. $(-\infty, 2]$

13. $(-7, 6)$
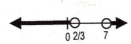

14. $(-\frac{1}{2}, \frac{1}{2}]$

15. $(-\infty, \frac{2}{3}) \cup (7, \infty)$
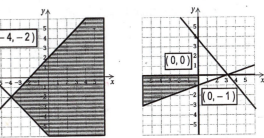

16. $(-\infty, -4) \cup (-3, \infty)$

17. $[8, 15)$

18. $\{x \mid x = \pm 2\}$
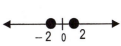

19. $(-\infty, -9) \cup (9, \infty)$
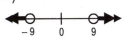

20. $(-\frac{3}{4}, \frac{7}{4})$

21. $(-\infty, -\frac{3}{2}] \cup [\frac{7}{6}, \infty)$

22. $\emptyset$

23. $(-\infty, -1) \cup (1, \infty)$

24. $\{x \mid x = 1\}$

25.

26.

27. $F(3, 0) = 71$; $F(7, 3) = 24$

28. 4 manicures, 46 haircuts; $822

29. $[-\frac{4}{3}, 0]$

30. $(\frac{3}{14}, \frac{11}{14})$

31. ~~$|2x+9| \leq 7$~~
    $|2x+7| \leq 9$

Answers for Chapter 4 Tests: FORM B

1. $\{x \mid x < 5\}; (-\infty, 5)$

2. $\{y \mid y > -50\}; (-50, \infty)$
   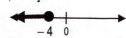

3. $\{y \mid y \leq -4\}; (-\infty, -4]$

4. $\{a \mid a \leq \frac{6}{7}\}; (-\infty, \frac{6}{7}]$
   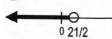

5. $\{x \mid x < \frac{21}{2}\}; (-\infty, \frac{21}{2})$
   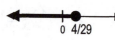

6. $\{x \mid x \leq \frac{4}{29}\}; (-\infty, \frac{4}{29}]$

7. $\{x \mid x > -\frac{5}{2}\}$

8. at least 140 miles

9. at most 3.5 hours

10. $\{2, 4\}$

11. $\{1, 2, 3, 4, 6, 8, 10\}$

12. $(-\infty, \frac{3}{2}]$

13. $(-6, 5)$
    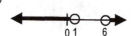

14. $(-2, -1]$
    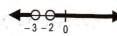

15. $(-\infty, 1) \cup (6, \infty)$

16. $(-\infty, -3) \cup (-2, \infty)$

17. $[10, 15)$

18. $\{x \mid x = \pm 2.5\}$
    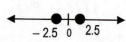

19. $(-\infty, -8) \cup (8, \infty)$
    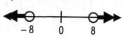

20. $(-\frac{1}{2}, \frac{5}{2})$
    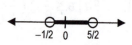

21. $(-\infty, -\frac{19}{7}] \cup [\frac{1}{7}, \infty)$

22. $\emptyset$
    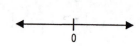

23. $(-\infty, \frac{7}{9}) \cup (\frac{17}{9}, \infty)$

24. $\{x \mid x = -2\}$

25.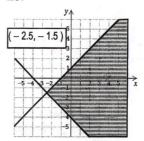

26.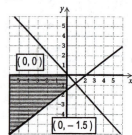

27. $F(10, 2) = 90; F(5, 0) = 40$

28. 27 pedicures, 23 massages; $661

29. $[-\frac{1}{2}, 0]$

30. $(\frac{5}{12}, 1)$

31. $|2x - 1| \leq 11$

Answers for Chapter 4 Tests: FORM C

1. $\{x \mid x < 4\}; (-\infty, 4)$
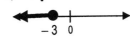

2. $\{y \mid y > -30\}; (-30, \infty)$

3. $\{y \mid y \leq -3\}; (-\infty, -3]$
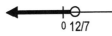

4. $\{a \mid a \leq \frac{8}{11}\}; (-\infty, \frac{8}{11}]$
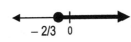

5. $\{x \mid x < \frac{12}{7}\}; (-\infty, \frac{12}{7})$

6. $\{x \mid x \geq -\frac{2}{3}\}; [-\frac{2}{3}, \infty)$

7. $\{x \mid x > -\frac{1}{2}\}$

8. at least 170 miles

9. at most 6.5 hours

10. $\{1, 3\}$

11. $\{1, 2, 3, 4, 5, 7, 9\}$

12. $(-\infty, 3]$

13. $(-5, 4)$
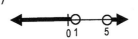

14. $(-4.5, -4]$
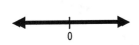

15. $(-\infty, 1) \cup (5, \infty)$

16. $(-\infty, \infty)$

17. $[6, 12)$

18. $\{x \mid x = \pm 3\}$
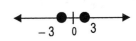

19. $(-\infty, -7) \cup (7, \infty)$
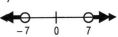

20. $(-\frac{5}{6}, \frac{3}{2})$
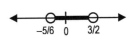

21. $(-\infty, -\frac{3}{4}] \cup [\frac{1}{2}, \infty)$
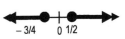

22. $\varnothing$
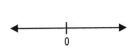

23. $(-\infty, 1) \cup (\frac{7}{3}, \infty)$
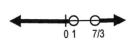

24. $\{x \mid x = -5\}$

25. 26.
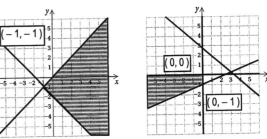

27. $F(3, 2) = 34; F(0, 0) = 0$

28. 6 pedicures, 24 massages; $468

29. $\varnothing$

30. $(\frac{7}{10}, \frac{13}{10})$

31. $|2x - 13| \leq 7$

317

Answers for Chapter 4 Tests: FORM D

1. $\{x \mid x < 3\}; (-\infty, 3)$
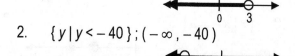

2. $\{y \mid y < -40\}; (-\infty, -40)$

3. $\{y \mid y \geq -2\}; [-2, \infty)$

4. $\{a \mid a \geq \frac{5}{8}\}; [\frac{5}{8}, \infty)$
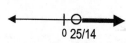

5. $\{x \mid x > \frac{25}{14}\}; (\frac{25}{14}, \infty)$
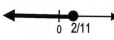

6. $\{x \mid x \leq \frac{2}{11}\}; (-\infty, \frac{2}{11}]$
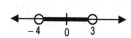

7. $\{x \mid x > -\frac{4}{3}\}$

8. at least 200 miles

9. at most 6.5 hours

10. $\{0, 2\}$

11. $\{0, 1, 2, 3, 4, 5, 6\}$

12. $(-\infty, \frac{4}{3}]$

13. $(-4, 3)$

14. $(-16, -13]$
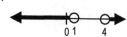

15. $(-\infty, 1) \cup (4, \infty)$
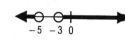

16. $(-\infty, -5) \cup (-3, \infty)$

17. $[4, 7.5)$
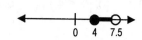

18. $\{x \mid x = \pm 3.5\}$
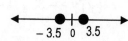

19. $(-\infty, -6) \cup (6, \infty)$
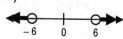

20. $(-\frac{2}{3}, \frac{2}{3})$
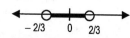

21. $(-\infty, -\frac{8}{9}] \cup [\frac{4}{9}, \infty)$
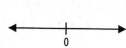

22. $\varnothing$

23. $(-\infty, \frac{5}{3}) \cup (\frac{11}{3}, \infty)$

24. $\{x \mid x = -6\}$

25.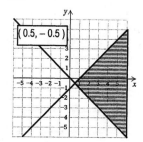

26.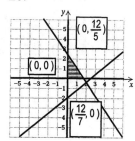

27. $F(5, 0) = 25; F(0, 5) = -40$

28. 24 brushes, 16 washes; $712

29. $[-\frac{2}{3}, 0]$

30. $(\frac{9}{8}, \frac{7}{4})$

31. $|2x - 45| \leq 15$

Answers for Chapter 4 Tests: FORM E

1. $\{x \mid x < 9\}; (-\infty, 9)$

2. $\{y \mid y < -30\}; (-\infty, -30)$

3. $\{y \mid y \geq -2\}; [-2, \infty)$

4. $\{a \mid a \geq \frac{14}{9}\}; [\frac{14}{9}, \infty)$

5. $\{x \mid x > \frac{12}{7}\}; (\frac{12}{7}, \infty)$

6. $\{x \mid x \geq -\frac{1}{5}\}; [-\frac{1}{5}, \infty)$

7. $\{x \mid x > 2\}$

8. at least 200 miles

9. at most 6.5 hours

10. $\{10, 20\}$

11. $\{0, 5, 10, 15, 20, 30, 40\}$

12. $(-\infty, -5]$

13. $(-5, 0)$

14. $(-9.5, -9]$

15. $(-\infty, 2) \cup (3, \infty)$

16. $(-\infty, -4) \cup (-4, \infty)$

17. $[2, 6)$
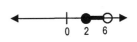

18. $\{x \mid x = \pm 4\}$
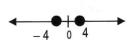

19. $(-\infty, -5) \cup (5, \infty)$
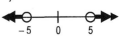

20. $(-\frac{9}{10}, \frac{13}{10})$

21. $(-\infty, -11] \cup [4, \infty)$

22. $\varnothing$
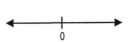

23. $(-\infty, 4) \cup (8, \infty)$

24. $\{x \mid x = -4\}$

25. 26.
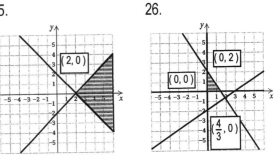

27. $F(6, 1) = 22; F(2, 5) = -30$

28. 2 brushes, 18 washes; $426

29. $[-\frac{2}{7}, 0]$

30. $(2, \frac{8}{3})$

31. $|2x + 32| \leq 4$

Answers for Chapter 4 Tests: FORM F

1. $\{x \mid x < 7\}; (-\infty, 7)$

2. $\{y \mid y < -30\}; (-\infty, -30)$

3. $\{y \mid y \geq -2\}; [-2, \infty)$

4. $\{a \mid a \geq 1\}; [1, \infty)$

5. $\{x \mid x > \frac{7}{2}\}; (\frac{7}{2}, \infty)$

6. $\{x \mid x \leq \frac{1}{9}\}; (-\infty, \frac{1}{9}]$

7. $\{x \mid x > 2\}$

8. at least 250 miles

9. at most 7.5 hours

10. $\{9\}$

11. $\{3, 6, 9, 12, 15, 18, 27, 36\}$

12. $(-\infty, -\frac{5}{2}]$

13. $(-4, -1)$

14. $(-\frac{10}{3}, -3]$

15. $(-\infty, 2) \cup (2, \infty)$

16. $(-\infty, -3) \cup (-2, \infty)$

17. $[6, \frac{32}{3})$

18. $\{x \mid x = \pm 4.5\}$

19. $(-\infty, -4) \cup (4, \infty)$

20. $(-\frac{4}{5}, \frac{8}{5})$

21. $(-\infty, -6] \cup [2, \infty)$

22. $\varnothing$

23. $(-\infty, \frac{3}{2}) \cup (\frac{5}{2}, \infty)$

24. $\{x \mid x = -2\}$

25.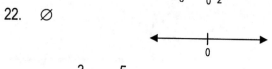

26.

27. $F(8, -2) = 56; F(-2, 2) = -26$

28. 45 brushes, 30 washes; $1335

29. $\varnothing$

30. $(5, 6)$

31. $|2x - 6| \leq 10$

Answers for Chapter 4 Tests: FORM G

1. d
2. d
3. c
4. a
5. b
6. a
7. a
8. b
9. c
10. c
11. d
12. b
13. d
14. a
15. c
16. b
17. b
18. a
19. d
20. d
21. d
22. c

Answers for Chapter 4 Tests: FORM H

1. d
2. a
3. a
4. a
5. b
6. c
7. c
8. b
9. b
10. d
11. c
12. c
13. c
14. c
15. d
16. a
17. b
18. a
19. b
20. d
21. c
22. a

Answers for Chapter 5 Tests: FORM A

1. 6
2. $-7x^5y - 2x^3y + 9x^2y^3 + 5xy^3$
3. $-1 - a^3$
4. $P(0) = 7; P(-2) = -17$
5. $h^2 - 3h + 2ah$
6. $xy + xy^2$
7. $6x^3 + 7x^2 - 12y - 5y^2$
8. $14m^3 - 6m^2n - 2mn^2 - 10n^3$
9. $-5a - 8b$
10. $3y^2 - y - 2y^3$
11. $45x^3y^3$
12. $10a^2 + 14ab - 12b^2$
13. $x^3 - y^3$
14. $4x^6 + 12x^3 + 9$
15. $9y^2 - 48y + 64$
16. $25x^2 - 64y^2$
17. $7x(2 - x^2)$
18. $(y^2 + 3)(y - 6)$
19. $(p + 1)(p - 17)$
20. $(2m + 1)(3m + 2)$
21. $(9y - 1)(9y + 1)$
22. $5(r - 1)(r^2 + r + 1)$
23. $(3x - 1)^2$
24. $(x - y)(x + y)(x^2 + y^2)$
25. $(y + 5 - 12t)(y + 5 + 12t)$
26. $3(12a - b)(12a + b)$
27. $2(2x - 3)(4x - 1)$
28. $4ab(4a^4 + 13b^4)$
29. $3xy(y^3 + 9x + 6y^2x - 4)$
30. $-4, 6$
31. ± 2
32. $-\frac{3}{2}, -1, 0$
33. $-\frac{1}{2}, 0$
34. $0, 6$
35. $\{x \mid x \in \mathbb{R}, x \neq -2\}$
36. length = 12 cm, width = 10 cm
37. 4 seconds
38. a) $-x^5 - 2x^4 - 2x^3 + x + 1$
    b) $(x^2 + x + 1)(-x^3 - x^2 + 1)$
39. $(2x^n - 3)(3x^n + 4)$

Answers for Chapter 5 Tests: FORM B

1. 6
2. $x^4y^2 - 8x^3y - 3x^2y^3 + 6xy^2$
3. -10
4. $P(0) = 6; P(-2) = 12$
5. $h^2 - 4h + 2ah$
6. $-xy - xy^2$
7. $6x^3 + 8x^2 - 10y - 6y^2$
8. $14m^3 - 6m^2n - 2mn^2 - 10n^3$
9. $-5a - 10b$
10. $-5y^2 - 2y + y^3$
11. $65x^3y^3$
12. $3a^2 - 7ab - 20b^2$
13. $-x^3 + 2x^2y - y^3$
14. $9x^6 + 24x^3 + 16$
15. $25y^2 - 20y + 4$
16. $36x^2 - 25y^2$
17. $8x(3 - x^2)$
18. $(y^2 + 4)(y - 4)$
19. $(p + 2)(p - 18)$
20. $(2m + 1)(4m + 3)$
21. $(10y - 11)(10y + 11)$
22. $(2r - 1)(4r^2 + 2r + 1)$
23. $(3x - 2)^2$
24. $5(x - y)(x + y)(x^2 + y^2)(x^4 + y^4)$
25. $(y + 6 - 11t)(y + 6 + 11t)$
26. $5(10a - b)(10a + b)$
27. $3(2x - 3)(5x - 1)$
28. $5ab(3a^4 + 7b^4)$
29. $4xy(2y^3 + x + 5yx - 3)$
30. $-4, 7$
31. ± 3
32. $-2, -\dfrac{1}{2}, 0$
33. $-\dfrac{1}{4}, 0$
34. $0, 7$
35. $\{x \mid x \in \mathbb{R}, x \neq -3\}$
36. length = 11 cm, width = 8 cm
37. 4 seconds
38. a) $-x^5 - x - 1$
    b) $(x^2 + x + 1)(-x^3 + x^2 - 1)$
39. $(2x^n + 3)(3x^n - 4)$

Answers for Chapter 5 Tests: FORM C

1. 6
2. $-4x^4y + 2x^3y^3 - 9x^2y + 7xy^4$
3. -7
4. $P(0) = 5; P(-2) = -1$
5. $h^2 - 41h + 2ah$
6. $19xy - 19xy^2$
7. $6x^3 + 9x^2 - 8y - 7y^2$
8. $8m^3 + 2m^2n - 2mn^2 - 6n^3$
9. $-5a - 12b$
10. $-5y^2 - 3y - 6y^3$
11. $77x^3y^3$
12. $8a^2 - 18ab - b^2$
13. $-x^3 - 2x^2y - y^3$
14. $16x^6 + 40x^3 + 25$
15. $49y^2 - 56y + 16$
16. $49x^2 - 9y^2$
17. $9x(4 - x^2)$
18. $(y^2 + 5)(y - 2)$
19. $(p + 3)(p - 16)$
20. $(2m + 1)(5m + 5)$
21. $(11y - 3)(11y + 3)$
22. $6(r - 1)(r^2 + r + 1)$
23. $(3x - 4)^2$
24. $(x - 3y)(x + 3y)(x^2 + 9y^2)$
25. $(y + 7 - 10t)(y + 7 + 10t)$
26. $7(8a - b)(8a + b)$
27. $5(3x - 2)(4x - 1)$
28. $6ab(9a^4 + 2b^4)$
29. $5xy(5y^3 + 3x + yx - 2)$
30. $-3, 7$
31. ± 4
32. $-3, -\dfrac{3}{2}, 0$
33. $-\dfrac{1}{5}, 0$
34. $0, 8$
35. $\{x \mid x \in \mathbb{R}, x \neq -4\}$
36. length = 10 cm, width = 6 cm
37. 8 seconds
38. a) $-x^5 + 2x^3 + x - 1$
    b) $(x^2 + x - 1)(-x^3 + x^2 + 1)$
39. $6(x^n - 2)(x^n + 1)$

Answers for Chapter 5 Tests: FORM D

1. 7
2. $-x^5y + 3x^3y^4 - 5x^2y + 8xy^3$
3. -6
4. $P(0) = 4;\ P(-2) = -22$
5. $h^2 - 7h - 2ah + 2a^2$
6. $-5xy + 9xy^2$
7. $2x^3 + 5x^2 - 9y - 14y^2$
8. $m^3 - 7m^2n - mn^2 + 3n^3$
9. $-5a - 14b$
10. $5y^2 - 4y - 3y^3$
11. $80x^4y^4$
12. $15a^2 + ab - 2b^2$
13. $x^3 + 2x^2y + 2xy^2 + y^3$
14. $4x^8 + 12x^4 + 9$
15. $y^2 - 12y + 36$
16. $64x^2 - 9y^2$
17. $10x(5 - x^2)$
18. $(y^2 - 3)(y + 6)$
19. $(p - 1)(p + 16)$
20. $(5m + 1)(2m + 2)$
21. $(8y - 5)(8y + 5)$
22. $(r - 2)(r^2 + 2r + 4)$
23. $(2x - 1)^2$
24. $(2x - y)(2x + y)(4x^2 + y^2)$
25. $(y + 8 - 9t)(y + 8 + 9t)$
26. $9(6a - b)(6a + b)$
27. $2(3x - 2)(5x - 1)$
28. $7ab^2(3a^3 + 4b^3)$
29. $3xy(5y^3 + 3x + yx - 2)$
30. $-2, 7$
31. ± 6
32. $-4, -\dfrac{1}{3}, 0$
33. $-\dfrac{1}{6}, 0$
34. $0, -9$
35. $\{x \mid x \in \mathbb{R},\ x \neq 5\}$
36. length = 2.5 in, width = 2 in
37. 6 seconds
38. a) $-x^5 + 2x^4 - 2x^3 + 2x^2 - x + 1$
    b) $(x^2 - x + 1)(-x^3 + x^2 + 1)$
39. $6(x^n + 2)(x^n - 1)$

Answers for Chapter 5 Tests: FORM E

1. 7
2. $4x^4y^3 - 2x^3y - 6x^2y + 9xy^2$
3. -6
4. $P(0) = 3; P(-2) = 13$
5. $h^2 - 8h - 2ah + 2a^2$
6. $14xy + 9xy^2$
7. $x^3 + 7x^2 - 8y - 14y^2$
8. $3m^3 - 5m^2n - 3mn^2 + 3n^3$
9. $4a - 16b$
10. $5y^2 + 3y$
11. $74x^4y^4$
12. $4a^2 + 10ab - 6b^2$
13. $-x^3 + y^3$
14. $9x^8 + 24x^4 + 16$
15. $4y^2 - 20y + 25$
16. $81x^2 - 25y^2$
17. $11x^2(4 - x^2)$
18. $(y^2 - 4)(y + 2)$
19. $(p - 2)(p + 18)$
20. $(4m + 1)(2m + 3)$
21. $(12y - 7)(12y + 7)$
22. $(r - 3)(r^2 + 3r + 9)$
23. $(2x - 5)^2$
24. $4(x - 2y)(x + 2y)(x^2 + 4y^2)$
25. $(y + 9 - 8t)(y + 9 + 8t)$
26. $11(4a - b)(4a + b)$
27. $3(3x - 1)(5x - 2)$
28. $2ab^2(5a^3 + 6b^3)$
29. $4xy(y^3 + 9x + 6yx - 4)$
30. $-3, 8$
31. ± 10
32. $-5, -\dfrac{2}{3}, 0$
33. $-\dfrac{1}{7}, 0$
34. $0, -2$
35. $\{x \mid x \in \mathbb{R}, x \neq 6\}$
36. length = 8 in, width = 5 in
37. 5 seconds
38. a) $-x^5 + 2x^4 + 2x^3 - x - 1$
    b) $(x^2 - x - 1)(-x^3 + x^2 + 1)$
39. $3(4x^n + 5)(x^n - 1)$

Answers for Chapter 5 Tests: FORM F

1. 9
2. $5x^4y^5 - 7x^3y - 3x^2y + xy^4$
3. -1
4. $P(0) = 2; P(-2) = 2$
5. $h^2 - 37h - 2ah + 2a^2$
6. $-4xy + 4xy^2$
7. $9x^3 + 9x^2 - 7y - 5y^2$
8. $5m^3 - 3m^2n + 3mn^2 - 3n^3$
9. $4a - 10b$
10. $-3y^2 + 2y + 3y^3$
11. $96x^4y^4$
12. $25a^2 - 5ab - 6b^2$
13. $-x^3 - 2x^2y + y^3$
14. $16x^8 + 40x^4 + 25$
15. $16y^2 - 24y + 9$
16. $4x^2 - 64y^2$
17. $12x^3(3 - x^2)$
18. $(y^2 - 5)(y + 4)$
19. $(p - 3)(p + 17)$
20. $(3m + 1)(2m + 5)$
21. $(6y - 5)(6y + 5)$
22. $7(r - 1)(r^2 + r + 1)$
23. $(2x - 3)^2$
24. $(x - 2y)(x + 2y)(x^2 + 4y^2)(x^4 + 16y^4)$
25. $(y + 10 - 7t)(y + 10 + 7t)$
26. $13(2a - b)(2a + b)$
27. $5(2x - 1)(4x - 3)$
28. $3ab^2(10a^3 + 3b^3)$
29. $5xy(2y^3 + x + 5yx - 3)$
30. $-6, 10$
31. $\pm \dfrac{3}{2}$
32. $-6, -\dfrac{2}{3}, 0$
33. $-\dfrac{1}{8}, 0$
34. $0, -3$
35. $\{x \mid x \in \mathbb{R}, x \neq 7\}$
36. length = 14 in, width = 9 in
37. 7 seconds
38. a) $x^5 - 2x^3 - x - 1$
    b) $(x^2 - x - 1)(x^3 + x^2 + 1)$
39. $3(4x^n - 5)(x^n + 1)$

Answers for Chapter 5 Tests: FORM G

1. c
2. d
3. b
4. b
5. a
6. c
7. d
8. d
9. b
10. b
11. b
12. c
13. a
14. c
15. b
16. d
17. a
18. b
19. c
20. d
21. d
22. d
23. a
24. b
25. d
26. c

Answers for Chapter 5 Tests: FORM H

1. d
2. a
3. b
4. c
5. c
6. a
7. b
8. c
9. d
10. a
11. c
12. c
13. b
14. d
15. a
16. d
17. c
18. c
19. b
20. a
21. a
22. c
23. b
24. b
25. c
26. a

Answers for Chapter 6 Tests: FORM A

1. $\dfrac{5}{6(t+2)}$

2. $\dfrac{x^2-x+1}{x+3}$

3. $(x-2)(x+3)(x+4)$

4. $\dfrac{x^3+16x}{x+4}$

5. $5(a-b)$

6. $\dfrac{ab+a^3-a^2b+ab^2-b^3}{a^2-b^2}$

7. $\dfrac{-(x^2+7x+70)}{(x-10)(x+10)(x^2+10x+100)}$

8. $\dfrac{y-12}{(y-7)(y+1)}$

9. $\dfrac{a^2(6a+5b)}{11a+b}$

10. $\dfrac{(x-10)(x-2)}{(x-8)(x+10)}$

11. $\dfrac{x^2-7x-4}{3x^2+13x-42}$

12. 2

13. $\dfrac{89}{3}$

14. $f(2)=-\dfrac{10}{3};\ f(-3)=-\dfrac{5}{8}$

15. $\dfrac{53}{8}$

16. $\dfrac{15}{32}$ hours, or $28\dfrac{1}{8}$ minutes = .46875 hr

17. $\dfrac{6b^2c}{a}-\dfrac{5bc^2}{3a}+4bc$

18. $y-5-\dfrac{20}{y-5}$

19. $3x^2+5+\dfrac{4x-4}{x^2+1}$

20. $x^2+6x+27+\dfrac{98}{x-4}$

21. $f(3)=216$

22. $R=\dfrac{R_1R_2}{R_1+R_2}$

23. $-5,-4$ or $4,5$

24. 10 km/h

25. $-\dfrac{16}{3}$

26. $\{x\mid x\in\mathbb{R}\}$ & $x\neq 0$ & $x\neq 8$

27. x–intercept: $(-11,0)$

    y–intercept: $\left(0,\dfrac{11}{30}\right)$

28. Tawana: 75

    Jeff: 45

Answers for Chapter 6 Tests: FORM B

1. $\dfrac{2}{3(t+3)}$

2. $\dfrac{x^2+x+1}{x-3}$

3. $(x-4)(x+2)(x+3)$

4. $\dfrac{x^3+9x}{x+4}$

5. $\dfrac{4(a+b)^2}{(a-b)}$

6. $\dfrac{2ab+a+b}{a^2-b^2}$

7. $\dfrac{-2(x^2+2x+10)}{(x-5)(x+5)(x^2+5x+25)}$

8. $\dfrac{y-17}{(y-6)(y+2)}$

9. $\dfrac{a^2(7a+4b)}{11a+3b}$

10. $\dfrac{(x-9)(x-3)}{(x-7)(x+9)}$

11. $\dfrac{2x^2-11x+7}{3x^2+10x-53}$

12. $-\dfrac{27}{8}$

13. $-\dfrac{33}{4}$

14. $f(3) = -\dfrac{11}{2}$; $f(-4) = -\dfrac{4}{9}$

15. 18

16. $\dfrac{35}{48}$ hours, or $43\dfrac{3}{4}$ minutes

17. $\dfrac{5b^2c}{a} - \dfrac{3bc^2}{4a} + 8bc$

18. $y - 26 - \dfrac{134}{y+4}$

19. $4x^2 - 1 + \dfrac{3x+3}{x^2+2}$

20. $x^2 + 6x + 23 + \dfrac{60}{x-3}$

21. $f(4) = 835$

22. $g = \dfrac{bde}{ad-bc}$

23. $-9, -8$ or $8, 9$

24. 9 km/h

25. $-\dfrac{33}{4}$

26. $\{x \mid x \in \mathbb{R}\}$

27. x–intercept: $(-18, 0)$
    y–intercept: $\left(0, -\dfrac{3}{2}\right)$

28. Xavier: 70
    Phyllis: 56

Answers for Chapter 6 Tests: FORM C

1. $\dfrac{3}{4(t+4)}$

2. $\dfrac{x^2 - 2x + 4}{x + 5}$

3. $(x-3)(x+2)(x+4)$

4. $\dfrac{x^3 + 64x}{x + 8}$

5. $6(a-b)$

6. $\dfrac{3ab + a^3 - a^2b + ab^2 - b^3}{a^2 - b^2}$

7. $\dfrac{-3(x^2 - 5x + 10)}{(x-2)(x+2)(x^2 - 2x + 4)}$ [16 crossed out, 10 written above]

8. $\dfrac{y - 11}{(y-5)(y+3)}$

9. $\dfrac{a^2(8a + 3b)}{11a + 5b}$ [2 corrected to 8]

10. $\dfrac{(x-8)(x-4)}{(x-6)(x+8)}$

11. $\dfrac{x^2 + x - 4}{3x^2 - 11x + 38}$

12. $-\dfrac{49}{19}$

13. $-\dfrac{31}{2}$

14. $f(4) = -12$; $f(-5) = -\dfrac{3}{10}$

15. $\dfrac{23}{2}$

16. $\dfrac{33}{34}$ hours

17. $\dfrac{7b^2c}{a} - \dfrac{7bc^2}{5a} + 2bc$

18. $y + 22 + \dfrac{100}{y - 4}$

19. $5x^2 - 9 + \dfrac{2x + 30}{x^2 + 3}$ [12 corrected to 30]

20. $x^2 + 6x + 19 + \dfrac{30}{x - 2}$

21. $f(5) = 2256$

22. $q_0 = \dfrac{F d_1 d_2}{k(q_2 d_2 + q_2 d_1)}$ [subscript 1 corrected to 2]

23. $-7, -6$ or $6, 7$

24. 8 km/h

25. $-\dfrac{56}{5}$

26. $\{x \mid x \in \mathbb{R}\ \&\ x \neq 0\ \&\ x \neq 10\}$

27. x–intercept: $(-25, 0)$

    y–intercept: $\left(0, \dfrac{25}{18}\right)$

28. Hosni: 55

    Oumy: 66

Answers for Chapter 6 Tests: FORM D

1. $\dfrac{4}{5(t+5)}$

2. $\dfrac{x^2+2x+4}{x-5}$

3. $(x-3)(x-2)(x+4)$

4. $\dfrac{x^3+49x}{x+7}$

5. $\dfrac{3(a+b)^2}{(a-b)}$

6. $\dfrac{4ab+a+b}{a^2-b^2}$

7. $\dfrac{-2(2x^2-7x+21)}{(x-3)(x+3)(x^2-3x+9)}$

8. $\dfrac{y-12}{(y-4)(y+4)}$

9. $\dfrac{a^2(9a+2b)}{11a+7b}$

10. $\dfrac{(x-7)(x-5)}{(x-4)(x+7)}$

11. $\dfrac{x^2+x-4}{3x^2+10x-53}$

12. $\dfrac{53}{22}$

13. 32

14. $f(6)=14$; $f(-7)=-\dfrac{1}{12}$

15. $\dfrac{28}{3}$

16. $\dfrac{35}{34}$ hours

17. $\dfrac{6b^2c}{a}-\dfrac{3bc^2}{2a}+2bc$

18. $y+12-\dfrac{26}{y+3}$

19. $3x^2+4+\dfrac{x}{x^2+1}$

20. $x^2+9x+45+\dfrac{173}{x-4}$

21. $f(2)=21$

22. $t_1=\dfrac{m(x_1-x_2)+pt_0}{p}$

23. $-12,-11$ or $11, 12$

24. 7 km/h

25. $-\dfrac{5}{2}$

26. $\{x \mid x \in \mathbb{R}\}$

27. x–intercept: $\left(\dfrac{16}{3},0\right)$

    y–intercept: $\left(0,\dfrac{2}{3}\right)$

28. Erin: 60

    Julia: 84

Answers for Chapter 6 Tests: FORM E

1. $\dfrac{6}{7(t+5)}$

2. $\dfrac{x^2-3x+9}{x+2}$

3. $(x-4)(x-2)(x+3)$

4. $\dfrac{x^3+4x}{x+2}$

5. $7(a-b)$

6. $\dfrac{4ab+a^3-a^2b+ab^2-b^3}{a^2-b^2}$

7. $\dfrac{-(5x^2+4x+8)}{(x-2)(x+2)(x^2+2x+4)}$

8. $\dfrac{y-11}{(y-3)(y+5)}$

9. $\dfrac{a^2(9a+5b)}{14a+4b}$

10. $\dfrac{(x-6)(x-4)}{(x-5)(x+6)}$

11. $\dfrac{2x^2+3x-28}{3x^2+13x-42}$

12. $\dfrac{8}{3}$

13. $\dfrac{27}{2}$

14. $f(7)=\dfrac{15}{2};\; f(-8)=0$

15. $\dfrac{33}{4}$

16. $\dfrac{26}{21}$ hours

17. $\dfrac{2b^2c}{a}-\dfrac{5bc^2}{7a}+bc$

18. $y-23-\dfrac{60}{y+3}$

19. $4x^2+\dfrac{x+5}{x^2+2}$

20. $x^2+9x+38+\dfrac{108}{x-3}$

21. $f(3)=-68$

22. $h_0=\dfrac{Fgh_1-aW}{Fg}$

23. $-10, -9$ or $9, 10$

24. 7 km/h

25. $-\dfrac{85}{6}$

26. $\{x\,|\,x\in\mathbb{R}\}$

27. x–intercept: $(-14, 0)$
    y–intercept: $\left(0, \dfrac{7}{10}\right)$

28. Robin: 68
    Segal: 51

Answers for Chapter 6 Tests: FORM F

1. $\dfrac{1}{2(t+4)}$

2. $\dfrac{x^2+3x+9}{x-2}$

3. $(x-4)(x-3)(x+2)$

4. $\dfrac{x^3+25x}{x+5}$

5. $\dfrac{2(a+b)^2}{(a-b)}$

6. $\dfrac{3ab+a+b}{a^2-b^2}$

7. $\dfrac{-2(3x^2-35x+350)}{(x-10)(x+10)(x^2-10x+100)}$

8. $\dfrac{y-8}{(y-2)(y+6)}$

9. $\dfrac{a^2(2a+b)}{3a+b}$

10. $\dfrac{(x-5)(x-3)}{(x-6)(x+5)}$

11. $\dfrac{x^2-11}{3x^2-11x+38}$

12. $-\dfrac{43}{31}$

13. $-\dfrac{106}{5}$

14. $f(8)=\dfrac{16}{3};\ f(-2)=-\dfrac{6}{7}$

15. $\dfrac{38}{5}$

16. $\dfrac{9}{7}$ hours

17. $\dfrac{b^2c}{a}-\dfrac{3bc^2}{4a}+5bc$

18. $y-44+\dfrac{364}{y+8}$

19. $5x^2-9+\dfrac{4x+15}{x^2+3}$

20. $x^2+9x+31+\dfrac{57}{x-2}$

21. $f(4)=-45$

22. $v_1=\dfrac{F(t_1-t_0)+mv_0}{m}$

23. $-4, -3$ or $3, 4$

24. 8 km/h

25. $\dfrac{1}{3}$

26. $\{x\,|\,x\in\mathbb{R}\}$

27. x–intercept: $(-23, 0)$

    y–intercept: $\left(0, -\dfrac{69}{50}\right)$

28. Oscar: 64

    Luis: 80

Answers for Chapter 6 Tests: FORM G

1. b
2. c
3. d
4. a
5. b
6. c
7. d
8. b
9. d
10. c
11. a
12. d
13. c
14. a
15. b
16. d
17. c
18. a
19. b
20. a
21. d

Answers for Chapter 6 Tests: FORM H

1. d
2. d
3. c
4. c
5. c
6. a
7. b
8. b
9. a
10. c
11. d
12. c
13. c
14. a
15. a
16. b
17. a
18. d
19. c
20. b
21. a

Answers for Chapter 7 Tests: FORM A

1. $4\sqrt{2}$
2. $-\dfrac{1}{x^3}$ $-\dfrac{3}{x^4}$
3. $|5a|$
4. $|x-7|$
5. $x\sqrt[5]{xy}$
6. $\left|\dfrac{4x^3}{5y}\right|$
7. $\sqrt[3]{15xy}$
8. $\sqrt[5]{xy^4}$
9. $|xy|\sqrt[4]{y}$
10. $\sqrt[20]{a}$
11. $4\sqrt{3}$
12. $\sqrt{xy}(x^2+4y)$
13. $16 - 4\sqrt{x} - 6x$
14. $\sqrt[5]{81a^4b^4}$
15. $(5x^3y)^{1/2}$
16. $(-\infty, 2]$
17. $14 + 6\sqrt{5}$
18. $\dfrac{\sqrt{70} - 2\sqrt{10}}{3}$
19. 5
20. $\varnothing$
21. 5 cm and $5\sqrt{2}\text{ cm} \approx 7.071\text{ cm}$
22. $\sqrt{29}\text{ km} \approx 5.385\text{ km}$
23. $9i\sqrt{2}$
24. $11 - 5i$
25. -24
26. $3 - 4i$
27. $-\dfrac{22}{25} - \dfrac{21}{25}i$
28. $-i$
29. 4
30. $-\dfrac{5}{2}i$

336

Answers for Chapter 7 Tests: FORM B

1. $5\sqrt{3}$
2. $-\dfrac{1}{x^3}$
3. $6a$
4. $x-8$
5. $x\sqrt[5]{x^2y^2}$
6. $\dfrac{9x}{5y^2}$
7. $\sqrt[3]{15x^2y}$
8. $\sqrt[5]{x^2y^2}$
9. $y\sqrt[4]{x^3}$
10. $\sqrt[10]{a}$
11. $4\sqrt{2}$
12. $\sqrt{x}(y+2x)$
13. $21-8\sqrt{x}-5x$
14. $2a^2b\sqrt[6]{16a^3b^4}$
15. $(3xy^2)^{1/2}$
16. $(-\infty, 3]$
17. $22+8\sqrt{6}$
18. $\dfrac{3\sqrt{70}+\sqrt{42}}{3}$
19. 6
20. $\varnothing$
21. 8 cm and $8\sqrt{2}$ cm $\approx$ 11.314 cm
22. $150\sqrt{89}$ m $\approx$ 1415.097 m
23. $8i\sqrt{2}$
24. $10-7i$
25. -12
26. $8+6i$
27. $-i$
28. i
29. 2
30. $-\dfrac{13}{3}i$

Answers for Chapter 7 Tests: FORM C

1. $6\sqrt{5}$
2. $-\dfrac{10}{x^2}$
3. $7a$
4. $x - 9$
5. $x\sqrt[5]{x^3 y^3}$
6. $\dfrac{5x^2}{7y^3}$
7. $\sqrt[3]{15x^2 y^2}$
8. $\sqrt[5]{x^3 y}$
9. $xy\sqrt[4]{xy^3}$
10. $\sqrt[20]{a^{19}}$
11. $4\sqrt{5}$
12. $\sqrt{y}\,(x + 5xy)$
13. $24 - 10\sqrt{x} - 4x$
14. $a\sqrt[5]{625a^3 b^4}$
15. $(2x^2 y)^{1/2}$
16. $(-\infty, 5]$
17. $19 - 10\sqrt{6}$
18. $\dfrac{5\sqrt{6} + \sqrt{30}}{20}$
19. 7
20. 9
21. 11 cm and $11\sqrt{2}$ cm ≈ 15.556 cm
22. $0.15\sqrt{41}$ m ≈ 0.960 m
23. $7i\sqrt{2}$
24. $9 - 6i$
25. -18
26. $5 + 12i$
27. $-\dfrac{39}{17} - \dfrac{3}{17}i$
28. 1
29. 3
30. $-\dfrac{25}{4}i$

Answers for Chapter 7 Tests: FORM D

1. $2\sqrt{6}$
2. $-\dfrac{2}{x}$
3. $8a$
4. $x-5$
5. $x\sqrt[5]{x^4 y^4}$
6. $\dfrac{x^4}{3y}$
7. $\sqrt[3]{15xy^2}$
8. $\sqrt[5]{x^4}$
9. $x\sqrt[4]{y^3}$
10. $\sqrt[10]{a^7}$
11. $4\sqrt{6}$
12. $\sqrt{xy}\,(y^2+3x)$
13. $25-10\sqrt{x}-3x$
14. $b\sqrt[6]{32a^5 b^4}$
15. $(7x^2 y)^{1/3}$
16. $[-4,-\infty)$
17. $31-12\sqrt{5}$
18. $\dfrac{6\sqrt{5}-\sqrt{15}}{33}$
19. 8
20. $\dfrac{1}{4}$
21. 14 cm and $14\sqrt{2}$ cm ≈ 19.799 cm
22. $100\sqrt{10}$ m ≈ 316.228 m
23. $6i\sqrt{3}$
24. $4+5i$
25. -30
26. $7-24i$
27. $-\dfrac{3}{5}+\dfrac{1}{5}i$
28. $-i$
29. 2
30. $-\dfrac{25}{3}i$

Answers for Chapter 7 Tests: FORM E

1. $3\sqrt{7}$
2. $-\dfrac{6}{x^2}$
3. $9a$
4. $x - 6$
5. $x^2 y$
6. $\dfrac{x^2}{3y^5}$
7. $\sqrt[3]{15xy^2}$
8. $\sqrt[5]{x^3 y}$
9. $x\sqrt[12]{x^2 y^7}$
10. $\sqrt[10]{a^3}$
11. $3\sqrt{3}$
12. $\sqrt{y}(xy + 2y)$
13. $24 - 8\sqrt{x} - 2x$
14. $2a^2 \sqrt[5]{2a^2 b^4}$
15. $(2xy)^{1/3}$
16. $[-2, \infty)$
17. $29 + 10\sqrt{7}$
18. $\dfrac{7\sqrt{3} + \sqrt{6}}{47}$
19. 7
20. $\emptyset$
21. 17 cm and $17\sqrt{2}$ cm ≈ 24.042 cm
22. $\sqrt{0.85}$ km ≈ 0.922 km
23. $5i\sqrt{3}$
24. $3 + 7i$
25. -32
26. $-5 + 12i$
27. $-\dfrac{13}{5} - \dfrac{19}{5}i$
28. $-i$
29. 7
30. $-\dfrac{13}{2}i$

Answers for Chapter 7 Tests: FORM F

1. $9\sqrt{10}$
2. $-\dfrac{4}{x^4}$
3. $11a$
4. $x-5$
5. $xy\sqrt[5]{y}$
6. $\dfrac{3x^2}{7y^4}$
7. $\sqrt[3]{15x^2y^2}$
8. $\sqrt[5]{xy^3}$
9. $x\sqrt[12]{xy^{11}}$
10. $\sqrt[20]{a^9}$
11. $\sqrt{2}$
12. $\sqrt{x}(1+6xy)$
13. $15-2\sqrt{x}-x$
14. $2ab\sqrt[3]{4a^2b^2}$
15. $(5x^2y^2)^{1/3}$
16. $[-3,\infty)$
17. $13-8\sqrt{3}$
18. $\dfrac{10\sqrt{2}+\sqrt{6}}{97}$
19. 6
20. $\dfrac{81}{16}$
21. 20 cm and $20\sqrt{2}$ cm ≈ 28.284 cm
22. $\sqrt{74}$ km ≈ 8.602 km
23. $4i\sqrt{3}$
24. $8+6i$
25. -45
26. $12+16i$
27. $-\dfrac{11}{5}+\dfrac{13}{5}i$
28. 1
29. 5
30. $-5i$

Answers for Chapter 7 Tests: FORM G

1. a
2. c
3. c
4. d
5. d
6. d
7. b
8. c
9. a
10. b
11. a
12. c
13. d
14. b
15. c
16. a
17. b
18. b
19. a
20. c
21. d
22. d
23. b
24. c
25. a
26. d
27. b

Answers for Chapter 7 Tests: FORM H

1. c
2. d
3. b
4. a
5. a
6. b
7. d
8. c
9. b
10. b
11. a
12. c
13. c
14. b
15. b
16. a
17. d
18. c
19. c
20. b
21. d
22. a
23. d
24. d
25. c
26. a
27. d

Answers for Chapter 8 Tests: FORM A

1. $\pm\sqrt{\dfrac{17}{2}}$

2. 10, 36

3. $\dfrac{-1 \pm i\sqrt{7}}{2}$

4. $\dfrac{1 \pm \sqrt{13}}{2}$

5. $\dfrac{-3 \pm \sqrt{33}}{4}$

6. $-2.302776, 1.302776$

7. $-\dfrac{5}{3}, \dfrac{3}{2}$

8. $x^2 + 12x + 36, (x+6)^2$

9. $x^2 - \dfrac{10}{9}x + \dfrac{25}{81}, (x - \dfrac{5}{9})^2$

10. $-6 \pm 2\sqrt{3}$

11. 10 km/h

12. 1 hour, or 60 minutes

13. two distinct rational solutions

14. $x^2 + \dfrac{19}{4}x - \dfrac{5}{4}$

15. $(-7, 0), (-3, 0), (-2, 0), (2, 0)$

16. 452.2 cm^2

17.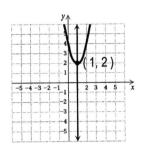
    min: $f(1) = 2$

18. x–intercepts: $(10, 0), (-9, 0)$
    y–intercept: $(0, -90)$

19. \$154 ; 300

20. $r = \sqrt{\dfrac{A}{\pi}}$

21. $f(x) = \dfrac{1}{6}x^2 - \dfrac{5}{6}x + \dfrac{2}{3}$

22. vertex: $\left(-\dfrac{1}{4}, \dfrac{7}{8}\right)$; axis: $x = -\dfrac{1}{4}$

    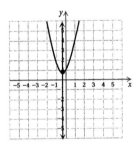

23. $[-3, -1]$

24. $(-\sqrt{2}, 0) \cup (\sqrt{2}, \infty)$

25. $\dfrac{1}{2}$

26. $x^4 - 8x^3 + 23x^2 - 28x + 10 = 0$

27. $x^6 - 14x^5 + 42x^4 + 98x^3 - 333x^2 - 140x + 490 = 0$

Answers for Chapter 8 Tests: FORM B

1. $\pm\sqrt{\dfrac{10}{3}}$

2. 1, 50

3. -1

4. $1 \pm 2\sqrt{2}$

5. $\dfrac{-1 \pm \sqrt{7}}{3}$

6. $-4.5413813, 1.5413813$

7. $-\dfrac{4}{5}, \dfrac{5}{2}$

8. $x^2 - 8x + 16, (x-4)^2$

9. $x^2 + \dfrac{8}{7}x + \dfrac{16}{49}, \left(x + \dfrac{4}{7}\right)^2$

10. $-5 \pm 2\sqrt{5}$

11. 10 km/h

12. 5 hours

13. two irrational solutions

14. $x^2 + \dfrac{10}{3}x - \dfrac{8}{3}$

15. $(-3, 0), (-1, 0), (1, 0)$

16. 615.4 cm^2

17.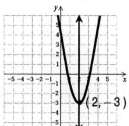
    min: $f(2) = -3$

18. x-intercepts: $(-6, 0), (5, 0)$
    y-intercept: $(0, -30)$

19. $146 ; 330

20. $v = \sqrt{\dfrac{2K}{m}}$

21. $f(x) = \dfrac{1}{2}x^2 - \dfrac{5}{2}x + 3$

22. vertex: $\left(-\dfrac{1}{4}, \dfrac{15}{8}\right)$; axis: $x = -\dfrac{1}{4}$

    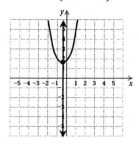

23. $[-2, -1]$

24. $(0, \infty)$

25. $\dfrac{1}{3}$

26. $x^4 - 8x^3 + 22x^2 - 24x + 5 = 0$

27. $x^6 - 4x^5 - 3x^4 + 28x^3 - 18x^2 - 40x + \ldots$
    $\ldots + 40 = 0$

Answers for Chapter 8 Tests: FORM C

1. $\pm\sqrt{\dfrac{15}{4}}$

2. $7, 20$

3. $-1 \pm i$

4. $\dfrac{3 \pm 2\sqrt{29}}{2}$

5. $\dfrac{-3 \pm \sqrt{57}}{8}$

6. $-5.85410197, 0.85410197$

7. $-\dfrac{1}{2}, \dfrac{7}{3}$

8. $x^2 + 20x + 100, (x+10)^2$

9. $x^2 - \dfrac{6}{5}x + \dfrac{9}{25}, \left(x - \dfrac{3}{5}\right)^2$

10. $-6, -2$

11. 20 km/h

12. 78 minutes

13. two complex solutions

14. $x^2 + \dfrac{9}{4}x - \dfrac{9}{4}$

15. $(-3, 0), \left(\dfrac{-3-\sqrt{13}}{2}, 0\right),$
    $(-1, 0), \left(\dfrac{-3+\sqrt{13}}{2}, 0\right)$

16. 803.8 cm²

17.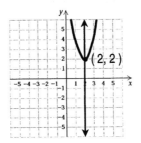
    min: $f(2) = 2$

18. x–intercepts: $(-4, 0), (5, 0)$
    y–intercept: $(0, -20)$

19. $138; 360

20. $r = \sqrt{\dfrac{GMm}{F}}$

21. $f(x) = \dfrac{4}{3}x^2 - \dfrac{32}{3}x + 203$

22. vertex: $\left(-\dfrac{1}{4}, \dfrac{23}{8}\right)$; axis: $x = -\dfrac{1}{4}$

    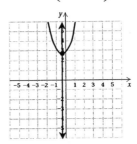

23. $(-\infty, -5] \cup [1, \infty)$

24. $(-2, 0) \cup (2, \infty)$

25. $\dfrac{1}{2}$

26. $x^4 - 8x^3 + 20x^2 - 16x - 5 = 0$

27. $x^6 - 6x^5 + 2x^4 + 42x^3 - 53x^2 - 60x + \ldots$
    $\ldots + 90 = 0$

Answers for Chapter 8 Tests: FORM D

1. $\pm\sqrt{\dfrac{8}{5}}$

2. 10, 11

3. $\dfrac{-3\pm\sqrt{5}}{2}$

4. $2\pm\sqrt{5}$

5. $\dfrac{-3\pm\sqrt{69}}{10}$

6. $-7.14005495, 0.14005495$

7. $-\dfrac{6}{5}, \dfrac{2}{3}$

8. $x^2-18x+81, (x-9)^2$

9. $x^2+\dfrac{4}{3}x+\dfrac{4}{9}, \left(x+\dfrac{2}{3}\right)^2$

10. $-3\pm\sqrt{6}$

11. 20 km/h

12. 20 hours

13. two irrational solutions

14. $x^2+\dfrac{9}{7}x-\dfrac{10}{7}$

15. $(-5,0),(-3,0),(-1,0),(1,0)$

16. 1017.4 cm²

17.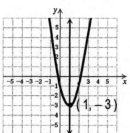
    min: $f(1)=-3$

18. x–intercepts: $(-11,0),(10,0)$
    y–intercept: $(0,-110)$

19. $130 ; 390

20. $b=\sqrt{c^2-a^2}$

21. $f(x)=\dfrac{1}{8}x^2-\dfrac{3}{4}x$

22. vertex: $\left(-\dfrac{1}{2},\dfrac{1}{2}\right)$; axis: $x=-\dfrac{1}{2}$

    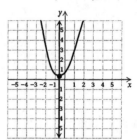

23. $(-\infty,-4]\cup[-1,\infty)$

24. $(0,\infty)$

25. $\dfrac{1}{2}$

26. $x^4-8x^3+19x^2-12x-10=0$

27. $x^6-8x^5+9x^4+56x^3-102x^2-80x+\ldots$
    $\ldots+160=0$

Answers for Chapter 8 Tests: FORM E

1. $\pm\sqrt{\dfrac{13}{6}}$

2. 4, 10

3. $-2, -1$

4. $\dfrac{3 \pm 3\sqrt{5}}{2}$

5. $\dfrac{-5 \pm \sqrt{85}}{6}$

6. $-3.1925824, 2.1925824$

7. $-2, \dfrac{7}{2}$

8. $x^2 + 6x + 9, (x+3)^2$

9. $x^2 - \dfrac{4}{9}x + \dfrac{4}{81}, \left(x - \dfrac{2}{9}\right)^2$

10. $-2 \pm \sqrt{3}$

11. 10 km/h

12. $4\dfrac{4}{5}$ hours

13. two distinct rational solutions

14. $x^2 + \dfrac{7}{4}x - \dfrac{1}{2}$

15. $(-5, 0), (-2+\sqrt{2}, 0),$
    $(-2-\sqrt{2}, 0), (1, 0)$

16. 1256 cm²

17.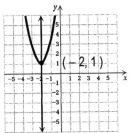
    min: $f(-2) = 1$

18. x–intercepts: $(-6, 0), (7, 0)$
    y–intercept: $(0, -42)$

19. \$122 ; 400

20. $v = \sqrt{\dfrac{3kT}{m}}$

21. $f(x) = -\dfrac{1}{5}x^2 + \dfrac{1}{5}x + \dfrac{12}{5}$

22. vertex: $\left(-\dfrac{1}{2}, \dfrac{3}{2}\right)$; axis: $x = -\dfrac{1}{2}$

    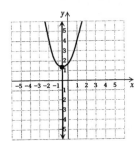

23. $[-4, -2]$

24. $(-\infty, -\sqrt{5}) \cup (0, \sqrt{5})$

25. $\dfrac{1}{3}$

26. $x^4 - 16x^3 + 93x^2 - 226x + 182 = 0$

27. $x^6 + 8x^5 + 9x^4 - 56x^3 - 102x^2 + 80x + \ldots$
    $\ldots + 160 = 0$

Answers for Chapter 8 Tests: FORM F

1. $\pm\sqrt{\dfrac{20}{7}}$

2. 4, 15

3. $\dfrac{-3 \pm i\sqrt{3}}{2}$

4. $1 \pm 2\sqrt{2}$

5. $\dfrac{-5 \pm \sqrt{105}}{8}$

6. $-3.79128785, 0.79128785$

7. $-\dfrac{4}{5}, \dfrac{3}{2}$

8. $x^2 - 2x + 1, (x-1)^2$

9. $x^2 + \dfrac{6}{7}x + \dfrac{9}{49}, \left(x + \dfrac{3}{7}\right)^2$

10. $-7 \pm \sqrt{13}$

11. 10 km/h

12. 1 hour

13. two distinct rational solutions

14. $x^2 + \dfrac{7}{3}x - 2$

15. $(-4, 0), (-2, 0), (-1, 0), (1, 0)$

16. 201 cm^2

17.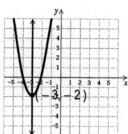
min: $f(-3) = -2$

18. x–intercepts: $(-8, 0), (7, 0)$
y–intercept: $(0, -56)$

19. $114 ; 430

20. $R = v\sqrt{\dfrac{l}{2K - mv^2}}$

21. $f(x) = \dfrac{1}{2}x^2 + \dfrac{3}{2}x - 5$

22. vertex: $\left(-\dfrac{1}{2}, \dfrac{5}{2}\right)$; axis: $x = -\dfrac{1}{2}$

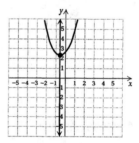

23. $(-\infty, -8] \cup [1, \infty)$

24. $(-\infty, 0)$

25. $\dfrac{1}{4}$

26. $x^4 - 16x^3 + 92x^2 - 216x + 156 = 0$

27. $x^6 + 6x^5 + 2x^4 - 42x^3 - 53x^2 + 60x + \ldots$
    $\ldots + 90 = 0$

Answers for Chapter 8 Tests: FORM G

1. a
2. c
3. d
4. a
5. d
6. b
7. b
8. a
9. d
10. c
11. d
12. a
13. d
14. b
15. c
16. b
17. a
18. a
19. c
20. a
21. c
22. c

Answers for Chapter 8 Tests: FORM H

1. d
2. a
3. c
4. d
5. c
6. d
7. c
8. b
9. c
10. a
11. b
12. b
13. b
14. c
15. d
16. a
17. b
18. b
19. a
20. d
21. a
22. d

Answers for Chapter 9 Tests: FORM A

1. $(f \circ g)(x) = 9x^2 + 15x + 6$
   $(g \circ f)(x) = 3x^2 + 3x + 2$
2. yes
3. $f^{-1}(x) = \dfrac{x+4}{3}$
4. $g^{-1}(x) = \sqrt[3]{x} - 4$
5.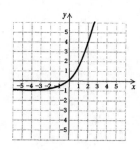
6.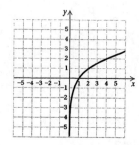
7. 4
8. $\dfrac{1}{2}$
9. 25
10. $\log_5 \dfrac{1}{125} = -3$
11. $\log_{121} 11 = \dfrac{1}{2}$
12. $2^m = 4$
13. $2^7 = 128$
14. $2\log a + 3\log b - \log c$
15. $\log_a x^{1/4} z^5$
16. 1
17. 5
18. 0
19. 0.470
20. 3.688
21. 3.807
22. domain: $(-\infty, \infty)$ range: $(2, \infty)$
23. domain: $(3, \infty)$ range: $(-\infty, \infty)$
24. -3
25. 2
26. 9
27. $\dfrac{1}{10{,}000}$
28. $\dfrac{1}{7}$
29. $\dfrac{\log 3.2}{\log 6} \approx 0.6492$
30. $e^{1/8} \approx 1.1331$
31. $\sqrt{6} \approx 2.4495$
32. a) 1.40 ft / sec
    b) 254,816
33. a) $k = 0.044$; $M(t) = 5e^{0.044t}$
    b) $M(10) = 7.8$ grams
    c) 9AM Tuesday
34. 3.15%
35. $-116, 127$
36. -1

Answers for Chapter 9 Tests: FORM B

1. $(f \circ g)(x) = -4x^2 - 10x - 6$
   $(g \circ f)(x) = -2x^2 + 2x + 3$
2. no
3. $f^{-1}(x) = \dfrac{x-5}{3}$
4. $g^{-1}(x) = \sqrt[3]{x} + 2$
5.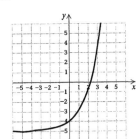
6.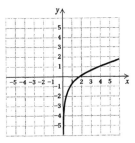
7. 4
8. $\dfrac{1}{2}$
9. 32
10. $\log_4 \dfrac{1}{256} = -4$
11. $\log_{64} 8 = \dfrac{1}{2}$
12. $3^m = 27$
13. $3^2 = 9$
14. $3 \log a + \log b - 2 \log c$
15. $\log_a x^{1/2} z^4$
16. 1
17. 19
18. 0
19. 0.980
20. 3.178
21. 3.218
22. domain: $(-\infty, \infty)$ range: $(-2, \infty)$
23. domain: $(-3, \infty)$ range: $(-\infty, \infty)$
24. -2
25. 8
26. 8
27. $\dfrac{1}{1000}$
28. 1
29. $\dfrac{\log 4.2}{\log 7} \approx 0.7375$
30. $e^{3/8} \approx 1.4550$
31. 5
32. a) 2.35 ft / sec
    b) 437,502
33. a) $k = 0.163$; $M(t) = 1.5 e^{0.163 t}$
    b) $M(10) = 7.6$ grams
    c) 10PM Monday
34. 3.46%
35. $-12, 4$
36. $-\dfrac{1}{3}$

351

Answers for Chapter 9 Tests: FORM C

1. $(f \circ g)(x) = 8x^2 - 22x + 15$
   $(g \circ f)(x) = 4x^2 + 2x - 3$
2. yes
3. $f^{-1}(x) = \dfrac{x+6}{2}$
4. $g^{-1}(x) = 2\sqrt[3]{x-1}$
5.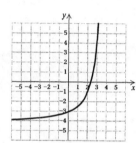
6.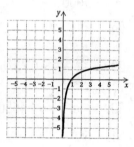
7. 2
8. $\dfrac{1}{2}$
9. 12
10. $\log_3 \dfrac{1}{243} = -5$
11. $\log_{144} 12 = \dfrac{1}{2}$
12. $4^m = 16$
13. $4^3 = 64$
14. $\log a + 2\log b - 3\log c$
15. $\log_a \dfrac{x^{1/3}}{z^3}$
16. 1
17. 0.052
18. 0
19. 0.510
20. 2.708
21. 4.158
22. domain: $(-\infty, \infty)$ range: $(3, \infty)$
23. domain: $(1, \infty)$ range: $(-\infty, \infty)$
24. -3
25. 5
26. 7
27. $\dfrac{1}{100}$
28. $\dfrac{2}{9}$
29. $\dfrac{\log 5.2}{\log 8} \approx 0.7928$
30. $e^{5/8} \approx 1.8682$
31. $2\sqrt{5} \approx 4.4721$
32. a) 2.39 ft / sec
    b) 751,162
33. a) $k = 0.384$; $M(t) = 0.5\,e^{0.384\,t}$
    b) $M(10) = 23.3$ grams
    c) 5PM Monday
34. 3.85%
35. $-61, 64$
36. $-\dfrac{7}{2}$

Answers for Chapter 9 Tests: FORM D

1. $(f \circ g)(x) = -18x^2 + 27x - 10$
   $(g \circ f)(x) = -6x^2 + 3x - 2$
2. no
3. $f^{-1}(x) = x - 7$
4. $g^{-1}(x) = 2\sqrt[3]{x} + 3$
5.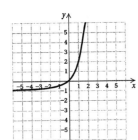
6.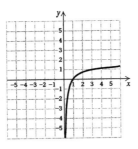
7. 1
8. $\dfrac{1}{2}$
9. 19
10. $\log_2 \dfrac{1}{64} = -6$
11. $\log_4 2 = \dfrac{1}{2}$
12. $5^m = 25$
13. $6^2 = 36$
14. $\dfrac{1}{2} \log a + \dfrac{1}{3} \log b - \log c$
15. $\log_a \dfrac{x^{1/2}}{z^2}$
16. 1
17. 38
18. 0
19. -0.470
20. 4.78
21. 2.198
22. domain: $(-\infty, \infty)$ range: $(-3, \infty)$
23. domain: $(-1, \infty)$ range: $(-\infty, \infty)$
24. -6
25. 2
26. 6
27. $\dfrac{1}{10}$
28. $\dfrac{7}{2}$
29. $\dfrac{\log 6.2}{\log 9} \approx 0.8304$
30. $e^{7/8} \approx 2.3989$
31. 6
32. a) 2.26 ft / sec
    b) 1,289,696
33. a) $k = 0.102$; $M(t) = 2e^{0.102t}$
    b) $M(10) = 5.5$ grams
    c) 4AM Tuesday
34. 4.33%
35. $-106, 110$
36. $\dfrac{17}{3}$

Answers for Chapter 9 Tests: FORM E

1. $(f \circ g)(x) = 9x^2 - 18x + 8$
   $(g \circ f)(x) = -3x^2 - 6x + 2$
2. yes
3. $f^{-1}(x) = \dfrac{x+2}{5}$
4. $g^{-1}(x) = 2\sqrt[3]{x} - 4$
5.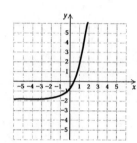
6.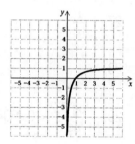
7. 4
8. $\dfrac{1}{2}$
9. 10
10. $\log_3 \dfrac{1}{81} = -4$
11. $\log_{169} 13 = \dfrac{1}{2}$
12. $6^m = 36$
13. $5^3 = 125$
14. $\dfrac{1}{3} \log a + \dfrac{1}{2} \log b - \log c$
15. $\log_a x^{1/3} z^2$
16. 1
17. 1012
18. 0
19. -0.406
20. 2.995
21. 2.772
22. domain: $(-\infty, \infty)$ range: $(-\infty, 1)$
23. domain: $\left(\dfrac{1}{2}, \infty\right)$ range: $(-\infty, \infty)$
24. -4
25. 6
26. 5
27. 1
28. $-\dfrac{1}{3}$
29. $\dfrac{\log 2.6}{\log 2} \approx 1.3785$
30. $e^{9/8} \approx 3.0802$
31. $\sqrt{15} \approx 3.8730$
32. a) 1.50 ft / sec
    b) 2,214,324
33. a) $k = 0.109$; $M(t) = 2.5\, e^{0.109\, t}$
    b) $M(10) = 7.4$ grams
    c) 12 Midnight Tuesday
34. 4.95%
35. $-317{,}308$
36. $\dfrac{7}{6}$

Answers for Chapter 9 Tests: FORM F

1. $(f \circ g)(x) = 4x^2 - 8x + 3$
   $(g \circ f)(x) = -2x^2 + 4x + 3$
2. no
3. $f^{-1}(x) = \dfrac{x-7}{6}$
4. $g^{-1}(x) = 4\sqrt[3]{x} + 3$
5.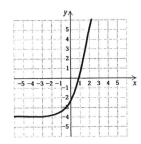
6.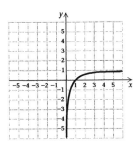
7. 3
8. $\dfrac{1}{2}$
9. 28
10. $\log_5 \dfrac{1}{625} = -4$
11. $\log_{81} 9 = \dfrac{1}{2}$
12. $7^m = 49$
13. $4^4 = 256$
14. $3 \log a + 2 \log b - \log c$
15. $\log_a x^{1/4} z^{3/2}$
16. 1
17. 3.14
18. 0
19. 0.406
20. 4.787
21. 4.381
22. domain: $(-\infty, \infty)$ range: $(-\infty, 2)$
23. domain: $\left(-\dfrac{1}{2}, \infty\right)$ range: $(-\infty, \infty)$
24. -7
25. 3
26. 4
27. 10
28. $\dfrac{9}{4}$
29. $\dfrac{\log 2.5}{\log 3} \approx 0.8340$
30. $e^{11/8} \approx 3.9551$
31. 5
32. a) 1.19 ft / sec
    b) 148,413
33. a) $k = 0.204$; $M(t) = 1.7 e^{0.204 t}$
    b) $M(10) = 13.1$ grams
    c) 7PM Monday
34. 5.78%
35. -125, 131
36. -1

355

Answers for Chapter 9 Tests: FORM G

1. d
2. b
3. c
4. d
5. c
6. a
7. a
8. c
9. c
10. b
11. c
12. d
13. c
14. a
15. b
16. c
17. c
18. b
19. d
20. b
21. d
22. d
23. b
24. b
25. a
26. b
27. c

Answers for Chapter 9 Tests: FORM H

1. c
2. d
3. b
4. c
5. b
6. c
7. b
8. a
9. b
10. d
11. b
12. a
13. a
14. b
15. c
16. b
17. b
18. c
19. b
20. a
21. b
22. a
23. a
24. a
25. c
26. c
27. a

Answers for Chapter 10 Tests: FORM A

1. $\sqrt{202} \approx 14.213$
2. $2\sqrt{64 + a^2}$
3. $(-3.5, 2.5)$
4. $(0, 0)$
5. center $(-2, 9)$; radius = 3
6. center $(-1, 4)$; radius = 4

7. parabola
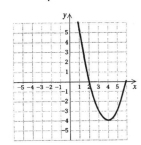

8. circle
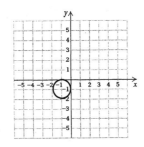

9. hyperbola
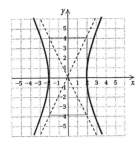

10. ellipse
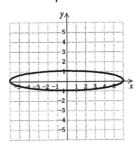

11. hyperbola
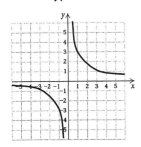

12. parabola
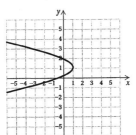

13. $(0, 2), (3, 0)$
14. $(-5, 0), (5, 0)$
15. $(4, 3), (-4, -3)$
16. $(\sqrt{5}, \sqrt{2}), (\sqrt{5}, -\sqrt{2}),$ $(-\sqrt{5}, \sqrt{2}), (-\sqrt{5}, -\sqrt{2})$
17. length = 7; width = 3
18. $\sqrt{7}$ m; $\sqrt{2}$ m
19. length = 20 m; width = 15 m
20. $1000; 5%
21. $\dfrac{(x-2)^2}{9} + \dfrac{(y-3)^2}{16} = 1$
22. $\left(0, \dfrac{33}{16}\right)$
23. 2

Answers for Chapter 10 Tests: FORM B

1. $4\sqrt{13} \approx 14.422$
2. $2\sqrt{49 + a^2}$
3. $(-2, 1)$
4. $(0, 0)$
5. center $(2, -8)$; radius = 4
6. center $(1, -3)$; radius = $\sqrt{8}$

7. parabola 8. circle

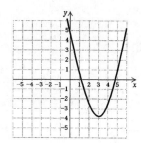 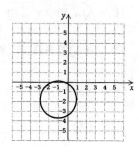

9. hyperbola 10. ellipse

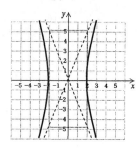 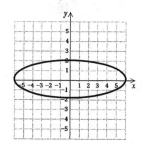

11. hyperbola 12. parabola

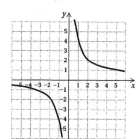 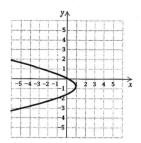

13. $(0, 5), (4, 0)$
14. $(-4, 0), (4, 0)$
15. $(7, 3), (-7, -3)$
16. $(2\sqrt{2}, 2), (2\sqrt{2}, -2),$ $(-2\sqrt{2}, 2), (-2\sqrt{2}, -2)$
17. length = 10; width = 4
18. $\sqrt{11}$ m; $\sqrt{5}$ m
19. length = 12 m; width = 5 m
20. $1000; 6%
21. $\dfrac{(x-3)^2}{9} + \dfrac{(y-2)^2}{16} = 1$
22. $\left(0, \dfrac{31}{14}\right)$
23. 4

Answers for Chapter 10 Tests: FORM C

1. $\sqrt{218} \approx 14.765$
2. $2\sqrt{36 + a^2}$
3. $(-0.5, -0.5)$
4. $(0, 0)$
5. center $(-2, 7)$; radius $= 5$
6. center $(-2, 3)$; radius $= \sqrt{5}$

7. parabola
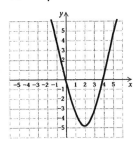

8. circle
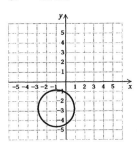

9. hyperbola
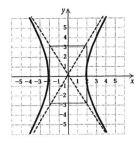

10. ellipse
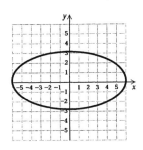

11. hyperbola
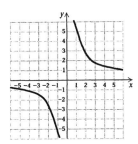

12. parabola
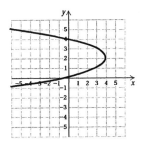

13. $(0, 4), (2, 0)$
14. $(-3, 0), (3, 0)$
15. $(6, 3), (-6, -3)$
16. $(\sqrt{3}, \sqrt{2}), (\sqrt{3}, -\sqrt{2}), (-\sqrt{3}, \sqrt{2}), (-\sqrt{3}, -\sqrt{2})$
17. length $= 9$; width $= 5$
18. 3 m; $\sqrt{2}$ m
19. length $= 24$ m; width $= 7$ m
20. $1100; 5%
21. $\dfrac{(x-4)^2}{25} + \dfrac{(y-1)^2}{4} = 1$
22. $(0, 2)$
23. 5

Answers for Chapter 10 Tests: FORM D

1. $2\sqrt{58} \approx 15.232$
2. $2\sqrt{25 + a^2}$
3. $(1, -2)$
4. $(0, 0)$
5. center $(2, -6)$; radius = 6
6. center $(2, -1)$; radius = 1

7. parabola 8. circle

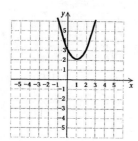 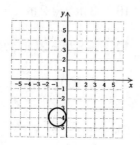

9. hyperbola 10. ellipse

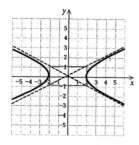 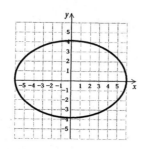

11. hyperbola 12. parabola

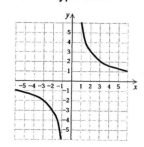 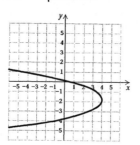

13. $(0, 3), (5, 0)$
14. $(-2, 0), (2, 0)$
15. $(8, 3), (-8, -3)$
16. $(\sqrt{7}, 1), (\sqrt{7}, -1),$
    $(-\sqrt{7}, 1), (-\sqrt{7}, -1)$
17. length = 13; width = 6
18. $\sqrt{5}$ m; 1 m
19. length = 24 m; width = 10 m
20. $1100; 6%
21. $\dfrac{(x-1)^2}{25} + \dfrac{(y-4)^2}{4} = 1$
22. $\left(0, \dfrac{3}{7}\right)$
23. 8

Answers for Chapter 10 Tests: FORM E

1. $\sqrt{218} \approx 14.765$
2. $2\sqrt{16 + a^2}$
3. $(1.5, -2.5)$
4. $(0, 0)$
5. center $(-7, 4)$; radius = 6
6. center $(-3, 2)$; radius = 3

7. parabola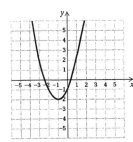
8. circle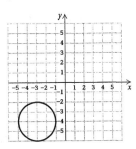

9. hyperbola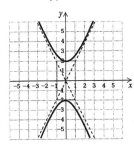
10. ellipse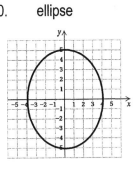

11. hyperbola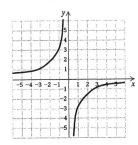
12. parabola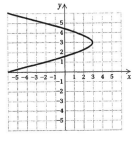

13. $(0, 5), (2, 0)$
14. $(-5, 0), (5, 0)$
15. $(5, 4), (-5, -4)$
16. $(\sqrt{7}, \sqrt{2}), (\sqrt{7}, -\sqrt{2}),$
    $(-\sqrt{7}, \sqrt{2}), (-\sqrt{7}, -\sqrt{2})$
17. length = 11 ; width = 6
18. $\sqrt{7}$ m ; 1 m
19. length = 8 m ; width = 6 m
20. $1300 ; 5%
21. $\dfrac{(x-1)^2}{4} + \dfrac{(y-4)^2}{25} = 1$
22. $\left(0, -\dfrac{26}{9}\right)$
23. 3

Answers for Chapter 10 Tests: FORM F

1. $10\sqrt{2} \approx 14.142$
2. $2\sqrt{9+a^2}$
3. $(1,-2)$
4. $(0,0)$
5. center $(7,-5)$; radius = 5
6. center $(3,-3)$; radius = $\sqrt{15}$

7. parabola   8. circle

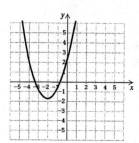 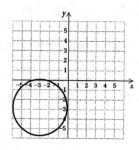

9. hyperbola   10. ellipse

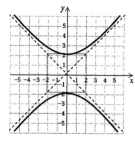 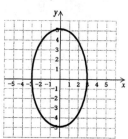

11. hyperbola   12. parabola

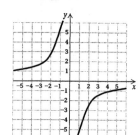 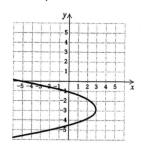

13. $(0,4),(3,0)$
14. $(-4,0),(4,0)$
15. $(5,3),(-5,-3)$
16. $(\sqrt{11},\sqrt{5}),(\sqrt{11},-\sqrt{5}),$
    $(-\sqrt{11},\sqrt{5}),(-\sqrt{11},-\sqrt{5})$
17. length = 12; width = 5
18. $\sqrt{3}$ m; $\sqrt{2}$ m
19. length = 15 m; width = 8 m
20. $1300; 6%
21. $\dfrac{(x-4)^2}{4} + \dfrac{(y-1)^2}{25} = 1$
22. $(0,1)$
23. 5

Answers for Chapter 10 Tests: FORM G

1. d
2. a
3. c
4. a
5. d
6. b
7. a
8. c
9. c
10. c
11. d
12. c
13. d
14. d
15. a
16. c
17. c
18. d
19. a
20. d
21. c

Answers for Chapter 10 Tests: FORM H

1. c
2. d
3. d
4. d
5. c
6. d
7. b
8. b
9. b
10. a
11. c
12. d
13. b
14. a
15. b
16. a
17. b
18. b
19. c
20. a
21. a

Answers for Chapter 11 Tests: FORM A

1. $3, 8, 13, 18, 23$; $a_{16} = 78$

2. $a_n = \left(\dfrac{1}{4}\right)^n$

3. $-1 - 7 - 25 - 79 - 241 = -353$

4. $\displaystyle\sum_{k=1}^{5}(-k)^k$

5. $a_{12} = -23$

6. $d = \dfrac{5}{8}$

7. $a_1 = 15$; $d = -3$

8. 1276

9. $g_6 = \dfrac{25}{256}$

10. $r = \dfrac{2}{3}$

11. $g_n = (-1)^n 2^{n+1}$

12. $S_9 = 9841 + 9841\,x$

13. 2.5

14. no limit

15. $2000

16. $\dfrac{5}{33}$

17. 72

18. $2934

19. $12,604.99

20. 4 m

21. 330

22. $x^{10} - 10x^8 y + 40x^6 y^2 - 80x^4 y^3 + \ldots$
    $\ldots + 80x^2 y^4 - 32y^5$

23. $210 a^6 x^4$

24. $\dfrac{n^2 + n}{4}$

25. $S_n = \dfrac{x^n - 2^n}{x^{n-1}(x-2)}$

Answers for Chapter 11 Tests: FORM B

1. 1, 6, 11, 16, 21 ; $a_{16} = 76$

2. $a_n = 2\left(\dfrac{1}{5}\right)^n$

3. $38 + 36 + 32 + 24 + 8 = 138$

4. $\displaystyle\sum_{k=1}^{5} \dfrac{k}{2}$

5. $a_{12} = 49$

6. $d = \dfrac{3}{4}$

7. $a_1 = \dfrac{4}{9}$; $d = -\dfrac{4}{3}$

8. 4095

9. $g_6 = \dfrac{32}{81}$

10. $r = \dfrac{3}{4}$

11. $g_n = (-1)^{n+1} 2^n$

12. $S_9 = 1533 + 1533x$

13. 20

14. no limit

15. $5000

16. $\dfrac{47}{99}$

17. 53

18. $2870

19. $15,756.24

20. $4\dfrac{2}{3}$ m

21. 462

22. $x^8 + 8x^6y + 24x^4y^2 + 32x^2y^3 + 16y^4$

23. $252a^5 x^5$

24. $\dfrac{n^2 + n}{6}$

25. $S_n = \dfrac{x^{2n} - 3^n}{x^{2n-2}(x^2 - 3)}$

Answers for Chapter 11 Tests: FORM C

1. $3, 9, 15, 21, 27$; $a_{16} = 93$

2. $a_n = 3\left(\dfrac{1}{2}\right)^n$

3. $3 + 1 - 3 - 11 - 27 = -37$

4. $\displaystyle\sum_{k=1}^{5}(-1)^k 2k$

5. $a_{12} = -45$

6. $d = \dfrac{5}{6}$

7. $a_1 = -\dfrac{1}{2}$; $d = -5$

8. 2730

9. $g_6 = \dfrac{320}{243}$

10. $r = \dfrac{2}{3}$

11. $g_n = 5\left(\dfrac{2}{3}\right)^n$

12. $S_9 = 39364 + 39364\, x$

13. 1

14. no limit

15. $2500

16. $\dfrac{7}{11}$

17. 124

18. $2535

19. $20,950.12

20. $5\dfrac{1}{3}$ m

21. 165

22. $x^5 - 10x^4 y^2 + 40x^3 y^4 - 80x^2 y^6 + \ldots$
    $\ldots + 80x\, y^8 - 32 y^{10}$

23. $210 a^4 x^6$

24. n^2

25. $S_n = \dfrac{x^n - 3^n}{x^{n-1}(x - 3)}$

Answers for Chapter 11 Tests: FORM D

1. $2, 8, 14, 20, 26$; $a_{16} = 92$

2. $a_n = 5\left(\dfrac{1}{3}\right)^n$

3. $5 + 16 + 27 + 16 + 1 = 65$

4. $\displaystyle\sum_{k=1}^{5}\left(\dfrac{1}{k}\right)^2$

5. $a_{12} = 71$

6. $d = 3$

7. $a_1 = -8$; $d = -\dfrac{7}{8}$

8. 35035

9. $g_6 = \dfrac{1}{2}$

10. $r = \dfrac{3}{4}$

11. $g_n = (-1)^n \left(\dfrac{3}{2}\right)^{n-1}$

12. $S_9 = 1022 + 511x$

13. 30

14. no limit

15. $7500

16. $\dfrac{226}{99}$

17. 74

18. $2465

19. $24,441.80

20. 6 m

21. 11

22. $x^4 + 8x^3y^2 + 24x^2y^4 + 32xy^6 + 16y^8$

23. $120a^3x^7$

24. $3(n^2 + n)$

25. $S_n = \dfrac{(2x)^n - 1}{(2x)^{n-1}(x-1)}$

Answers for Chapter 11 Tests: FORM E

1. $3, 7, 11, 15, 19$; $a_{16} = 63$

2. $a_n = \left(\dfrac{5}{2}\right)^n$

3. $1 + 0 - 1 + 16 - 243 = -227$

4. $\displaystyle\sum_{k=1}^{5} (-1)^{k+1} k^2$

5. $a_{12} = -67$

6. $d = \dfrac{1}{4}$

7. $a_1 = \dfrac{7}{4}$; $d = -7$

8. 1320

9. $g_6 = \dfrac{1}{625}$

10. $r = \dfrac{3}{2}$

11. $g_n = 3\left(\dfrac{1}{2}\right)^{n+2}$

12. $S_9 = 511 + 2555x$

13. 10

14. no limit

15. $4000

16. $\dfrac{118}{99}$

17. 51

18. $1869

19. $25,209.98

20. 9 m

21. 91

22. $32x^{10} - 80x^8y + 80x^6y^2 - 40x^4y^3 + \ldots$
    $\ldots + 10x^2y^4 - y^5$

23. $715a^9x^4$

24. $\dfrac{3n^2 - n}{2}$

25. $S_n = \dfrac{x^{3n} - 2^n}{x^{3n-3}(x^3 - 2)}$

Answers for Chapter 11 Tests: FORM F

1. $5, 8, 11, 14, 17 \,;\, a_{16} = 61$

2. $a_n = \left(\dfrac{3}{4}\right)^n$

3. $0 + 1 - 2 + 9 - 64 = -56$

4. $\displaystyle\sum_{k=1}^{5}(-1)^{k+1}\,2^k$

5. $a_{12} = -17$

6. $d = \dfrac{4}{3}$

7. $a_1 = 100 \,;\, d = -\dfrac{5}{4}$

8. 1863

9. $g_6 = \dfrac{1}{5}$

10. $r = \dfrac{4}{3}$

11. $g_n = \left(\dfrac{2}{5}\right)^{n-2}$

12. $S_9 = 29523 + 511x$

13. 3

14. no limit

15. $2500

16. $\dfrac{16}{37}$

17. 83

18. $1380

19. $18,214.39

20. $10\dfrac{1}{2}$ m

21. 1001

22. $16x^8 + 32x^6 y + 24x^4 y^2 + 8x^2 y^3 + y^4$

23. $1287 a^8 x^5$

24. $\dfrac{5n^2 - 3}{2}$

25. $S_n = \dfrac{3^n + x^n}{3^{n-1}(3 + x)}$

Answers for Chapter 11 Tests: FORM G

1. a
2. a
3. d
4. b
5. d
6. b
7. b
8. d
9. d
10. c
11. d
12. a
13. d
14. c
15. c
16. c
17. b
18. b
19. a
20. d
21. c
22. d
23. d
24. b
25. b

Answers for Chapter 11 Tests: FORM H

1. d
2. d
3. a
4. d
5. c
6. d
7. c
8. c
9. c
10. a
11. a
12. c
13. b
14. b
15. a
16. d
17. c
18. d
19. d
20. c
21. a
22. b
23. b
24. c
25. d

Answers for Final Tests: FORM A

1. 4.62
2. $-\dfrac{11}{10}$
3. $3a^2b + 3ab^2 + 1$
4. 3
5. 3.02×10^{-6}
6.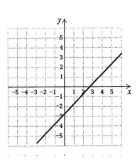
7. $\dfrac{9}{2}$; $(0, -18)$
8. -7
9. parallel
10. $y = -\dfrac{5}{6}x - \dfrac{13}{2}$
11. $(9, -4)$
12. $\left(\dfrac{1}{3}, \dfrac{5}{4}\right)$
13. $100,000 mortgage; $10,000 car loan; $5000 credit card
14. $(3, 5, 1)$
15. $(-2, 4, -3)$
16. $\{a \mid a \geq \dfrac{5}{8}\}$; $[\dfrac{5}{8}, \infty)$

    ⟵|●━━⟶
        0 5/8

17. at least 200 miles
18. $\{0, 2\}$
19. $(-\infty, 1) \cup (4, \infty)$

    ⟵━━○─○━━⟶
       0 1   4

20. $\varnothing$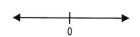
21. $h^2 - 7h - 2ah + 2a^2$
22. $5y^2 - 4y - 3y^3$
23. $80x^4y^4$
24. $10x(5 - x^2)$
25. $(8y - 5)(8y + 5)$
26. $\{x \mid x \in \mathbb{R}, x \neq 5\}$
27. $\dfrac{4}{5(t+5)}$
28. $\dfrac{3(a+b)^2}{(a-b)}$
29. 32
30. $3x^2 + 4 + \dfrac{x}{x^2+1}$
31. $f(2) = 21$
32. 7 km/h
33. $x - 5$
34. $25 - 10\sqrt{x} - 3x$
35. $\dfrac{6\sqrt{5} - \sqrt{15}}{33}$
36. $4 + 5i$
37. $-\dfrac{3}{5} + \dfrac{1}{5}i$
38. $-i$
39. $\dfrac{-3 \pm \sqrt{5}}{2}$

Answers for Final Tests: FORM A
continued

40. $-7.14005495,\ 0.14005495$

41. 20 hours

42. two irrational solutions

43. vertex: $\left(-\dfrac{1}{2},\dfrac{1}{2}\right)$; axis: $x=-\dfrac{1}{2}$

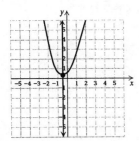

44. $(0,\infty)$

45. $(f\circ g)(x)=-18x^2+27x-10$
    $(g\circ f)(x)=-6x^2+3x-2$

46. $g^{-1}(x)=2\sqrt[3]{x}+3$

47. $\log_2\dfrac{1}{64}=-6$

48. $\dfrac{1}{2}\log a+\dfrac{1}{3}\log b-\log c$

49. -0.470

50. 4.78

51. 2

52. $(1,-2)$

53. circle

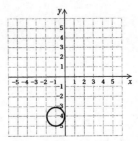

54. $(8,3),(-8,-3)$

55. length = 13; width = 6

56. $5+16+27+16+1=65$

57. $r=\dfrac{3}{4}$

58. no limit

59. $\dfrac{226}{99}$

60. 11

Answers for Final Tests: FORM B

1. 4.62
2. $\frac{19}{10}$
3. $7a^2b - 7ab^2 - 2$
4. -1
5. 3.06×10^{-6}
6.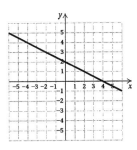
7. $\frac{11}{4}$; $\left(0, \frac{11}{2}\right)$
8. undefined
9. parallel
10. $y = -\frac{6}{5}x + 8$
11. $(10, -3)$
12. $\left(-\frac{2}{5}, \frac{1}{4}\right)$
13. $10,000 mortgage; $12,000 car loan; $6000 credit card
14. $(3, 0, -1)$
15. $(1, 0, -1)$
16. $\{a \mid a \geq \frac{14}{9}\}$; $[\frac{14}{9}, \infty)$
18. at least 200 miles
18. $\{10, 20\}$
19. $(-\infty, 2) \cup (3, \infty)$
20. $\varnothing$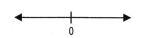
21. $h^2 - 8h - 2ah + 2a^2$
22. $5y^2 + 3y$
23. $74x^4y^4$
24. $11x^2(4 - x^2)$
25. $(12y - 7)(12y + 7)$
26. $\{x \mid x \in \mathbb{R}, x \neq 6\}$
27. $\frac{6}{7(t+5)}$
28. $7(a - b)$
29. $\frac{27}{2}$
30. $4x^2 + \frac{x+5}{x^2+2}$
31. $f(3) = -68$
32. 7 km/h
33. $x - 6$
35. $24 - 8\sqrt{x} - 2x$
35. $\frac{7\sqrt{3} + \sqrt{6}}{47}$
36. $3 + 7i$
37. $-\frac{13}{5} - \frac{19}{5}i$
38. $-i$
40. $-2, -1$

373

Answers for Final Tests: FORM B
continued

40. $-3.1925824, 2.1925824$

41. $4\frac{4}{5}$ hours

42. two distinct rational solutions

43. vertex: $\left(-\frac{1}{2}, \frac{3}{2}\right)$; axis: $x = -\frac{1}{2}$

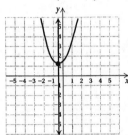

44. $(-\infty, -\sqrt{5}) \cup (0, \sqrt{5})$

45. $(f \circ g)(x) = 9x^2 - 18x + 8$
    $(g \circ f)(x) = -3x^2 - 6x + 2$

46. $g^{-1}(x) = 2\sqrt[3]{x} - 4$

47. $\log_3 \frac{1}{81} = -4$

48. $\frac{1}{3} \log a + \frac{1}{2} \log b - \log c$

49. -0.406

50. 2.995

51. 6

53. $(1.5, -2.5)$

53. circle

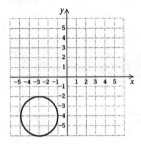

54. $(5, 4), (-5, -4)$

55. length = 11 ; width = 6

56. $1 + 0 - 1 + 16 - 243 = -227$

57. $r = \frac{3}{2}$

58. no limit

59. $\frac{118}{99}$

60. 91

Answers for Final Tests: FORM C

1. -5.28
2. $\dfrac{43}{14}$
3. $4a^2b - 11ab^2 - 3$
4. 4
5. 2.92×10^{-6}
6.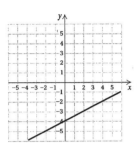
7. $\dfrac{9}{2}$; $(0, -18)$
8. 0
9. neither
10. $y = -\dfrac{7}{4}x + \dfrac{27}{4}$
11. $(11, -2)$
12. $\left(-\dfrac{3}{4}, \dfrac{3}{5}\right)$
13. $200,000 mortgage; $7000 car loan; $5000 credit card
14. $(2, -4, 3)$
15. $(2, -1, 2)$
16. $\{a \mid a \geq 1\}$; $[1, \infty)$
19. at least 250 miles
18. $\{9\}$
19. $(-\infty, 2) \cup (2, \infty)$
20. $\varnothing$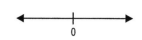
21. $h^2 - 37h - 2ah + 2a^2$
22. $-3y^2 + 2y + 3y^3$
23. $96x^4y^4$
24. $12x^3(3 - x^2)$
25. $(6y - 5)(6y + 5)$
26. $\{x \mid x \in \mathbb{R}, x \neq 7\}$
27. $\dfrac{1}{2(t+4)}$
28. $\dfrac{2(a+b)^2}{(a-b)}$
29. $-\dfrac{106}{5}$
30. $5x^2 - 9 + \dfrac{4x + 15}{x^2 + 3}$
31. $f(4) = -45$
32. 8 km/h
33. $x - 5$
36. $15 - 2\sqrt{x} - x$
35. $\dfrac{10\sqrt{2} + \sqrt{6}}{97}$
36. $8 + 6i$
37. $-\dfrac{11}{5} + \dfrac{13}{5}i$
38. 1
41. $\dfrac{-3 \pm i\sqrt{3}}{2}$

Answers for Final Tests: FORM C
continued

40. $-3.79128785, 0.79128785$

41. 1 hour

42. two distinct rational solutions

43. vertex: $\left(-\frac{1}{2}, \frac{5}{2}\right)$; axis: $x = -\frac{1}{2}$

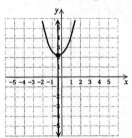

44. $(-\infty, 0)$

45. $(f \circ g)(x) = 4x^2 - 8x + 3$
    $(g \circ f)(x) = -2x^2 + 4x + 3$

46. $g^{-1}(x) = 4\sqrt[3]{x} + 3$

47. $\log_5 \frac{1}{625} = -4$

48. $3 \log a + 2 \log b - \log c$

49. 0.406

50. 4.787

51. 3

54. $(1, -2)$

53. circle

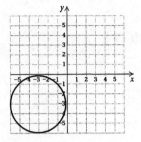

54. $(5, 3), (-5, -3)$

55. length = 12; width = 5

56. $0 + 1 - 2 + 9 - 64 = -56$

57. $r = \frac{4}{3}$

58. no limit

59. $\frac{16}{37}$

60. 1001

Answers for Final Tests: FORM D

1. $d = \frac{5}{8}$

2. $\sqrt{202} \approx 14.213$

3. $f^{-1}(x) = \frac{x+4}{3}$

4. $\frac{-1 \pm i\sqrt{7}}{2}$

5. $\sqrt[20]{a}$

6. $\frac{5}{6(t+2)}$

7. $h^2 - 3h + 2ah$

8. $\{x \mid x < \frac{8}{7}\} ; (-\infty, \frac{8}{7})$

9. $(-1, -3)$

10. $r = \frac{2}{3}$

11. 5.72

12.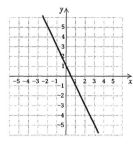

13.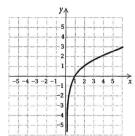

14. parabola

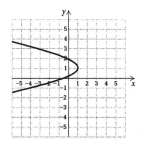

15. $-2.302776, 1.302776$

16. $(-\infty, 2]$

20. $(x-2)(x+3)(x+4)$

18. $3y^2 - y - 2y^3$

19. at least 135 miles

20. width = 5; length = 7

21. $-\frac{3}{8} ; \left(0, -\frac{9}{4}\right)$

22. $\frac{4}{15}$

23. 2.5

24. $(-5, 0), (5, 0)$

25. $2^7 = 128$

26. 1 hour, or 60 minutes

27. 5

28. $\frac{(x-10)(x-2)}{(x-8)(x+10)}$

29. $10a^2 + 14ab - 12b^2$

30. $\{0, 1, 2, 3, 4, 5\}$

31. $\left(\frac{1}{2}, -\frac{1}{2}, \frac{1}{2}\right)$

32. $\frac{3}{2}$

33. $9a^2 b + 9ab^2 + 1$

34. 4 m

35. length = 7 ; width = 3

36. -3

37. two distinct rational solutions

38. $3 - 4i$

377

Answers for Final Tests: FORM D
continued

39. $\frac{15}{32}$ hours, or $28\frac{1}{8}$ minutes

40. $7x(2-x^2)$

41. $(9y-1)(9y+1)$

42. $(1, 2)$

43. $f(x) = 2x + 1$

44. -3

45. $210a^6 x^4$

46. center $(-2, 9)$; radius = 3

47. $\frac{\log 3.2}{\log 6} \approx 0.6492$

48. vertex: $\left(-\frac{1}{4}, \frac{7}{8}\right)$; axis: $x = -\frac{1}{4}$

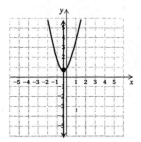

49. $-i$

50. $3x^2 + 5 + \frac{4x-4}{x^2+1}$

51. ± 2

52. $\left(-\frac{3}{4}, \frac{7}{4}\right)$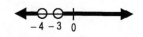

53. ($2, 62)

54. perpendicular

55. $\frac{y^4}{81x^2}$

56. $\frac{5}{33}$

57. 3.15%

58. $\frac{\sqrt{70} - 2\sqrt{10}}{3}$

59. $(-\infty, -4) \cup (-3, \infty)$

60. a) $C(x) = 15{,}000 + 10x$
    b) $R(x) = 35x$
    c) $P(x) = 25x - 15{,}000$
    d) $P(300) = -7500$; $P(900) = 7500$
    e) $x = 600$

Answers for Final Tests: FORM E

1. $d = \dfrac{3}{4}$

2. $4\sqrt{13} \approx 14.422$

3. $f^{-1}(x) = \dfrac{x-5}{3}$

4. -1

5. $\sqrt[10]{a}$

6. $\dfrac{2}{3(t+3)}$

7. $h^2 - 4h + 2ah$

8. $\{x \mid x < \dfrac{21}{2}\}; (-\infty, \dfrac{21}{2})$

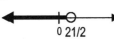

9. $\left(-\dfrac{15}{14}, -\dfrac{9}{7}\right)$

10. $r = \dfrac{3}{4}$

11. 5.72

13.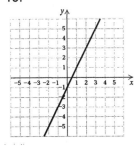

13.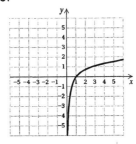

14. parabola

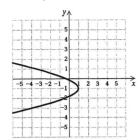

15. $-4.5413813, 1.5413813$

16. $(-\infty, 3]$

21. $(x-4)(x+2)(x+3)$

18. $-5y^2 - 2y + y^3$

19. at least 140 miles

20. width = 8; length = 24

21. $\dfrac{5}{6}; \left(0, \dfrac{1}{2}\right)$

22. $-\dfrac{5}{21}$

23. 20

24. $(-4, 0), (4, 0)$

25. $3^2 = 9$

26. 5 hours

27. 6

28. $\dfrac{(x-9)(x-3)}{(x-7)(x+9)}$

29. $3a^2 - 7ab - 20b^2$

30. $\{1, 2, 3, 4, 6, 8, 10\}$

31. $(2, 3, 4)$

32. 2

33. $2a^2b + 5ab^2 + 2$

34. $4\dfrac{2}{3}$ m

35. length = 10; width = 4

36. -2

37. two irrational solutions

38. $8 + 6i$

Answers for Final Tests: FORM E continued

39. $\frac{35}{48}$ hours, or $43\frac{3}{4}$ minutes

40. $8x(3-x^2)$

41. $(10y-11)(10y+11)$

42. $(2,1)$

43. $f(x)=-2x+2$

44. 2

45. $252a^5x^5$

46. center $(2,-8)$; radius $=4$

47. $\frac{\log 4.2}{\log 7} \approx 0.7375$

48. vertex: $\left(-\frac{1}{4}, \frac{15}{8}\right)$; axis: $x=-\frac{1}{4}$

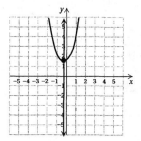

49. $-i$

50. $4x^2-1+\frac{3x+3}{x^2+2}$

51. ± 3

52. $\left(-\frac{1}{2}, \frac{5}{2}\right)$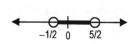

53. $(\$3, 63)$

54. neither

55. $-\frac{y^9}{27x^6}$

56. $\frac{47}{99}$

57. 3.46%

58. $\frac{3\sqrt{70}+\sqrt{42}}{3}$

59. $(-\infty, -3) \cup (-2, \infty)$

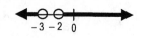

60. a) $C(x) = 24{,}000 + 15x$
    b) $R(x) = 45x$
    c) $P(x) = 30x - 24{,}000$
    d) $P(250) = -16{,}500$; $P(1000) = 6000$
    e) $x = 800$

380

Answers for Final Tests: FORM F

1. $d = \dfrac{5}{6}$

2. $\sqrt{218} \approx 14.765$

3. $f^{-1}(x) = \dfrac{x+6}{2}$

4. $-1 \pm i$

5. $\sqrt[20]{a^{19}}$

6. $\dfrac{3}{4(t+4)}$

7. $h^2 - 41h + 2ah$

8. $\{x \mid x < \dfrac{12}{7}\}; (-\infty, \dfrac{12}{7})$

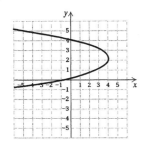

9. $\left(\dfrac{30}{11}, -\dfrac{5}{11}\right)$

10. $r = \dfrac{2}{3}$

11. 4.72

14. 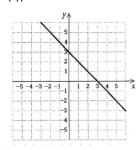 13.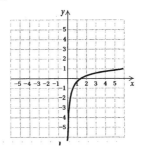

14. parabola

15. $-5.85410197, 0.85410197$

16. $(-\infty, 5]$

22. $(x-3)(x+2)(x+4)$

18. $-5y^2 - 3y - 6y^3$

19. at least 170 miles

20. width = 11; length = 31

21. $-\dfrac{7}{4}; \left(0, \dfrac{5}{4}\right)$

22. $\dfrac{1}{20}$

23. 1

24. $(-3, 0), (3, 0)$

25. $4^3 = 64$

26. 78 minutes

27. 7

28. $\dfrac{(x-8)(x-4)}{(x-6)(x+8)}$

29. $8a^2 - 18ab - b^2$

30. $\{1, 2, 3, 4, 5, 7, 9\}$

31. $\left(\dfrac{1}{2}, 1, \dfrac{3}{2}\right)$

32. 3

33. $8a^2b - ab^2 + 1$

34. $5\dfrac{1}{3}$ m

35. length = 9 ; width = 5

36. -3

37. two complex solutions

38. $5 + 12i$

Answers for Final Tests: FORM F
continued

39. $\frac{33}{34}$ hours

40. $9x(4-x^2)$

41. $(11y-3)(11y+3)$

42. $(2,3)$

43. $f(x) = -2x - 2$

44. -2

45. $210a^4 x^6$

46. center $(-2,7)$; radius $= 5$

47. $\frac{\log 5.2}{\log 8} \approx 0.7928$

48. vertex: $\left(-\frac{1}{4}, \frac{23}{8}\right)$; axis: $x = -\frac{1}{4}$

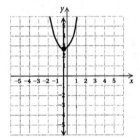

49. 1

50. $5x^2 - 9 + \frac{2x+12}{x^2+3}$

51. ± 4

52. $\left(-\frac{5}{6}, \frac{3}{2}\right)$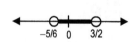

53. $(\$3, 71)$

54. perpendicular

55. $\frac{y^4}{64x^6}$

56. $\frac{7}{11}$

57. 3.85%

58. $\frac{5\sqrt{6} + \sqrt{30}}{20}$

59. $(-\infty, \infty)$

60. a) $C(x) = 32{,}250 + 22x$
    b) $R(x) = 57x$
    c) $P(x) = 35x - 32{,}250$
    d) $P(575) = -13{,}125$; $P(1500) = 19{,}250$
    e) $x = 950$

Answers for Final Tests: FORM G

1. a
2. d
3. c
4. c
5. a
6. d
7. b
8. a
9. a
10. d
11. d
12. b
13. c
14. c
15. d
16. b
17. c
18. d
19. c
20. c
21. a
22. a
23. c
24. a
25. b
26. b
27. c
28. d
29. a
30. c
31. b
32. b
33. d
34. b
35. b
36. c
37. c
38. a
39. a
40. a
41. b
42. c
43. a
44. a
45. c
46. b
47. a
48. c
49. c
50. a

Answers for Final Tests: FORM H

1. c
2. c
3. d
4. d
5. d
6. b
7. b
8. c
9. a
10. d
11. a
12. c
13. d
14. d
15. b
16. b
17. b
18. b
19. a
20. d
21. a
22. d
23. d
24. c
25. a
26. d
27. b
28. a
29. d
30. a

31. c
32. b
33. b
34. d
35. c
36. a
37. b
38. d
39. a
40. d
41. c
42. d
43. a
44. b
45. d
46. a
47. c
48. a
49. b
50. b

EXTRA PRACTICE 1
Addition and Subtraction of Real Numbers
Use After Section 1.2   Name_____

Examples: Add.
a) $-15+(-7)=-22$
b) $3.45+(-0.7)=2.75$
c) $-\dfrac{15}{16}+\dfrac{3}{8}=-\dfrac{15}{16}+\dfrac{3}{8}\cdot\dfrac{2}{2}=-\dfrac{15}{16}+\dfrac{6}{16}=-\dfrac{9}{16}$

Add.

1. $-47+15=$ _____
2. $3.2+(-9.4)=$ _____
3. $-\dfrac{5}{8}+\dfrac{3}{8}=$ _____

4. $-18.7+(-5.9)=$ _____
5. $-\dfrac{3}{4}+\left(-\dfrac{2}{5}\right)=$ _____
6. $-17+(-15)=$ _____

7. $\dfrac{17}{30}+\left(-\dfrac{5}{6}\right)=$ _____
8. $45+(-79)=$ _____
9. $-5.23+18.45=$ _____

10. $-6+(-6)=$ _____
11. $\dfrac{5}{7}+\left(-\dfrac{5}{7}\right)=$ _____
12. $-19+37=$ _____

13. $-1.9+0.7=$ _____
14. $\dfrac{3}{5}+\left(-\dfrac{4}{5}\right)=$ _____
15. $5.29+(-5.29)=$ _____

16. $-\dfrac{11}{12}+\dfrac{3}{4}=$ _____
17. $-81+(-5)=$ _____
18. $6.3+(-4.21)=$ _____

19. $75+(-4)=$ _____
20. $\dfrac{2}{3}+\left(-\dfrac{1}{15}\right)=$ _____
21. $16+(-16)=$ _____

22. $-2.5+(-17.9)=$ _____
23. $-72+15=$ _____
24. $-\dfrac{17}{24}+\left(-\dfrac{1}{6}\right)=$ _____

25. $\dfrac{7}{18}+\left(-\dfrac{1}{3}\right)=$ _____
26. $18.2+(-8.5)=$ _____
27. $-\dfrac{1}{7}+\dfrac{4}{5}=$ _____

28. $-72+72=$ _____
29. $-\dfrac{1}{8}+\left(-\dfrac{1}{8}\right)=$ _____
30. $93+(-9)=$ _____

31. $\dfrac{5}{8}+\left(-\dfrac{8}{5}\right)=$ _____
32. $-32+6=$ _____
33. $-23.17+2.3=$ _____

34. $-3.012+(-2.103)=$ _____
35. $-\dfrac{1}{4}+\left(-\dfrac{3}{4}\right)=$ _____
36. $-\dfrac{4}{9}+\dfrac{2}{5}=$ _____

37. $-24+31=$ _____
38. $0.9+(-0.73)=$ _____
39. $0+(-29)=$ _____

40. $\dfrac{3}{8}+\left(-\dfrac{7}{24}\right)=$ _____
41. $-71+(-1)=$ _____
42. $-0.003+0.1=$ _____

EXTRA PRACTICE 1
Addition and Subtraction of Real Numbers
Use after Section 1.2

Examples: Subtract.
a) $54 - (-13) = 54 + 13 = 67$
b) $-0.25 - 14.8 = -0.25 + (-14.8) = -15.05$
c) $-\frac{7}{12} - \left(-\frac{1}{12}\right) = -\frac{7}{12} + \frac{1}{12} = -\frac{6}{12} = -\frac{1}{2}$

Subtract.

43. $8 - 11 =$ _____
44. $8 - (-11) =$ _____
45. $-8 - 11 =$ _____

46. $-8 - (-11) =$ _____
47. $19 - 19 =$ _____
48. $19 - (-19) =$ _____

49. $-19 - 19 =$ _____
50. $-19 - (-19) =$ _____
51. $15 - (-3) =$ _____

52. $-93 - 14 =$ _____
53. $46 - (-12) =$ _____
54. $-8 - 32 =$ _____

55. $-\frac{4}{5} - \frac{1}{2} =$ _____
56. $3.7 - (-1.5) =$ _____
57. $86 - (-3) =$ _____

58. $-15.2 + 1.5 =$ _____
59. $-\frac{7}{10} - \left(-\frac{7}{10}\right) =$ _____
60. $-18 - 32 =$ _____

61. $1 - (-43) =$ _____
62. $-1.8 + (-19.3) =$ _____
63. $\frac{5}{8} - \left(-\frac{1}{3}\right) =$ _____

64. $-17 - 38 =$ _____
65. $9 - \left(-\frac{1}{3}\right) =$ _____
66. $4.2 - (-0.4) =$ _____

67. $-23.6 - 0.43 =$ _____
68. $-50 - (-17) =$ _____
69. $-\frac{5}{12} - \frac{1}{4} =$ _____

70. $\frac{2}{5} - \frac{2}{3} =$ _____
71. $-19 - 3.5 =$ _____
72. $16 - 40 =$ _____

73. $-85 - 49 =$ _____
74. $-\frac{8}{15} - \left(-\frac{1}{3}\right) =$ _____
75. $-0.3 - (-5.12) =$ _____

76. $-9.37 - (-1.7) =$ _____
77. $6 - (-25) =$ _____
78. $\frac{5}{6} - \frac{6}{7} =$ _____

79. $75 - (-10) =$ _____
80. $-\frac{4}{3} - \frac{3}{4} =$ _____
81. $-72 - (-19) =$ _____

82. $\frac{7}{19} - \left(-\frac{1}{2}\right) =$ _____
83. $4.6 - 7 =$ _____
84. $\frac{5}{8} - \left(-\frac{5}{8}\right) =$ _____

EXTRA PRACTICE 2
Solving Problems
Use after Section 1.4   Name_____

Example: Six plus five times a number is 10 more than three times the number. What is the number?

$$\underbrace{\text{Six}}_{6} \underbrace{\text{plus}}_{+} \underbrace{\text{five}}_{5} \underbrace{\text{times}}_{\cdot} \underbrace{\text{a number}}_{x} \underbrace{\text{is}}_{=} \underbrace{10}_{10} \underbrace{\text{more than}}_{+} \underbrace{\text{three}}_{3} \underbrace{\text{times}}_{\cdot} \underbrace{\text{the number.}}_{x}$$

Solve: $6 + 5x = 10 + 3x$
$6 + 2x = 10$
$2x = 4$
$x = 2$

The value of 2 checks in the original problem. The number is 2.

Solve.

1. Seven plus twice a number is five less than four times the number. What is the number?
   _____

2. The first angle of a triangle is three times the second. The third angle is eight times the second. Find the measure of the second angle. _____

3. The perimeter of a rectangle is 30 inches. The length is seven inches more than the width. What is the length? _____

4. Find two consecutive odd integers such that three times the first plus two times the second is 29. _____

5. Four minus nine times a number is twenty-six more than twice the number. What is the number? _____

6. Eight plus three times a number is 18 more than four times the number. What is the number? _____

EXTRA PRACTICE 2
Solving Problems
Use after Section 1.4

---

7. Nine minus five times a number is five minus three times the number. What is the number? _____

8. The sum of four consecutive integers is 78. Find the integers. _____

9. Five less than three times a number is fifteen more than half the number. What is the number? _____

10. A piece of ribbon 48 in. long is cut into two pieces so that one piece is one-third as long as the other. Find the length of each piece. _____

11. The perimeter of a rectangle is 84 ft. The length is 6 ft less than twice the width. What are the dimensions? _____

12. Elise pays $630.70 for a printer. If the price paid includes a 6% sales tax, what is the price of the printer itself? _____

13. Four more than three times a number is seventeen less than six times the number. Find the number. _____

14. The second angle of a triangle is 10° more than the first, and the third is 5° less than three times the first. Find the measures of the angles. _____

15. Find three consecutive even integers such that the sum of the first, twice the second, and three times the third is 124. _____

16. A paddleboat moves at a rate of 12 km/h in still water. How long will it take the boat to travel 63 km downriver if the river's current moves at a rate of 6 km/h? _____

EXTRA PRACTICE 3
Properties of Exponents
Use after Section 1.6   NAME_____

Examples: Multiply and Simplify
a) $7^2 \cdot 7^5 = 7^{2+5} = 7^7$
b) $(-2x^2 y) \cdot (4xy^3) = -8x^{2+1} y^{1+3} = -8x^3 y^4$

Divide and Simplify
a) $\dfrac{8^5}{8^2} = 8^{5-2} = 8^3$
b) $\dfrac{-18x^2 y^4}{-3xy^2} = 6x^{2-1} y^{4-2} = 6xy^2$

Multiply and Simplify (Leave the answer in exponential notation).

1. $3^4 \cdot 3^2$ _____
2. $4^6 \cdot 4$ _____
3. $9^3 \cdot 9^2$ _____
4. $f^4 \cdot f^5$ _____
5. $r^2 \cdot r^6$ _____
6. $(-e^3)(e^5)$ _____
7. $(z^2)(-8r^2)(z^3 r)$ _____
8. $(5x^2)(2xy)$ _____
9. $(-3n^3)(-4n^2)$ _____
10. $(6x^4)(3x^2)$ _____
11. $(4a^7)(-3a^2 b^3)$ _____
12. $(8x^3 y^6)(-5y^3 x)$ _____
13. $(4n^2 y)(3xy^2)$ _____
14. $(x^2 y^3)(-5y^2 x^3)$ _____

Divide and Simplify.

15. $\dfrac{b^3}{b}$ _____
16. $\dfrac{4x^2}{3x}$ _____
17. $\dfrac{5x^4}{x^2}$ _____
18. $\dfrac{13n^5}{7n^2}$ _____
19. $\dfrac{6n^3}{4n^2}$ _____
20. $\dfrac{4y^4}{2y^3}$ _____
21. $\dfrac{-5a^8}{6a^2}$ _____
22. $\dfrac{-15n^6}{5n^2}$ _____
23. $\dfrac{22x^5}{18x^3}$ _____
24. $\dfrac{14x^3 y^2 z^4}{-2xy^2 z^3}$ _____
25. $\dfrac{18a^4 b^8 c^6}{6a^2 b^3 c}$ _____
26. $\dfrac{16a^4 b^7 c^6}{5a^2 b^5 c}$ _____

Simplify.

27. $(1)^{-8}$ _____
28. $(-4)^5$ _____
29. $(-5)^3$ _____
30. $(6)^{-3}$ _____
31. $(-4)^{-2}$ _____
32. $(-2)^9$ _____
33. $(-6)^{-3}$ _____
34. $(-9)^4$ _____
35. $(-3)^6$ _____
36. $(4)^{-5}$ _____
37. $(3)^{-7}$ _____
38. $(-1)^{-8}$ _____

EXTRA PRACTICE 3
Properties of Exponents
Use after Section 1.6

Write an equivalent expression without negative exponents.

39. a^{-3} _____

40. $\dfrac{x^{-3}}{5xy^2}$ _____

41. $x^{-4}y^{-2}$ _____

42. $\dfrac{(4x)^{-3}}{y^4}$ _____

43. $(4c^5x^2)^{-2}$ _____

44. $\dfrac{2y^2}{4x^2y^3}$ _____

45. $(df^3)^{-4}$ _____

46. $\dfrac{x^3y^2z^5}{z^4x^5}$ _____

47. $3f^3n^{-7}$ _____

48. $(m^4y^6)^{-3}$ _____

49. $5m^4c^2x^{-3}y^2$ _____

Write an equivalent expression with negative exponents.

50. $\dfrac{1}{2^4}$ _____

51. $\dfrac{3}{2^2}$ _____

52. $\dfrac{1}{(-3)^3}$ _____

53. $5x^{-2}$ _____

54. n^5 _____

55. $\dfrac{1}{y^2}$ _____

56. $\dfrac{-3}{(2y)^3}$ _____

57. $\dfrac{1}{(2x)^2}$ _____

Simplify. Use only positive exponents.

58. $5^3 \cdot 5^{-4}$ _____

59. $6^4(-6)^2$ _____

60. $x^4 \cdot x^{-5}$ _____

61. $(4x^2)(3x^2y^3)$ _____

62. $(8a^{-3}b^2)(b^2a^3c)$ _____

63. $(5xy^2)(3y^3z)$ _____

64. $\dfrac{3x^2y^3}{2yx^3}$ _____

65. $\dfrac{4a^2bc^4}{3a^4cm^3}$ _____

66. $\dfrac{((4x)^{-3})(3x^2)}{xy^2}$ _____

67. $\dfrac{8x^3y^2z^4}{16x^5y^3z}$ _____

68. $4(a^2b^3)^4$ _____

69. $(6x^3y^2)^{-2}$ _____

70. $8(x^3y^2)^3$ _____

71. $(4^{-3})^{-3}$ _____

72. $\dfrac{-4x^3y^5z^2}{16x^5y^7z}$ _____

73. $\dfrac{4a^2bc^3}{-8a^8b^4c^3d^{-5}}$ _____

74. $\left(\dfrac{3x^3y^{-4}}{4x^2y^3}\right)^{-3}$ _____

75. $\left(\dfrac{14x^3y^2z^5}{7xy^3z^3}\right)^2$ _____

76. $\left(\dfrac{3a^2b^3c^5}{4ab^2c^2}\right)^{-3}$ _____

77. $(4x^2)^6$ _____

78. $(4^{x-2})(4^2)$ _____

79. $(5y^2x^3)^{-6}$ _____

80. $\left(\dfrac{(2^{-3}a^2b^3c)(4a^2xb)}{(5ab^2c^2)(x^2ab^2)}\right)^{-3}$ _____

EXTRA PRACTICE 4
Graphing Linear Equations
Use after Section 2.1   Name_____

Examples: Graph.

a) $y = -\dfrac{3}{4}x + 1$

x	y
-4	4
0	1
4	-2

b) $2y - 3x = 6$

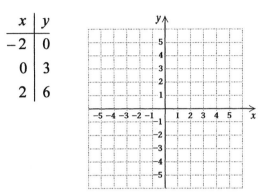

x	y
-2	0
0	3
2	6

Graph.

1. $y = 2x - 1$

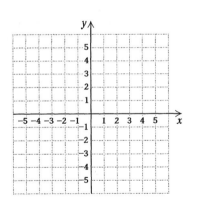

2. $y = 3x$

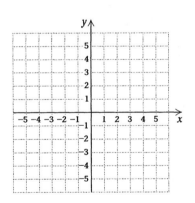

3. $y = 5x - 3$

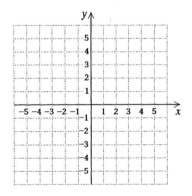

EXTRA PRACTICE 4
Graphing Linear Equations
Use after Section 2.1

4. $y = -1.5x$

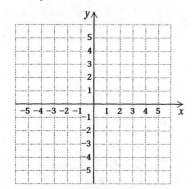

5. $2x + 5y = 10$

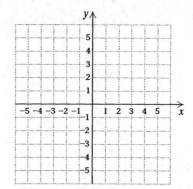

6. $y = -2x + 5$

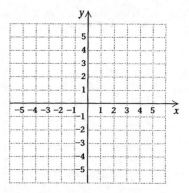

7. $y = -3x$

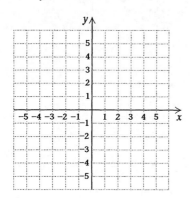

8. $-2x - 3y = 12$

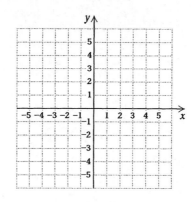

9. $y = \dfrac{5}{2}x$

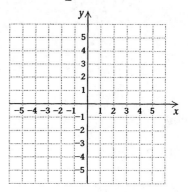

10. $y = \dfrac{1}{2}x + 2$

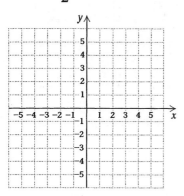

11. $3y + x = 0$

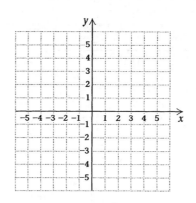

12. $y = -\dfrac{1}{2}x$

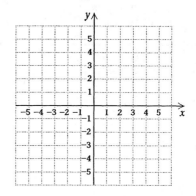

EXTRA PRACTICE 4
Graphing Linear Equations
Use after Section 2.1

Name_____

13. $y = \dfrac{3}{4}x$

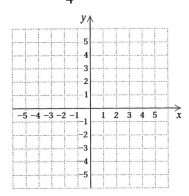

14. $y = -1 + x$

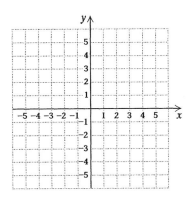

15. $3y + 15 = 5x$

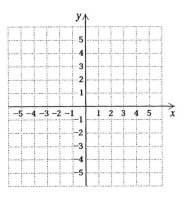

16. $3x + y = 3$

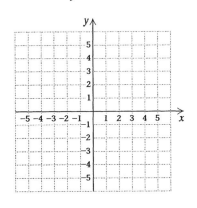

17. $y = -\dfrac{2}{3}x$

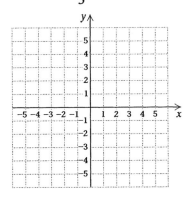

18. $y = x - 1$

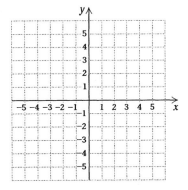

19. $y = -5x$

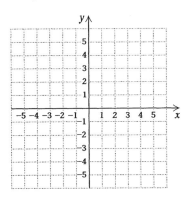

20. $2x + 10 = -5y$

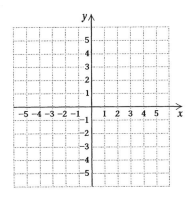

21. $y = -\dfrac{4}{5}x + 1$

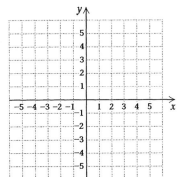

EXTRA PRACTICE 4
Graphing Linear Equations
Use after Section 2.1

22. $y = 6x - 2$

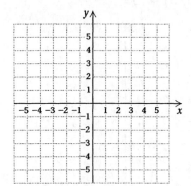

23. $3x - 3y = 9$

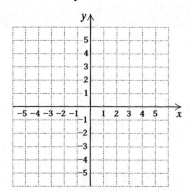

24. $y = -x$

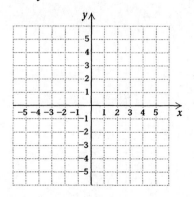

25. $y = -3x + 2$

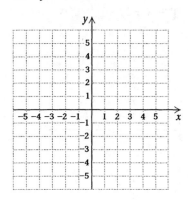

26. $y = \dfrac{3}{2}x$

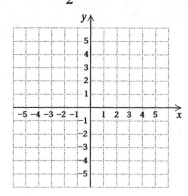

27. $x - 3y = 9$

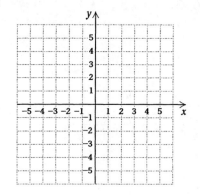

28. $y = -\dfrac{2}{3}x$

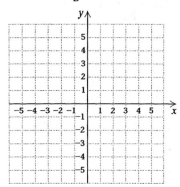

29. $y = 5x - 3$

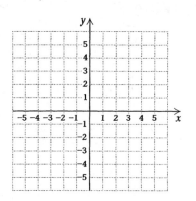

30. $y = 0.5x + 1$

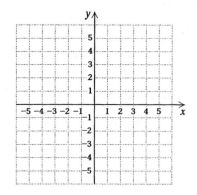

EXTRA PRACTICE 5
Solving Systems of Linear Equations
Use after Section 3.2   Name_____

Examples:
a)   Solve using the substitution method: $5x - 2y = 4$,
$$y = 5 - x.$$

Substitute $5 - x$ for y.

$$5x - 2y = 4$$
$$5x - 2(5 - x) = 4$$
$$5x - 10 + 2x = 4$$
$$7x = 14$$
$$x = 2$$

Then substitute 2 for x and solve for y.

$$y = 5 - x$$
$$y = 5 - 2$$
$$y = 3$$

The solution is $(2, 3)$.

b) Solve using the elimination method: $2x + 7y = -1$,
$$-x - 2y = 2.$$

Multiply the second equation by 2 and then add.

$$2x + 7y = -1$$
$$-2x - 4y = 4$$
$$\overline{\,3y = 3}$$
$$y = 1$$

Then substitute 1 for y and solve for x.

$$2x + 7y = -1$$
$$2x + 7 \cdot 1 = -1$$
$$2x + 7 = -1$$
$$2x = -8$$
$$x = -4$$

The solution is $(-4, 1)$.

Solve.

1. $4x + 3y = 1$,
    $x = 1 - y$ _____

2. $2x - y = 6$,
    $-x + y = -1$ _____

3. $6x - y = 3$,
    $4x - 2y = -2$ _____

4. $2x + 3y = 7$,
    $x = 1 - 4y$ _____

5. $2x + 3y = 6$,
    $x - 3y = -15$ _____

6. $7x - 5y = 4$,
    $y = 3x - 4$ _____

EXTRA PRACTICE 5
Solving Systems of Linear Equations
Use after Section 3.2

7. $2y - 5x = -1$,
$x = 2y + 5$

8. $4x + 3y = 1$,
$3x + 5y = -13$

9. $6x - 5y = 3$,
$4x + 3y = 21$

10. $x + y = 4$,
$3x + 4y = 10$

11. $-3x + y = 2$,
$7x - 8y = 1$

12. $7x + 2y = 2$,
$x - 2y = 14$

13. $9y - 2x = -7$,
$x - 3y = 5$

14. $3x - 5y = 8$,
$4x - 7y = 12$

15. $5x + 2y = 12$,
$3x - 4y = 2$

16. $x + 4y = 7$,
$3x + 7y = 6$

17. $5x - 8y = 25$,
$-x + 4y = -7$

18. $0.5x + 2y = 9$,
$4x - 1.5y = 2$

19. $8x - 6y = 0$,
$x + 9y = \dfrac{13}{4}$

20. $\dfrac{2}{3}x + \dfrac{1}{4}y = 18$,
$\dfrac{1}{6}x - \dfrac{3}{8}y = -6$

EXTRA PRACTICE 6
Solving Applications: Systems of Two Equations
Use after Section 3.3   Name_____

Example: The campus bookstore sells two kinds of sweatshirts. The hooded ones sell for $39.50 and the crewneck ones sell for $34.50. During the first week of school, a total of 250 sweatshirts were sold at a total value of $9185. How many of each kind were sold?

We let x represent the number of hooded sweatshirts sold and y represent the number of crewneck sweatshirts sold.

The total sold was 250, so we have $x + y = 250$. The total amount taken in was $9185, thus we have $39.50x + 34.50y = 9185$.

We solve the following system.

$$x + y = 250$$
$$39.5x + 34.5y = 9185$$

or

$$x + y = 250$$
$$395x + 345y = 91{,}850 \text{ (Multiplying by 10)}$$

The solution of the system is $x = 112$ and $y = 138$. These values check. Thus 112 hooded sweatshirts and 138 crewneck sweatshirts were sold.

Solve.

1. The sum of two numbers is -11. Twice the first number minus the second is 32. Find the numbers. _____

2. Two investments are made totaling $16,000. For a certain year these investments yield $970 in simple interest. Part of the $16,000 is invested at 5% and the rest at 7%. How much is invested at 7%? _____

3. A collection of nickels and dimes is worth $3.30. There are 42 coins in all. How many of each kind of coin are there? _____

4. Patrick is 4 years younger than his sister Alice. In five years, Patrick will be $\frac{3}{4}$ as old as Alice. How old is Patrick now? _____

5. The difference between two numbers is 14. Twice the smaller is 7 more than the larger. What are the numbers? _____

6. The perimeter of a lot is 84 ft. The length exceeds the width by 16 feet. Find the length and the width. _____

EXTRA PRACTICE 6
Solving Applications: Systems of Two Equations
Use after Section 3.3

---

7. One night a theater sold 548 movie tickets. An adult's ticket costs $6.50, and a child's ticket costs $3.50. In all, $2881 was taken in. How many of each kind of ticket were sold? _____

8. A train leaves Smithville and travels south at a speed of 60 mph. Three hours later, a second train leaves on a parallel track and travels south at 90 mph. How far from the station will they meet? _____

9. The sum of a certain number and a second number is 21. The second number minus the first number is -57. Find the numbers. _____

10. The perimeter of a rectangular field is 110 feet. The length is 7 feet more than twice the width. Find the dimensions. _____

11. A chemist has one solution that is 20% saline and a second that is 65% saline. How many gallons of each should be mixed together to get 120 gallons of a solution that is 50% saline? _____

12. Two investments are made totaling $23,000. For a certain year these investments yield $2095 in simple interest. Part of the $23,000 is invested at 8% and the rest at 11%. How much is invested at each rate? _____

13. Two angles are complementary. One angle is 10° less than three times the other. Find the measures of the angles. _____

14. A small boat took 2 hr to make a trip downstream with a 4-mph current. The return trip against the same current took 3 hr. Find the speed of the boat in still water. _____

EXTRA PRACTICE 7
Solving Systems of Equations in Three Variables
Use after Sections 3.4 and 3.5 NAME_____

See Sections 3.4 and 3.5 for examples of both solving systems of equations in three variables and applications.

Solve each system. If a system's equations are dependent or if there is no solution, state this.

1. $2x - 3y + z = -9,$
   $2x - 4y + 3z = -16,$
   $4x + y - 3z = 13$

2. $x + 4y + z = 11,$
   $3x + 4y = 7,$
   $5x + 2y + 2z = 19$

3. $3x + 2y + 2z = 29,$
   $9x + 8y + 9z = 116,$
   $x + 2y + 9z = 86$

4. $7x + 3y + 8z = 144,$
   $y + 2z = 21,$
   $2x + 8y + 7z = 105$

5. $x - 6y + 7z = -39,$
   $3x - 2y = 6,$
   $5x - 9y + 5z = -36$

6. $-5x + 5y + z = 8,$
   $-5x + 3y + z = -8,$
   $-9x + y + 8z = -31$

7. $7y - z = 50,$
   $-3x - 2z = -40,$
   $2x - 3y - z = 8$

8. $-3x + 8y - 9z = 29,$
   $-6x + 6y - 5z = 1,$
   $-2x + 5y - 5z = 18$

9. $-2x - 9y + 5z = 52,$
   $-8x - 4y + 4z = 48,$
   $-5x - 4y + 3z = 37$

10. $9x - 8y - 3z = -43,$
    $8x + y - z = 14,$
    $-8x - y - 6z = -63$

11. $y - 6z = 3,$
    $-x - y + z = -9,$
    $4x + 8y - 4z = 72$

12. $-3x - 7y - z = -43,$
    $6x + 3y - 9z = -24,$
    $-6x - 6y + 6z = 6$

13. $-9x + 8y - 2z = -8,$
    $5x - 4y - 2z = 4,$
    $7x - 6y = 6$

14. $-3y - 7z = -27,$
    $-6x + y - 5z = -45,$
    $-4x + 4y - 6z = 0$

15. $x + 8y + 7z = 34,$
    $6x + 7y + 4z = 49,$
    $2x + 16y + 14z = 68$

16. $18x - 16y + 15z = 158,$
    $-6x + 8y + 9z = 86,$
    $-36x + 32y - 30z = -79$

17. $3x - 14y + 8z = 27,$
    $-19x + 15y + 17z = 211,$
    $-2y + 4z = 32$

18. $-6x + 8y - 5z = -6,$
    $2x - 9y - 8z = -88,$
    $-8x - 8y + 9z = 80$

399

EXTRA PRACTICE 7
Solving Systems of Equations in Three Variables
Use After Sections 3.4 and 3.5

Solve.

19. The sum of 3 numbers is 22. The third is 2 more than twice the second. The first is twice the second. Find the numbers. _____

20. The sum of 3 numbers is 8. The second is 5 more than twice the third. The first is three times the sum of the second and third. Find the numbers. _____

21. The sum of 3 numbers is 12. Twice the first plus the second is twice the third. Three times the second plus the first is 3 times the third. Find the numbers. _____

22. The third of 3 numbers is 27 less than twice the second. The first and the third differ by 2 more than the second number. Twice the first is the same as the second minus the third. Find the numbers. _____

23. The sum of 3 numbers is 8. Twice the first is one more than the second. The third is 5 less than $\frac{1}{2}$ the first. Find the numbers. _____

24. In triangle ABC, angle C is 20° more than angle A. Angle B is twice the sum of angles A and C. Find the angles. _____

25. In triangle ABC, angle B is 2° less than twice angle A. Angle C is 7° more than twice angle A. Find the angles. _____

26. In triangle XYZ, angle Y is 11° less than twice angle Z. Angle Y is 23° less than three times angle X. _____

27. Albert, Beth, and Cathy can fold 460 napkins in an hour. Albert and Cathy can fold 247 in an hour. Cathy and Beth can fold 310 in an hour. How many napkins can each person fold individually in an hour? _____

28. John bought 3 types of donuts; apple, bluebery, and chocolate. He bought one for every person in a class of 30 people. John realized that most people like chocolate, so he bought as many chocolate as the other two types combined. Apple donuts cost 45¢ a piece, blueberry cost 48¢ a piece and chocolate cost 50¢ a piece. John spent a total of $14.46. How many of each type did he buy? _____

EXTRA PRACTICE 8
Solving Inequalities With Both Principles
Use after Section 4.1   Name_____

Examples: Solve.

a)
$$4 - 6x < -10$$
$$-4 + 4 - 6x < -4 + (-10)$$
$$-6x < -14$$
$$-\frac{1}{6} \cdot (-6x) > -\frac{1}{6} \cdot (-14)$$
$$x > \frac{14}{6}$$
$$x > \frac{7}{3}$$

The solution set is $\left\{ x \mid x > \frac{7}{3} \right\}$.

b)
$$7y - 4 \geq 5y - 12$$
$$7y - 4 + 4 \geq 5y - 12 + 4$$
$$7y \geq 5y - 8$$
$$-5y + 7y \geq -5y + 5y - 8$$
$$2y \geq -8$$
$$\frac{1}{2} \cdot 2y \geq \frac{1}{2} \cdot (-8)$$
$$y \geq -4$$

The solution set is $\{y \mid y \geq -4\}$.

Solve.

1. $4x + 7 \leq 23$ _____

2. $3y + 4 \geq 8y - 5$ _____

3. $12y + 5 < 3y - 7$ _____

4. $0.7x < 1.4 - 0.3x$ _____

5. $3x + 20 - 5x \geq 5 + x$ _____

6. $-5y + 3 < -7$ _____

7. $14 - 5x > 2x + 3$ _____

8. $3y - 10 \leq 8y - 5$ _____

9. $\frac{1}{2}y - 4 < 4$ _____

10. $2 - y > y + 2$ _____

EXTRA PRACTICE 8
Solving Inequalities With Both Principles
Use after Section 4.1

11. $7x - 4 \leq 6x + 4$ _____

12. $55 \geq 8y - 1$ _____

13. $-3x + 2 > -25$ _____

14. $y - 8 + \frac{1}{2}y \leq -29$ _____

15. $0.3y > 0.8y - 11$ _____

16. $2x - \frac{2}{3} < \frac{1}{3} - 2x$ _____

17. $8x + 0.9 \leq -0.1$ _____

18. $0.4y - 3 \geq 2$ _____

19. $6y - 4 > 16$ _____

20. $8 - 5x \leq x + 4$ _____

21. $7 - 0.3y \leq 0.3y - 2$ _____

22. $15x - 7 - 18x > 5$ _____

23. $3 \geq 7 - 2x$ _____

24. $\frac{11}{8} - 2x < \frac{3}{8} - 3x$ _____

25. $18 - 5x < 4x - 7$ _____

26. $\frac{1}{2} \geq 2 - \frac{3}{2}y$ _____

27. $29 > 11x - 4$ _____

28. $7 - y \leq y - 7$ _____

29. $-4.3y - 14 < 72$ _____

30. $\frac{3}{5}x - 4 \geq \frac{2}{5}x + 8$ _____

EXTRA PRACTICE 9
Solving Equations and Inequalities With Absolute Value
Use after Section 4.3   Name_____

Examples: Solve.

a) $|3x-5|=16$

    $3x-5=-16$ or $3x-5=16$

    $3x=-11$ or $\quad 3x=21$

    $x=-\dfrac{11}{3}$ or $\quad x=7$

    The solution set is $\left\{-\dfrac{11}{3},7\right\}$.

b) $|3x-5|\leq 16$

    $-16\leq 3x-5\leq 16$

    $-11\leq 3x\leq 21$

    $-\dfrac{11}{3}\leq x\leq 7$

    The solution set is $\left\{x\mid -\dfrac{11}{3}\leq x\leq 7\right\}$.

c) $|3x-5|>16$

    $3x-5<-16$ or $3x-5>16$

    $3x<-11$ or $\quad 3x>21$

    $x<-\dfrac{11}{3}$ or $\quad x>7$

    The solution set is $\left\{x\mid x<-\dfrac{11}{3} \text{ or } x>7\right\}$.

Solve.

1. $|8x-3|>21$ _____

2. $|y-2|\leq 7$ _____

3. $|5x+8|<23$ _____

4. $|9-2x|=5$ _____

5. $|x|=4$ _____

6. $\left|\dfrac{1}{2}y-3\right|\geq 3$ _____

7. $|y+9|\leq 2$ _____

8. $\left|y+\dfrac{1}{3}\right|>\dfrac{4}{3}$ _____

9. $|-4x+3|>13$ _____

10. $\left|\dfrac{5}{8}x\right|<10$ _____

EXTRA PRACTICE 9
Solving Equations and Inequalities With Absolute Value
Use after Section 4.3

---

11. $|10y - 1.3| = 4.7$ _____

12. $|9 - 4x| \geq 15$ _____

13. $|x + 9| > 17$ _____

14. $\left|\frac{3}{4} + x\right| = \frac{1}{4}$ _____

15. $|9 - y| > 11$ _____

16. $|y| \leq \frac{1}{5}$ _____

17. $\left|\frac{3}{7}y\right| > \frac{3}{7}$ _____

18. $|3 - x| = 2$ _____

19. $|5x - 2| \geq 15$ _____

20. $|17 - 4x| < 23$ _____

21. $|y - 3| = 51$ _____

22. $|19 - x| > 19$ _____

23. $|8x - 3| \leq 5$ _____

24. $|2y - 9| < 15$ _____

25. $|y| > 9$ _____

26. $\left|3y - \frac{5}{9}\right| \leq \frac{4}{9}$ _____

27. $|8 - 3y| < 35$ _____

28. $|0.2x + 0.5| \geq 0.9$ _____

29. $\left|x - \frac{2}{9}\right| \geq \frac{4}{9}$ _____

30. $|34 - 4y| \leq 14$ _____

EXTRA PRACTICE 10
Inequalities in Two Variables
Use After Section 4.4   NAME_____

Examples: Graph.
a) $y < 5$ b) $x + y \geq 3$

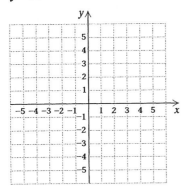 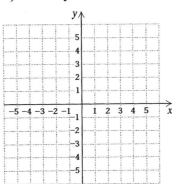

Determine if the given ordered pair is a solution to the inequality.

1. $(3,6)$; $2x + 3y > 8$ _____   2. $(-1,4)$; $2x + 6y < 19$ _____

3. $(3,-8)$; $4x - 8y > 20$ _____   4. $(1,3)$; $6x + 3y > 10$ _____

5. $(1,11)$; $4x - 3y < -29$ _____   6. $(6,7)$; $-3x + 9y > 25$ _____

Graph.
7. $y < 2$ 8. $x > 3$ 9. $y \geq 4$

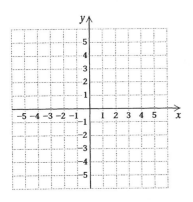 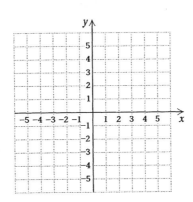 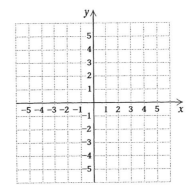

405

EXTRA PRACTICE 10
Inequalities in Two Variables
Use After Section 4.4

10. $x \leq 4$
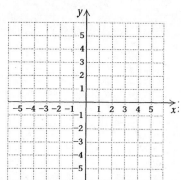

11. $y < x + 3$
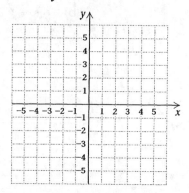

12. $y \leq x - 5$
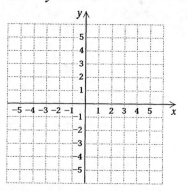

13. $y > 6 + x$
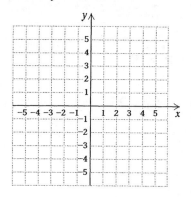

14. $x \leq 3y + 2$
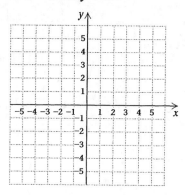

15. $2x + 3y \geq 5$
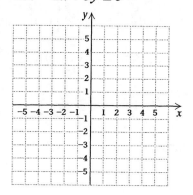

16. $2y - 5x < 13$
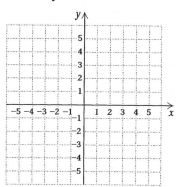

17. $-5 \leq x < 3$
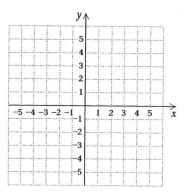

18. $x - 3y \geq 4$
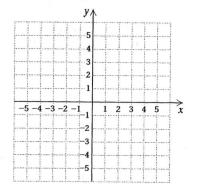

EXTRA PRACTICE 10
Inequalities in Two Variables
Use After Section 4.4

NAME

19. $4x + 3y \geq 7$

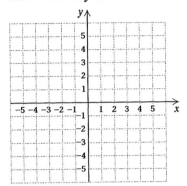

20. $2 < y \leq 4$

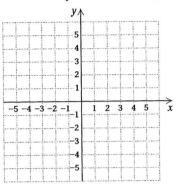

21. $4x + 2 \leq 3y + x$

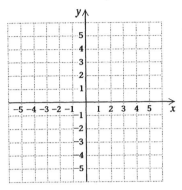

Graph each system.

22. $y \geq 4;$
    $y \leq 5 + x$

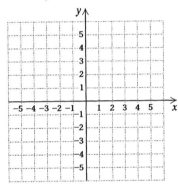

23. $y < x + 2;$
    $x < 3y$

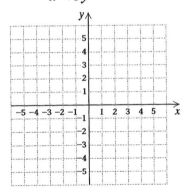

24. $y > 3x + 2;$
    $y < x$

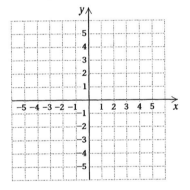

25. $5x + 2y > 3;$
    $2x - 3y \leq 4$

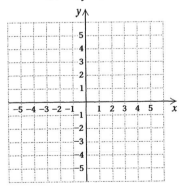

26. $3y \geq 4x;$
    $y > x - 2$

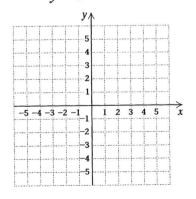

27. $4x + 3 < y;$
    $3y \geq 2$

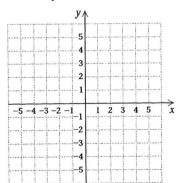

407

EXTRA PRACTICE 10
Inequalities in Two Variables
Use After Section 4.4

28. $6x + 3 \leq y + 2$;
    $x \geq 0$

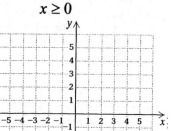

29. $x + y > 0$;
    $3y < x + 3$

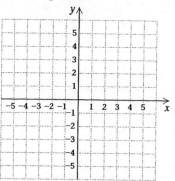

30. $4x + 3y < 5$;
    $5x \geq 3y$

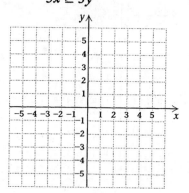

Graph the system. Find coordinates of any vertices formed.

31. $y \geq 2x + 3$;
    $y \leq 5x - 2$;
    $x \geq 3$

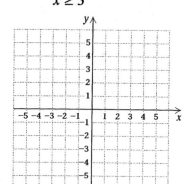

32. $y \leq 9x - 3$;
    $x \leq 5$;
    $x \geq y + 2$

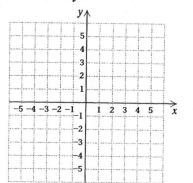

33. $x + 3y \leq 14$;
    $3y + 2x \geq 4$;
    $x + y \geq 4$

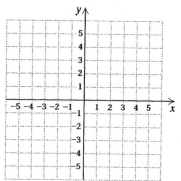

34. $3x + 2y \geq 4$;
    $3x + 2y \leq 7$;
    $0 \leq x \leq 1$

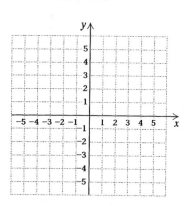

35. $5x + 2y \geq 3$;
    $2y - 5x \leq 4$;
    $x \geq 0$;
    $y \geq 0$

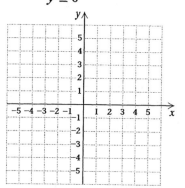

36. $3x + 2y \geq 6$;
    $3x + 2y \leq 10$;
    $x \geq 2$;
    $y \leq 5$

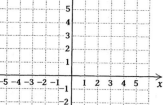

EXTRA PRACTICE 11
Factoring Polynomials
Use after Sections 5.3 - 5.7   Name_____

Examples. Factor completely.
a) $36x^2 - 25 = (6x)^2 - 5^2 = (6x+5)(6x-5)$
b) $8x^2 - 56x + 98 = 2(4x^2 - 28x + 49) = 2(2x-7)^2$
c) $y^3 + 64 = y^3 + 4^3 = (y+4)(y^2 - 4y + 16)$
d) $3x^2 - 10x - 8 = (3x+2)(x-4)$
e) $8x^3 - 27 = (2x)^3 - 3^3 = (2x-3)(4x^2 + 6x + 9)$
f) $x^3 - 5x^2 - 14x = x(x-7)(x+2)$

Factor completely.

1. $125x^3 - 1 =$ _____   2. $w^2 - 64 =$ _____

3. $y^2 - 12y + 36 =$ _____   4. $x^2 - 8x - 48 =$ _____

5. $a^3 - 7a^2 + 12a =$ _____   6. $25a^2 + 8b^2 =$ _____

7. $(x-3)(x+7) + (x-3)(x-4) =$ _____   8. $6x^2 + 12x + 6 =$ _____

9. $y^2 - 11y + 18 =$ _____   10. $40 + 3b - b^2 =$ _____

11. $3x^5 - 12x^2 =$ _____   12. $250x^3 + 2 =$ _____

13. $7xy^4 - 7xz^4 =$ _____   14. $2y^4 + 5y^3 - 12y^2 =$ _____

15. $24x^2 - 7x - 5 =$ _____   16. $y^2 + 14y - 32 =$ _____

17. $0.04w^2 + 0.28w + 0.49 =$ _____   18. $4x^3 + 40x^2 + 64x =$ _____

19. $64y^3 + 27 =$ _____   20. $\frac{1}{81} - x^2 =$ _____

EXTRA PRACTICE 11
Factoring Polynomials
Use after Sections 5.3 - 5.7

21. $5x^2 - 2x + 3 =$ _____
22. $x^3 - 343 =$ _____
23. $40y^2 + 28y - 48 =$ _____
24. $3ab - 5bc + bd =$ _____
25. $8c^6 - 125d^6 =$ _____
26. $81 - 18z + z^2 =$ _____
27. $x^4 + 10x^3 + 25x^2 =$ _____
28. $xz - xw - yz + yw =$ _____
29. $y^2 + 5y - 36 =$ _____
30. $x^2 - 11x - 42 =$ _____
31. $7a^2 - 7b^2 =$ _____
32. $216 - a^3 =$ _____
33. $81 + 18y + y^2 =$ _____
34. $b^2 - 5b - 14 =$ _____
35. $q^4 - 10q^3 + 21q^2 =$ _____
36. $9x^2y^2 - 25y^4 =$ _____
37. $105 + 8x - x^2 =$ _____
38. $x^2 - 3x - 2 =$ _____
39. $6y^3 + 48 =$ _____
40. $a^3 - 14a^2 + 49a =$ _____
41. $3y^2 - 34y - 24 =$ _____
42. $a^2 + 8a + 16 =$ _____
43. $y^2 - 121 =$ _____
44. $42 + a - a^2 =$ _____
45. $9x^3 - 24x^2 + 16x =$ _____
46. $x^3 - \dfrac{1}{8} =$ _____
47. $10w^2 + 29w - 21 =$ _____
48. $16x^2 + 54x - 7 =$ _____
49. $27x^2 - 30x - 8 =$ _____
50. $x^6 - 1 =$ _____
51. $x^2 - 0.6x + 0.09 =$ _____
52. $4x^2 - 13x - 35 =$ _____
53. $125x^6 - 81 =$ _____
54. $49x^3 - 14x^2 + x =$ _____
55. $40y^2 + 7y - 3 =$ _____
56. $15w^2 - 15w - 90 =$ _____
57. $0.04a^2 - 0.49b^2 =$ _____
58. $x^3y^2 + 7x^2y^2 - 18xy^2 =$ _____
59. $2x^6 - 54y^6 =$ _____
60. $\dfrac{1}{4}x^2 - 5x + 25 =$ _____

EXTRA PRACTICE 12
Applications of Polynomial Equations
Use After Section 5.8   NAME_____

Examples: Solve.
a) $x^2 - 2x = 15$
   $x^2 - 2x - 15 = 0$
   $(x-5)(x+3) = 0$
   $x = 5$ or $x = -3$
   $\{-3, 5\}$

b) Let $f(x) = x^2 - 4x - 10$.
   Find a so that $f(a) = 11$.
   $f(a) = a^2 - 4a - 10$. Since $f(a) = 11$,
   $11 = a^2 - 4a - 10$
   $0 = a^2 - 4a - 21$
   $0 = (a-7)(a+3)$
   $a = 7$ or $a = -3$
   $\{-3, 7\}$

Solve.

1. $x^2 - 1 = 0$ _____
2. $x^2 - x - 12 = 0$ _____
3. $x^2 - 5x - 24 = 0$ _____
4. $x^2 + 3x - 10 = 0$ _____
5. $x^2 + 12x + 27 = 0$ _____
6. $x^2 - 2x + 1 = 0$ _____
7. $x^2 + 2x - 8 = 0$ _____
8. $x^2 + 9x + 18 = 0$ _____
9. $x^2 - 9x + 14 = 0$ _____
10. $x^2 + 7x + 10 = 0$ _____
11. $x^2 - 9x - 10 = 0$ _____
12. $x^2 + 2x = 0$ _____
13. $x^2 - 13x + 30 = 0$ _____
14. $x^2 + 11x + 28 = 0$ _____
15. $6x^2 - 43x - 40 = 0$ _____
16. $3x^2 + 17x - 56 = 0$ _____
17. $x^2 - 7x - 8 = 0$ _____
18. $2x^2 + 10x + 12 = 0$ _____
19. $2x^2 + 13x + 18 = 0$ _____
20. $x^2 - 11x + 30 = 0$ _____
21. $4x^2 + 19x + 12 = 0$ _____
22. $3x^2 - 4x - 32 = 0$ _____
23. $x^2 - 5x - 24 = 0$ _____
24. $x^2 - 18x + 81 = 0$ _____

25. Let $f(x) = x^2 + x - 10$. Find a so that $f(a) = 2$. _____
26. Let $f(x) = x^2 - 3x - 6$. Find a so that $f(a) = 4$. _____
27. Let $f(x) = x^2 - 6x + 16$. Find a so that $f(a) = 7$. _____
28. Let $f(x) = x^2 + 12x + 38$. Find a so that $f(a) = 11$. _____
29. Let $f(x) = x^2 + 4x - 4$. Find a so that $f(a) = -4$. _____
30. Let $f(x) = x^2 - 5x - 11$. Find a so that $f(a) = 3$. _____

Find the domain of the function f given by each of the following.

31. $f(x) = \dfrac{10}{x^2 - x - 6}$ _____

32. $f(x) = \dfrac{x+3}{x^2 - x - 12}$ _____

33. $f(x) = \dfrac{2x}{x^2 + 11x + 24}$ _____

34. $f(x) = \dfrac{x-3}{x^2 + x - 20}$ _____

35. $f(x) = \dfrac{4}{x^2 + 5x - 24}$ _____

36. $f(x) = \dfrac{x^2 - 4x - 4}{x^2 + 4x - 12}$ _____

EXTRA PRACTICE 12
Applications of Polynomial Equations
Use After Section 5.8

37. A positive number plus its square is 210. What is the number? _____

38. A number plus its square is 72. What is the number? _____

39. The product of 2 consecutive positive integers is 156. What are the numbers? _____

40. The product of 2 consecutive even positive integers is 224. What are the numbers? _____

41. The length of a cover of a book is 2 in. more than the width. If the area is 99 in.^{2}, what are the dimensions of the cover? _____

42. A rectangular window has a surface area of 6 ft^2. One side is 1 ft longer than the other. What are the lengths of the sides? _____

43. Alex, the carpenter, adds 2 ft of height and 2 ft of width to a freestanding square wall. It now has an area of 64 ft^2. What was the area of the original wall? _____

44. Jenny wants a garden to be 7 ft longer than it is wide and to have an area of 60 sq. ft. What are the dimensions? _____

45. Three consecutive odd integers are such that the square of the third is 88 more than the square of the first. Find the integers. _____

46. Three consecutive integers are such that the square of the third is 47 more than the square of the second. Find the integers. _____

47. The Jones family has a square T.V. set. The diagonal of the T.V. is $14\sqrt{2}$ in. What is the area of the T.V.? _____

48. A room is 10 ft longer than it is wide. It has an area of 875 ft^2. What are the dimensions of the room? _____

EXTRA PRACTICE 13
Multiplying and Dividing Rational Expressions
Use After Section 6.1   NAME_____

Examples:
a) Simplify by removing a factor of 1.

$$\frac{x^2-x-6}{x^2-7x+12} = \frac{(x-3)(x+2)}{(x-3)(x-4)}$$
$$= \frac{x-3}{x-3} \cdot \frac{x+2}{x-4}$$
$$= 1 \cdot \frac{x+2}{x-4}$$
$$= \frac{x+2}{x-4}$$

b) Multiply and simplify.

$$\frac{x+5}{x-3} \cdot \frac{x^2-9}{x^2} = \frac{x+5}{x-3} \cdot \frac{(x-3)(x+3)}{x^2}$$
$$= \frac{x-3}{x-3} \cdot \frac{(x+5)(x+3)}{x^2}$$
$$= 1 \cdot \frac{x^2+8x+15}{x^2}$$
$$= \frac{x^2+8x+15}{x^2}$$

c) Divide and simplify.

$$\frac{x^2-25}{x+2} \div \frac{x+5}{x^2+3x+2} = \frac{x^2-25}{x+2} \cdot \frac{x^2+3x+2}{x+5}$$
$$= \frac{(x+5)(x-5)}{x+2} \cdot \frac{(x+2)(x+1)}{x+5}$$
$$= \frac{(x+5)(x-5)(x+2)(x+1)}{(x+2)(x+5)}$$
$$= (x-5)(x+1)$$
$$= x^2-4x-5$$

Simplify by removing a factor of 1.

1. $\dfrac{8x^3}{x^2}$ _____

2. $\dfrac{14x^2}{7x}$ _____

3. $\dfrac{4x^2+16}{2}$ _____

4. $\dfrac{6x^2+3x}{9x^3}$ _____

5. $\dfrac{13x+12x^2}{3x}$ _____

6. $\dfrac{5x^2}{15x^3+25x}$ _____

7. $\dfrac{x^2+7x+12}{x^2-16}$ _____

8. $\dfrac{x^2-x-12}{x^2-9x+20}$ _____

9. $\dfrac{x^2+8x+15}{x^2+14x+45}$ _____

10. $\dfrac{x^2-4x-12}{x^2-11x+30}$ _____

11. $\dfrac{2x^2+11x-21}{2x^2+3x-9}$ _____

12. $\dfrac{4x^2+23x+15}{2x^2+13x+15}$ _____

413

EXTRA PRACTICE 13
Multiplying and Dividing Rational Expressions
Use After Section 6.1

Multiply and simplify.

13. $\dfrac{4y^3}{5x^2} \cdot \dfrac{3x}{2y}$ _____

14. $\dfrac{4a^3}{3b^4} \cdot \dfrac{9b^6}{2a^2}$ _____

15. $\dfrac{x^2-1}{2x-2} \cdot \dfrac{x-1}{x+1}$ _____

16. $\dfrac{6x^2+13x+6}{2x+3} \cdot \dfrac{x}{3x+2}$ _____

17. $\dfrac{15x^2+29x+12}{x+1} \cdot \dfrac{x^3+x^2}{5x+3}$ _____

18. $\dfrac{12x^2+40x+25}{6x^2+25x+24} \cdot \dfrac{6x^2+31x+40}{6x^2+25x+25}$ _____

19. $\dfrac{x^2-x-72}{6x^2-7x-20} \cdot \dfrac{6x^2-11x-10}{2x^2+7x-72}$ _____

Divide and simplify.

20. $\dfrac{x^2-1}{x-5} \div \dfrac{x+1}{x^2-3x-10}$ _____

21. $\dfrac{x^2+x-6}{2x^2+3x-2} \div \dfrac{x^2+5x+6}{x^2+x-2}$ _____

22. $\dfrac{10x^2+29x+10}{3x^2+23x+30} \div \dfrac{15x^2+31x+10}{3x+5}$ _____

23. $\dfrac{6x^2+13x+6}{4x^2-15x-4} \div \dfrac{3x^2-4x-4}{4x^2-7x-2}$ _____

24. $\dfrac{x^2+3x+2}{x^2-4} \div \dfrac{x^2-1}{x+2}$ _____

25. $\dfrac{x^2-7x+10}{2x^2+3x-5} \div \dfrac{x^2+3x-10}{x^2+4x-5}$ _____

26. $\dfrac{x^2+2x-80}{x+5} \div \dfrac{x^2-13x+40}{x}$ _____

27. $\dfrac{x^2-3x+2}{x^2-7x+12} \div \dfrac{x^2-5x+6}{x^2-9x+20}$ _____

EXTRA PRACTICE 14
Addition And Subtraction of Rational Expressions
Use after Section 6.2   Name_____

Example: Do this calculation.

$$\frac{3}{4x+8} + \frac{5}{x+2} - \frac{2}{x-2}$$

$$= \frac{3}{4(x+2)} + \frac{5}{x+2} - \frac{2}{x-2}, \quad LCD = 4(x+2)(x-2)$$

$$= \frac{3}{4(x+2)} \cdot \frac{x-2}{x-2} + \frac{5}{x+2} \cdot \frac{4(x-2)}{4(x-2)} - \frac{2}{x-2} \cdot \frac{4(x+2)}{4(x+2)}$$

$$= \frac{3(x-2) + 20(x-2) - 8(x+2)}{4(x+2)(x-2)}$$

$$= \frac{3x - 6 + 20x - 40 - 8x - 16}{4(x+2)(x-2)}$$

$$= \frac{15x - 62}{4(x+2)(x-2)}$$

Do these calculations. Simplify by removing a factor of 1 when possible.

1. $\dfrac{3}{x-2} + \dfrac{x}{x^2-4} =$ _____

2. $\dfrac{2x}{5y} + \dfrac{x}{-5y} =$ _____

3. $\dfrac{2x}{x-y} + \dfrac{3}{y-x} =$ _____

4. $\dfrac{ab}{a+b} - \dfrac{b}{3} =$ _____

5. $\dfrac{1}{3x^2 - x - 2} - \dfrac{1}{2x^2 + 3x - 5} =$ _____

6. $\dfrac{5a^2}{a+b} + \dfrac{5b^2}{-b-a} =$ _____

EXTRA PRACTICE 14
Addition And Subtraction of Rational Expressions
Use after Section 6.2

7. $\dfrac{x+y}{x-3} + \dfrac{5}{y} =$ _____

8. $\dfrac{5}{y} + \dfrac{y}{x+y} =$ _____

9. $\dfrac{a-3}{a} + \dfrac{1}{b} =$ _____

10. $\dfrac{x+y}{4x-3} - \dfrac{x-y}{3-4x} =$ _____

11. $\dfrac{4}{2x^2 - 13x - 7} - \dfrac{4}{x^2 + x - 56} =$ _____

12. $\dfrac{2x+1}{x^2 - 4} + \dfrac{x}{5x^2 + 9x - 2} =$ _____

13. $\dfrac{2(3x-1)}{x^2 - 9} + \dfrac{x}{x+3} - \dfrac{4}{x-3} =$ _____

14. $\dfrac{3x}{3x^2 + 2x - 1} + \dfrac{x}{x^2 + 5x + 4} =$ _____

15. $\dfrac{x-y}{xy} + \dfrac{y-z}{yz} - \dfrac{x-z}{xz} =$ _____

16. $\dfrac{a}{a+5} + \dfrac{a}{a-5} - 2 =$ _____

17. $\dfrac{x}{5x-25} - \dfrac{5}{5x-x^2} - \dfrac{2}{x-5} =$ _____

18. $\dfrac{2b}{b^2 - 49} - \dfrac{1}{b-7} + \dfrac{1}{b+7} =$ _____

19. $\dfrac{1}{x^2 - 16} + x - \dfrac{1}{x-4} =$ _____

20. $\dfrac{1}{x^2 - 9} + \dfrac{1}{x^2 - x - 12} - \dfrac{1}{x^2 - 7x + 12} =$ _____

EXTRA PRACTICE 15
Simplifying Complex Rational Expressions
Use after Section 6.3   Name_____

Example: Simplify.

$$\frac{\frac{2}{x}-\frac{2}{y}}{\frac{x^2-y^2}{2xy}} = \frac{\frac{2}{x}-\frac{2}{y}}{\frac{x^2-y^2}{2xy}} \cdot \frac{2xy}{2xy} \quad \text{or} \quad \frac{\frac{2}{x}-\frac{2}{y}}{\frac{x^2-y^2}{2xy}} = \frac{\frac{2}{x}\cdot\frac{y}{y}-\frac{2}{y}\cdot\frac{x}{x}}{\frac{x^2-y^2}{2xy}}$$

$$= \frac{\left(\frac{2}{x}-\frac{2}{y}\right)\cdot 2xy}{\left(\frac{x^2-y^2}{2xy}\right)\cdot 2xy} \qquad\qquad = \frac{\frac{2y-2x}{xy}}{\frac{x^2-y^2}{2xy}}$$

$$= \frac{\frac{2}{x}\cdot 2xy - \frac{2}{y}\cdot 2xy}{\left(\frac{x^2-y^2}{2xy}\right)\cdot 2xy} \qquad = \frac{2(y-x)}{xy}\cdot\frac{2xy}{x^2-y^2}$$

$$= \frac{4y-4x}{x^2-y^2} \qquad\qquad\qquad = \frac{-2(x-y)}{xy}\cdot\frac{2xy}{(x+y)(x-y)}$$

$$= \frac{4(y-x)}{(x+y)(x-y)} = \frac{-4(x-y)}{(x+y)(x-y)} \qquad = \frac{-4xy(x-y)}{xy(x+y)(x-y)}$$

$$= -\frac{4}{x+y} \qquad\qquad\qquad\qquad = -\frac{4}{x+y}$$

Simplify.

1. $\dfrac{1+\dfrac{1}{x}}{x-\dfrac{1}{x}} = $ _____

2. $\dfrac{\dfrac{x}{y}-\dfrac{y}{x}}{\dfrac{1}{y}-\dfrac{1}{x}} = $ _____

3. $\dfrac{\dfrac{2}{a}-\dfrac{1}{b}}{\dfrac{5}{b}-\dfrac{3}{a}} = $ _____

4. $\dfrac{4}{\dfrac{1}{x}+\dfrac{1}{y}} = $ _____

5. $\dfrac{\dfrac{1}{x}-1}{\dfrac{1}{x}-2} = $ _____

6. $\dfrac{a-\dfrac{3a}{b}}{\dfrac{b}{a}-b} = $ _____

417

EXTRA PRACTICE 15
Simplifying Complex Rational Expressions
Use after Section 6.3

7. $\dfrac{x^2-y^2}{\dfrac{1}{x}-\dfrac{1}{y}} =$ _____

8. $\dfrac{\dfrac{x}{4}-2x}{x+\dfrac{1}{x}} =$ _____

9. $\dfrac{\dfrac{3}{a}+\dfrac{3}{c}}{\dfrac{a^2-c^2}{3ac}} =$ _____

10. $\dfrac{1+\dfrac{1}{x}}{1-\dfrac{1}{x^2}} =$ _____

11. $\dfrac{a^2-b^2}{\dfrac{1}{b}-\dfrac{1}{a}} =$ _____

12. $\dfrac{\dfrac{xy}{x^2-y^2}}{\dfrac{y}{x+y}} =$ _____

13. $\dfrac{\dfrac{x^2-36}{x^2-4}}{\dfrac{3x^2-20x+12}{x^2-3x+2}} =$ _____

14. $\dfrac{\dfrac{1}{y^2}-3}{\dfrac{1}{y^2}+9} =$ _____

15. $\dfrac{x-\dfrac{1}{4}}{x-\dfrac{1}{8}} =$ _____

16. $\dfrac{\dfrac{x^2-8x+15}{x^3+8}}{\dfrac{x^2-9}{x^2+5x+6}} =$ _____

EXTRA PRACTICE 16
Solving Rational Equations Including Problem Solving
Use after Sections 6.4 and 6.5   Name_____

Example: Solve.

$$\frac{y-1}{5y} - \frac{y+1}{10y} = \frac{1}{4}$$

The LCD is 20y.

$$20y\left(\frac{y-1}{5y} - \frac{y+1}{10y}\right) = 20y \cdot \frac{1}{4}$$
$$4(y-1) - 2(y+1) = 5y$$
$$4y - 4 - 2y - 2 = 5y$$
$$2y - 6 = 5y$$
$$-6 = 3y$$
$$-2 = y$$

Check:

$$\frac{y-1}{5y} - \frac{y+1}{10y} = \frac{1}{4}$$

$$\frac{-2-1}{5(-2)} - \frac{-2+1}{10(-2)} \Big| \frac{1}{4}$$

$$\frac{-3}{-10} - \frac{-1}{-20}$$

$$\frac{3}{10} - \frac{1}{20}$$

$$\frac{6}{20} - \frac{1}{20}$$

$$\frac{5}{20}$$

$$\frac{1}{4}$$

The solution is -2.

Solve.

1. $5x - \dfrac{2}{x} = -9$ _____

2. $\dfrac{y-7}{y+2} = \dfrac{3}{8}$ _____

3. $\dfrac{6}{3x+2} = \dfrac{4}{3x}$ _____

4. $3x + \dfrac{8}{x} = -14$ _____

5. $\dfrac{9}{x} - \dfrac{2}{x} = 4 - \dfrac{6}{x}$ _____

6. $\dfrac{10}{a} - \dfrac{10}{a+4} = \dfrac{5}{a}$ _____

7. $\dfrac{3}{y} + \dfrac{2}{y+2} = \dfrac{6}{y^2+2y}$ _____

8. $\dfrac{4}{x-3} + \dfrac{3x}{x^2-9} = \dfrac{6}{x+3}$ _____

9. $\dfrac{3}{x+1} + \dfrac{2}{x-2} = \dfrac{21}{x^2-x-2}$ _____

10. $\dfrac{6}{y^2+6y} - \dfrac{2}{y^2-6y} = \dfrac{1}{y^2-36}$ _____

EXTRA PRACTICE 16
Solving Rational Equations Including Problem Solving
Use after Sections 6.4 and 6.5

See Section 6.5 for examples of problem solving involving work and motion problems.

Solve.

11. The sum of a number and 4 times its reciprocal is -5. Find the number. _____

12. One machine takes 15 minutes to do a certain job. Another machine takes 20 minutes to do the same job. Find the time it would take both machines, working together, to do the same job. _____

13. On a river a boat travels 240 miles downstream in the same amount of time that it takes to travel 160 miles upstream. The speed of the boat in still water is 25 mph. Find the speed of the river. _____

14. A water pipe of a certain size can fill a pool in 5 hours. Another pipe can fill the same pool in 2 hours. How long will it take to fill the pool if both pipes are used together? _____

15. A train leaves Bensenville traveling north at 60 mph. Two hours later a second train leaves on a parallel track also traveling north but at 80 mph. How far from Bensenville will the trains be when the second train catches the first? _____

16. Speedy Press can print an order of booklets in 3.5 hr. Blue Star Printers can do the same job in 4.5 hr. How long would it take if both presses are used? _____

17. The wind is blowing 25 mph. An airplane can fly 90 miles upwind in the same amount of time it takes to fly 110 miles downwind. What is the speed of the airplane in still air? _____

18. A student working at a pizza parlor takes 2 times longer to make pizza than her manager does. Working together they can make 36 pizzas each hour. How long would it take the student to make 36 pizzas working by herself? _____

19. Sue can bicycle 8 mph with no wind. She bikes 24 mi against the wind and 24 mi with the wind in a total time of 8 hr. Find the wind speed. _____

20. A new machine takes one-third the time to complete a certain job that an old machine does. Working together they can complete the job in 8 minutes. How long would it take the old machine to complete the job if it were working alone? _____

EXTRA PRACTICE 17
Division of Polynomials
Use after Section 6.6   Name_____

Examples: Divide.

a) $(20x^8 - 12x^5 + 28x^3) \div 4x^3$

$$\frac{20x^8 - 12x^5 + 28x^3}{4x^3}$$

$$= \frac{20x^8}{4x^3} - \frac{12x^5}{4x^3} + \frac{28x^3}{4x^3}$$

$$= 5x^5 - 3x^2 + 7$$

Answer: $5x^5 - 3x^2 + 7$

b) $(x^4 - 3x^2 + x - 8) \div (x - 2)$

$$\begin{array}{r} x^3 + 2x^2 + x + 3 \\ x-2 \overline{\smash{)}x^4 + 0x^3 - 3x^2 + x - 8} \\ \underline{x^4 - 2x^3} \\ 2x^3 - 3x^2 \\ \underline{2x^3 - 4x^2} \\ x^2 + x \\ \underline{x^2 - 2x} \\ 3x - 8 \\ \underline{3x - 6} \\ -2 \end{array}$$

Answer: $x^3 + 2x^2 + x + 3$, R -2, or
$$x^3 + 2x^2 + x + 3 + \frac{-2}{x-2}$$

Divide.

1. $(15x^5 + 12x^4 - 21x^3) \div 3x^2 =$

_____

2. $(15x^2 - 22x - 15) \div (3x - 5) =$

_____

3. $(5x^2 + 44x - 9) \div (5x - 1) =$

_____

4. $(7x^3 + 9x^2 - 3x + 7) \div (x + 1) =$

_____

5. $(10y^3 - 16y^2 + 2y - 4) \div (5y + 2) =$

_____

6. $(36x^2 + 12x^5 - 18x^3) \div 6x^2 =$

_____

7. $(y^4 - 16) \div (y + 2) =$

_____

8. $(a^4 + 2a^2 - 15) \div (a - 3) =$

_____

EXTRA PRACTICE 17
Division of Polynomials
Use after Section 6.6

---

9. $(x^5 - x^4 + 3x^3 - 2) \div (x + 2) =$

10. $(x^3 + 216) \div (x + 6) =$

11. $(7x^4 + 8x^6 - 5x^3) \div x =$

12. $(x^8 - 1) \div (x^2 + 1) =$

13. $(3b^2 - 13b - 10) \div (b - 5) =$

14. $(15x^2 + x - 28) \div (3x - 4) =$

15. $(8a^2 - 2a - 1) \div (2a - 1) =$

16. $(5x^2 - 3x + 2) \div (x - 2) =$

17. $(32y^3 - 4y^2 + 15y - 9) \div (8y + 1) =$

18. $(4x^3 - 9x^2 - 9x) \div (x - 3) =$

19. $(b^3 - 2b^2 + b - 9) \div (b + 2) =$

20. $(10x^3 + 25x^2 - 15x) \div 5x =$

21. $(8x^3 + 11x - 8) \div (x + 3) =$

22. $(x^8 - x^6 + 4x^4 - 24) \div (x^2 - 2) =$

23. $(x^6 + 3x^3 - 5x^2) \div (x + 1) =$

24. $(x^{12} - x^6 + 4) \div (x^6 - 4) =$

EXTRA PRACTICE 18
Synthetic Division
Use After Section 6.7   NAME_____

Examples:
a) $(x^3 - 2x^2 + x - 15) \div (x + 5)$

$$\begin{array}{r|rrrr} -5 & 1 & -2 & 1 & -15 \\ & & -5 & 35 & -180 \\ \hline & 1 & -7 & 36 & | -195 \end{array}$$

The quotient is $x^2 - 7x + 36$.
The remainder is -195.

b) $(x^4 - x^2 + 3) \div x(x - 1)$

$$\begin{array}{r|rrrrr} 1 & 1 & 0 & -1 & 0 & 3 \\ & & 1 & 1 & 0 & 0 \\ \hline & 1 & 1 & 0 & 0 & | 3 \end{array}$$

The quotient is $x^3 + x^2$.
The remainder is 3.

c) Let $f(x) = 4x^5 - 2x^3 + x - 9$.
Use synthetic division to find $f(3)$.

$$\begin{array}{r|rrrrrr} 3 & 4 & 0 & -2 & 0 & 1 & -9 \\ & & 12 & 36 & 102 & 306 & 921 \\ \hline & 4 & 12 & 34 & 102 & 307 & | 912 \end{array}$$

The remainder tells us that $f(x) = 912$.

Use synthetic division to divide.

1. $(x^3 + 3x^2 - 2x + 5) \div (x - 3)$

    _____

2. $(x^3 - x^2 + 4x - 3) \div (x - 2)$

    _____

3. $(x^3 - 10x^2 + x + 120) \div (x - 5)$

    _____

4. $(2x^3 - 7x^2 - 19x + 60) \div (x + 3)$

    _____

5. $(2x^3 + 9x^2 - 53x + 27) \div (x + 8)$

    _____

6. $(x^3 - 3x^2 - 10x + 29) \div (x - 4)$

    _____

7. $(3x^3 + 5x^2 + 2x - 3) \div (x + 6)$

    _____

8. $(2x^3 - 4x + 3)(x - 3)$

    _____

9. $(x^4 - 16) \div (x - 2)$

    _____

10. $(x^3 - 2x^2 + 3) \div (x + 4)$

    _____

11. $(x + 4x^2 + 3x + 4) \div (x - 2)$

    _____

12. $(x^3 + 3x^2 + 2x - 4) \div (x + 1)$

    _____

EXTRA PRACTICE 18
Synthetic Division
Use After Section 6.7

---

13. $(2x^3 - 9x^2 + 9x - 2) \div (x - 7)$

    _____

14. $(x^3 + 6x^2 - 3x + 1) \div (x + 6)$

    _____

15. $(x^3 + 5x^2 - 7x - 4) \div (x - 2)$

    _____

16. $(x^3 - 5x^2 + 4x + 8) \div (x - 7)$

    _____

17. $(x^3 - 3x^2 - 2x - 4) \div (x + 3)$

    _____

18. $(x^3 + 6x + 5) \div (x + 8)$

    _____

19. $(x^3 + 2x^2 + 6x + 3) \div (x - 4)$

    _____

20. $(x^3 - 7x^2 + 3x - 6) \div (x + 5)$

    _____

Use synthetic division to find the indicated function value.

21. $f(x) = 2x^4 + 3x^3 - 2x^2 + x - 1$; $f(5)$ _____

22. $g(x) = 4x^4 - 2x^3 - 6x^2 + 3x + 4$; $g(-3)$ _____

23. $h(x) = 3x^4 + 8x^3 - 3x^2 - 2x + 5$; $h(2)$ _____

24. $f(x) = 4x^4 + 3x^3 + 6x^2 + 2x + 6$; $f(-6)$ _____

25. $f(x) = 2x^4 - 6x^3 - 2x^2 - 6x - 8$; $f(5)$ _____

26. $q(x) = 4x^4 + 4x^3 + 5x^2 + 4x + 5$; $q(3)$ _____

27. $r(x) = 5x^4 - 8x^3 - 3x^2 + 5x - 3$; $r(-4)$ _____

28. $f(x) = 3x^4 - 6x^3 - 9x^2 - 6x + 4$; $f(7)$ _____

29. $f(x) = 6x^4 + 4x^3 + 8x^2 - 6x + 3$; $f(4)$ _____

30. $f(x) = 7x^4 - 5x^3 + 6x^2 + 8x + 6$; $f(8)$ _____

EXTRA PRACTICE 19
Radical Expressions and Rational Numbers as Exponents
Use after Sections 7.1 and 7.2 Name_____

Examples: a) Given $f(x) = \sqrt{3x-16}$, find $f(7)$ and $f(3)$.
$$f(7) = \sqrt{3 \cdot 7 - 16} = \sqrt{21-16} = \sqrt{5}$$
$$f(3) = \sqrt{3 \cdot 3 - 16} = \sqrt{9-16} = \sqrt{-7}$$
Since $\sqrt{-7}$ is not a real number, we say $\sqrt{-7}$ does not exist.

b) Rewrite $(a^3 b^2)^{\frac{1}{4}}$ without fractional exponents.
$$(a^3 b^2)^{\frac{1}{4}} = \sqrt[4]{a^3 b^2}.$$

c) Simplify $x^{\frac{2}{3}} \cdot x^{\frac{1}{4}} = x^{\frac{8}{12}} \cdot x^{\frac{3}{12}} = x^{\frac{11}{12}} = \sqrt[12]{x^{11}}$.

For each function, find the function value if it exists.

1. $f(x) = \sqrt{3x+5}$; $f(2), f(6), f(9), f(-2)$ _____

2. $f(x) = \sqrt{6x-10}$; $f(4), f(-6), f(5), f(3)$ _____

3. $f(x) = \sqrt{4x+8}$; $f(2), f(-2), f(3), f(-3)$ _____

4. $f(x) = \sqrt[3]{5x+6}$; $f(2), f(-6), f(3), f(8)$ _____

5. $f(x) = \sqrt[3]{6x-4}$; $f(-1), f(3), f(4), f(-6)$ _____

6. $g(x) = \sqrt[4]{2x+3}$; $g(0), g(-2), g(1), g(2)$ _____

Simplify. Assume that variables can represent any real number.

7. $\sqrt{(-5)^2}$ _____ 8. $\sqrt{16x^2}$ _____ 9. $\sqrt{81x^3}$ _____

10. $\sqrt[3]{343x^3}$ _____ 11. $\sqrt[3]{64z^3}$ _____ 12. $\sqrt[3]{(-27)}$ _____

13. $\sqrt[7]{(-z)^7}$ _____ 14. $\sqrt{4x^2+12x+9}$ _____ 15. $\sqrt{(2x+3)^2}$ _____

16. $\sqrt[13]{(-6x)^{13}}$ _____ 17. $\sqrt[3]{\dfrac{216}{343}}$ _____ 18. $\sqrt[3]{-\dfrac{125}{729}}$ _____

19. $\sqrt[7]{\dfrac{128}{2187}}$ _____ 20. $\sqrt[5]{(4ab)^5}$ _____

EXTRA PRACTICE 19
Radical Expressions and Rational Numbers as Exponents
Use after Sections 7.1 and 7.2

Find the domain of each of the following functions.

21. $f(x) = \sqrt{x-3}$ _____
22. $f(x) = \sqrt{2x+5}$ _____
23. $f(x) = \sqrt{x-9}$ _____
24. $f(x) = \sqrt[3]{3x-5}$ _____
25. $f(x) = 3 + \sqrt[4]{x-5}$ _____
26. $f(x) = 2 + \sqrt{2x+3}$ _____

Assume for all exercises that even roots are of nonnegative quantities and that all denominators are nonzero.

Rewrite without fractional exponents.

27. $y^{1/6}$ _____
28. $4^{1/2}$ _____
29. $32^{1/5}$ _____
30. $16^{3/2}$ _____
31. $(xy)^{1/6}$ _____
32. $(169x^3)^{1/2}$ _____
33. $64^{5/4}$ _____
34. $(16xy^2z)^{1/3}$ _____
35. $(625)^{3/4}$ _____

Rewrite with fractional exponents.

36. $\sqrt[3]{16}$ _____
37. $\sqrt[5]{x^2}$ _____
38. $\sqrt[3]{x^2 y}$ _____
39. $\left(\sqrt{4xy}\right)^3$ _____
40. $\sqrt[4]{10ab^3}$ _____
41. $\left(\sqrt[5]{abc^2}\right)^3$ _____
42. $\sqrt[4]{x^2 y^2}$ _____
43. $\sqrt{3x^2 y^3}$ _____
44. $\sqrt[5]{x^3 y^2 z^5}$ _____

Simplify. Do not use negative exponents.

45. $y^{-3/4}$ _____
46. $(10x^2 y)^{-3/2}$ _____
47. $\left(\dfrac{1}{64}\right)^{-1/2}$ _____
48. $\left(\dfrac{5a}{6bc}\right)^{-2/3}$ _____
49. $\dfrac{x^{-3/2}}{x^{-2/5}}$ _____
50. $x^{3/2} \cdot x^{2/5}$ _____
51. $\left(x^{-5/2} \cdot y^{2/3}\right)^{1/3}$ _____
52. $\left(x^{5/6}\right)^{1/3}$ _____
53. $\left(a^{3/5} \cdot b^{1/2}\right)^{1/4}$ _____
54. $6^{3/5} \cdot 6^{1/5}$ _____
55. $4^{2/5} \cdot 4^{-3/2}$ _____

EXTRA PRACTICE 20
Multiplying, Dividing, and Simplifying Radical Expressions
Use after Sections 7.3 and 7.4   Name_____

Examples. Simplify. Assume that all variables represent positive numbers.

a) $\sqrt[3]{320x^6y^4z^2}$

$= \sqrt[3]{64 \cdot 5 \cdot x^6 \cdot y^3 \cdot y \cdot z^2}$

$= \sqrt[3]{64x^6y^3}\sqrt[3]{5yz^2}$

$= 4x^2y\sqrt[3]{5yz^2}$

b) $\sqrt[4]{(81a^8b^4)^2}$

$= \left(\sqrt[4]{3^4a^8b^4}\right)^2$

$= (3a^2b)^2$

$= 9a^4b^2$

c) $\sqrt{\dfrac{75y^5}{16x^2}}$

$= \dfrac{\sqrt{75y^5}}{\sqrt{16x^2}}$

$= \dfrac{\sqrt{25y^4 \cdot 3y}}{\sqrt{16x^2}}$

$= \dfrac{5y^2\sqrt{3y}}{4x}$

Simplify. Assume that all variables represent positive numbers.

1. $\sqrt{20x^3yz^2} =$ _____

2. $\sqrt[3]{128x^4y^2} =$ _____

3. $\sqrt[4]{a^{16}b^{12}} =$ _____

4. $\sqrt{\dfrac{49a^3}{b^4}} =$ _____

5. $\sqrt{45a^3bc^2} =$ _____

6. $\sqrt{16^3} =$ _____

7. $\sqrt[3]{\dfrac{16x^5}{y^6}} =$ _____

8. $\sqrt[4]{64a^7b^{12}} =$ _____

9. $\sqrt{50a^2b^5} =$ _____

10. $\sqrt[5]{(32x^{10})^3} =$ _____

11. $\sqrt{\dfrac{16x^3}{81}} =$ _____

12. $\sqrt{500x^2yz^{11}} =$ _____

13. $\sqrt[3]{216^2} =$ _____

14. $\sqrt[3]{\dfrac{64a^7}{27}} =$ _____

15. $\sqrt[3]{240x^4y^5} =$ _____

16. $\sqrt[4]{x^7y^9z^{12}} =$ _____

17. $\sqrt{\dfrac{24x^3}{25}} =$ _____

18. $\sqrt[4]{256^3} =$ _____

19. $\sqrt[5]{(32a^5b^{10})^3} =$ _____

20. $\sqrt[3]{(54a^3)^2} =$ _____

EXTRA PRACTICE 20
Multiplying, Dividing, and Simplifying Radical Expressions
Use after Sections 7.3 and 7.4

Examples. Assume that all variables represent positive numbers.

a) Multiply and simplify.

$$\sqrt{32xy^3}\sqrt{4x^2y^5}$$
$$= \sqrt{128x^3y^8}$$
$$= \sqrt{64 \cdot 2 \cdot x^2 \cdot x \cdot y^8}$$
$$= \sqrt{64x^2y^8}\sqrt{2x}$$
$$= 8xy^4\sqrt{2x}$$

b) Divide and simplify.

$$\frac{\sqrt[3]{56a^5b^{14}}}{\sqrt[3]{7ab^5}}$$
$$= \sqrt[3]{\frac{56a^5b^{14}}{7ab^5}}$$
$$= \sqrt[3]{8a^4b^9}$$
$$= \sqrt[3]{8 \cdot a^3 \cdot a \cdot b^9} = 2ab^3\sqrt[3]{a}$$

Multiply or divide and simplify. Assume that all variables represent positive numbers.

21. $\sqrt[3]{5(x+2)^2}\sqrt[3]{25(x+2)^2} =$ _____

22. $\dfrac{\sqrt{32a^5b^3}}{\sqrt{2ab^2}} =$ _____

23. $\dfrac{6\sqrt{45x^3}}{3\sqrt{5x}} =$ _____

24. $\sqrt[3]{x^7}\sqrt[3]{64xy^2} =$ _____

25. $\sqrt{8x^3y}\sqrt{3xy^2} =$ _____

26. $\dfrac{\sqrt[3]{81a^5b^8}}{\sqrt[3]{3ab^2}} =$ _____

27. $\dfrac{\sqrt[3]{625x^6y^4}}{\sqrt[3]{5xy}} =$ _____

28. $\sqrt{6(x+3)^3}\sqrt{3(x+3)} =$ _____

29. $\sqrt[3]{6^5a^2b}\sqrt[3]{6^2ab} =$ _____

30. $\dfrac{\sqrt[3]{27xy^7}}{\sqrt[3]{xy}} =$ _____

31. $\dfrac{9\sqrt[5]{160x^8y^{11}}}{3\sqrt[5]{5xy^2}} =$ _____

32. $\sqrt[3]{4(y-3)^2}\sqrt[3]{2(y-3)^5} =$ _____

EXTRA PRACTICE 21
Solving Radical Equations
Use after Section 7.6

Name_____

Example: Solve.
$\sqrt{x+19} - \sqrt{x-20} = 3$

$\sqrt{x+19} = \sqrt{x-20} + 3$
$(\sqrt{x+19})^2 = (\sqrt{x-20}+3)^2$
$x + 19 = x - 20 + 6\sqrt{x-20} + 9$
$30 = 6\sqrt{x-20}$
$5 = \sqrt{x-20}$
$5^2 = (\sqrt{x-20})^2$
$25 = x - 20$
$45 = x$

The solution is 45.

Check:
$\sqrt{x+19} - \sqrt{x-20} = 3$
$\sqrt{45+19} - \sqrt{45-20}$ | 3
$\sqrt{64} - \sqrt{25}$
$8 - 5$
3

Solve.

1. $x + 2 = \sqrt{7x+2}$ _____

2. $\sqrt{x} - 3 = 3$ _____

3. $\sqrt{x+9} + \sqrt{x+2} = 7$ _____

4. $y - 5 = \sqrt{y-3}$ _____

5. $\sqrt{-3x+4} = 2 - x$ _____

6. $1 - x = \sqrt{-5x+1}$ _____

7. $\sqrt{a} + 2 = 5$ _____

8. $\sqrt{x-5} + \sqrt{x+6} = 11$ _____

EXTRA PRACTICE 21
Solving Radical Equations
Use after Section 7.6

9. $\sqrt[3]{x-1} - 3 = 0$ _____

10. $\sqrt{y+3} - \sqrt{2y-8} = 1$ _____

11. $\sqrt{x+12} - \sqrt{x-12} = 12$ _____

12. $\sqrt{x+4} - \sqrt{2x+9} = -1$ _____

13. $\sqrt{x+7} + \sqrt{x-4} = 11$ _____

14. $5 - \sqrt{x} = 1$ _____

15. $\sqrt[3]{4x+3} + 2 = 5$ _____

16. $\sqrt{x+9} + \sqrt{x+4} = 5$ _____

17. $\sqrt{x+10} + \sqrt{x} = 3$ _____

18. $\sqrt{5x+3} = \sqrt{3x+7}$ _____

19. $\sqrt{7x+8} - \sqrt{41-2x} = 3$ _____

20. $\sqrt{10-2x} - \sqrt{5x+16} = 3$ _____

EXTRA PRACTICE 22
Geometric Applications
Use after Section 7.7   Name_____

See Section 7.7 for examples.

Find the length of the missing side. Give an exact answer.

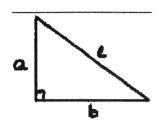

1. $a = 3$, $b = 4$ _____   2. $a = 5$, $b = 12$ _____
3. $a = 15$, $c = 17$ _____   4. $b = 12$, $c = 18$ _____
5. $a = 5$, $c = 5\sqrt{2}$ _____ 6. $a = 9$, $c = 17$ _____
7. $b = 3$, $c = 4$ _____   8. $a = 10$, $b = 10$ _____
9. $a = 8$, $c = 14$ _____   10. $b = 6$, $c = 12$ _____

Give an exact answer.

11. A wire reaches from the top of a 20-ft pole to a point 7 feet from its base. How long is the wire? _____

12. John is in one corner of a 12 ft x 15 ft room. His brother is in the opposite corner. John throws a ball to his brother. How far does he have to throw it? _____

13. Jim starts at post A in a corn field. He heads directly south 10 feet, then east 12 feet, to post C. He then turns south 18 feet from C and east another 4 feet, where he reaches post B. What is the shortest distance between posts A and B if he must pass C on the way?
_____

14. A 19" T.V. set (measured on the diagonal) has a width of 8 inches. What is its height?
_____

For each of the following, find the missing length(s). Give exact answers. (Drawings *not* to scale.)

15. 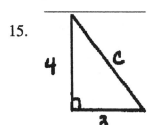 _____   16. 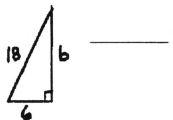 _____

431

EXTRA PRACTICE 22
Geometric Applications
Use after Section 7.7

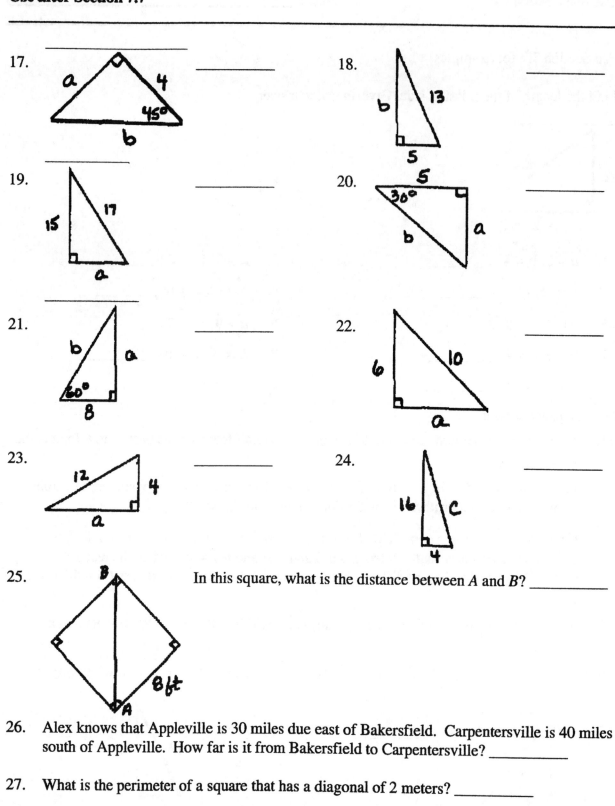

25. In this square, what is the distance between A and B? _____

26. Alex knows that Appleville is 30 miles due east of Bakersfield. Carpentersville is 40 miles south of Appleville. How far is it from Bakersfield to Carpentersville? _____

27. What is the perimeter of a square that has a diagonal of 2 meters? _____

28. A rectangle has one side 2 ft longer than another. The diagonal is 10 ft. What is the perimeter? _____

EXTRA PRACTICE 23
Solving Quadratic Equations Using the Quadratic Formula
Use after Section 8.2   Name_____

Example: Solve $5x^2 - 8x + 2 = 0$ using the quadratic formula.
$5x^2 - 8x + 2 = 0$
$a = 5 \quad b = -8 \quad c = 2$

$x = \dfrac{-(-8) \pm \sqrt{(-8)^2 - 4(5)(2)}}{2(5)}$

$\left[\begin{array}{l}\text{Quadratic Formula:}\\ x = \dfrac{-b \pm \sqrt{b^2 - 4ac}}{2a}\end{array}\right]$

$= \dfrac{8 \pm \sqrt{64 - 40}}{10} = \dfrac{8 \pm \sqrt{24}}{10}$

$= \dfrac{8 \pm 2\sqrt{6}}{10} = \dfrac{2(4 \pm \sqrt{6})}{2 \cdot 5} = \dfrac{4 \pm \sqrt{6}}{5}$

Solve.

1. $5x^2 + 4x + 8 = 0$ _____   2. $x^2 - 4x + 6 = 0$ _____

3. $3x^2 + x = 5$ _____   4. $x^2 - 7x + 5 = 0$ _____

5. $x^2 - 3x + 4 = 0$ _____   6. $5x^2 - 2 = x$ _____

7. $3x^2 - 7x + 3 = 0$ _____   8. $x^2 - 3x + 9 = 0$ _____

9. $4 + \dfrac{2}{x^2} = \dfrac{1}{x}$ _____   10. $2x^2 - 3x + 4 = 0$ _____

11. $x^2 - 2x - 2 = 0$ _____   12. $5 = x^2 - 6x$ _____

EXTRA PRACTICE 23
Solving Quadratic Equations Using the Quadratic Formula
Use after Section 8.2

13. $5 - \dfrac{1}{x} + \dfrac{3}{x^2} = 0$ _____ 14. $x^2 - 6x + 12 = 0$ _____

15. $x^2 - 3x = 6$ _____ 16. $8x^2 - 5 = -4x$ _____

17. $3 + \dfrac{2}{x} + \dfrac{5}{x^2} = 0$ _____ 18. $x^2 + 12 = 10x$ _____

19. $x^2 + 4 = 6x$ _____ 20. $6x^2 - x - 1 = 0$ _____

21. $10x^2 - 4x - 1 = 0$ _____ 22. $7 - \dfrac{2}{x} = \dfrac{2}{x^2}$ _____

23. $x^2 - 5x + 2 = 0$ _____ 24. $3x^2 + 8x = 10$ _____

25. $4x^2 = 3x + 2$ _____ 26. $x^2 + 9x + 5 = 0$ _____

27. $x^2 - 8x + 3 = 0$ _____ 28. $2x^2 = 8x + 2$ _____

29. $7x^2 - 2x + 3 = 0$ _____ 30. $15x^2 - 2x - 10 = 0$ _____

EXTRA PRACTICE 24
Solving Problems Using Quadratic Equations
Use after Sections 8.1 and 8.3   Name_____

Example: $3600 was invested at interest rate r, compounded annually. In 2 years, it grew to $3969. What was the interest rate?

We use the compound-interest formula, $A = P(1+r)^t$, to solve this problem.
$A = \$3969$, $P = \$3600$, $t = 2$, and we solve for r.

$$A = P(1+r)^t$$

$$3969 = 3600(1+r)^2$$

$$\frac{3969}{3600} = (1+r)^2$$

$$\pm\sqrt{\frac{3969}{3600}} = 1+r$$

$$\pm\frac{63}{60} = 1+r$$

$$-\frac{60}{63} \pm \frac{63}{60} = r$$

$$\frac{3}{60} = \frac{1}{20} = r \quad \text{or} \quad -\frac{123}{60} = -\frac{41}{20} = r$$

Since the interest rate cannot be negative, we need only check $\frac{1}{20}$, or 5%. 5% does check, so the interest rate was 5%.

Solve. Use a calculator and approximate answers to the nearest tenth of a percent or tenth of a second.

1. $2000 was invested at interest rate r, compounded annually. In 2 years, it grew to $2333. What was the interest rate? _____

2. The formula $s = 16t^2$ is used to approximate the distance s, in feet, that an object falls freely from rest in t seconds. Use the formula to find how long it would take an object to fall freely from the top of the 984 ft tall Eiffel Tower. _____

3. $5500 was invested at interest rate r, compounded annually. In 2 years, it grew to $6180. What was the interest rate? _____

4. Use the formula $s = 16t^2$ to find the approximate time t, that an object falls freely from rest from a height of 1325 ft. _____

5. $7000 was invested at interest rate r, compounded annually. In 2 years, it grew to $7836. What was the interest rate? _____

EXTRA PRACTICE 24
Solving Problems Using Quadratic Equations
Use after Sections 8.1 and 8.3

See Section 8.3 for examples of problem solving involving work and motion problems.

6. A boat travels 40 miles upstream and then turns around and travels 40 miles downstream. The total time for both trips is 6 hours. If the stream flows at 5 mph, how fast does the boat travel in still water? _____

7. It takes Jim 15 hours longer to build a wall than it does Corey. If they work together, they can build the wall in 18 hours. How long would it take Corey to build the wall alone?
_____

8. Jose's motorcycle traveled 270 mi at a certain speed. Had he gone 15 mph faster, the trip would have taken 3 hr less. Find the speed of the motorcycle. _____

9. Gary and Marsha work together to type a short story, and it takes them 6 hr. It would take Marsha 5 hr more than Gary to type the story alone. How long would each need to type the story if they worked alone? _____

10. Karen's Honda travels 432 mi at a certain speed. If the car had gone 6 mph slower, the trip would have taken 1 hr more. Find Karen's speed. _____

11. It takes Danielle 2 hours longer to deliver the papers than it does Stan. If they work together it takes them 1 hour. How long would it take Danielle to deliver the papers alone? Round the answer to the nearest tenth of an hour. _____

12. A boat travels 16 miles upstream and then turns around and travels 16 miles downstream. The total time for both trips is 4 hours. If the stream flows at 2 mph, how fast does the boat travel in still water? Round the answer to the nearest tenth. _____

EXTRA PRACTICE 25
Solving Equations Reducible to Quadratic
Use after Section 8.5   Name_____

Example. Solve: $(1+3\sqrt{x})^2 - 11(1+3\sqrt{x}) + 28 = 0$

Let $u = 1+3\sqrt{x}$ and substitute u for $1+3\sqrt{x}$.

$u^2 - 11u + 28 = 0$

$(u-7)(u-4) = 0$

$u - 7 = 0$ or $u - 4 = 0$

$u = 7$ $u = 4$

Substitute $1+3\sqrt{x}$ for u and solve for x.

$1+3\sqrt{x} = 7$ or $1+3\sqrt{x} = 4$

$3\sqrt{x} = 6$ or $3\sqrt{x} = 3$

$\sqrt{x} = 2$ or $\sqrt{x} = 1$

$x = 4$ or $x = 1$

Both values check. The solutions are 4 and 1.

Solve.

1. $a - 6\sqrt{a} - 27 = 0$ _____

2. $x^4 - 8x^2 + 12 = 0$ _____

3. $5x^{-2} - 5x^{-1} - 60 = 0$ _____

4. $(3x-1)^2 + 2(3x-1) - 15 = 0$ _____

5. $a - 10\sqrt{a} + 9 = 0$ _____

6. $(5-\sqrt{x})^2 + 5(5-\sqrt{x}) - 24 = 0$ _____

EXTRA PRACTICE 25
Solving Equations Reducible to Quadratic
Use after Section 8.5

---

7. $x^4 - 6x^2 + 8 = 0$ _____

8. $x - 13\sqrt{x} + 36 = 0$ _____

9. $(y^2 - 2y)^2 - 11(y^2 - 2y) + 24 = 0$ _____

10. $x^4 + 4x^2 - 21 = 0$ _____

11. $(x^2 - 5x)^2 - 2(x^2 - 5x) - 24 = 0$ _____

12. $a - 12\sqrt{a} + 20 = 0$ _____

13. $(4x + 2)^2 - 10(4x + 2) + 25 = 0$ _____

14. $(\sqrt{x} - 7)^2 - 13(\sqrt{x} - 7) + 40 = 0$ _____

15. $x^4 - 7x^2 + 12 = 0$ _____

16. $2y^{-2} + 7y^{-1} - 15 = 0$ _____

EXTRA PRACTICE 26
Graphing Quadratic Functions
Use after Section 8.6   Name_____

Example. Graph: $f(x) = -x^2 + 4x - 3$

$f(x) = -(x^2 - 4x) - 3$
$ = -[x^2 - 4x + (4 - 4)] - 3$
$ = -(x^2 - 4x + 4) + 4 - 3$
$ = -(x - 2)^2 + 1$

x	$f(x)$
2	1
1	0
3	0
0	-3
4	-3

Line of symmetry: $x = 2$
Vertex: $(2, 1)$

Graph.

1. $f(x) = 3x^2$

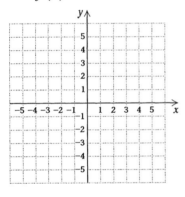

2. $f(x) = (x - 1)^2$

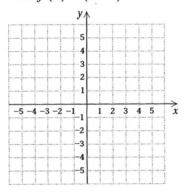

3. $f(x) = (x - 2)^2 + 3$

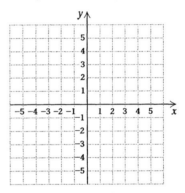

4. $f(x) = 2(x - 3)^2 + 1$

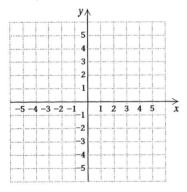

5. $f(x) = x^2 - 6x + 7$

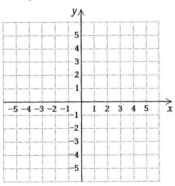

6. $f(x) = -4x^2$

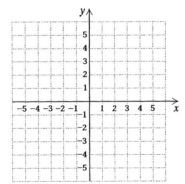

EXTRA PRACTICE 26
Graphing Quadratic Functions
Use after Section 8.6

7. $f(x) = x^2 + 4x + 2$

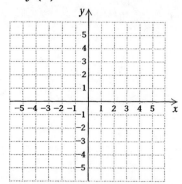

8. $f(x) = 3(x-1)^2$

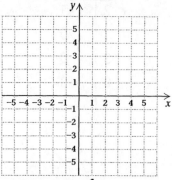

9. $f(x) = -2x^2 - 20x - 47$

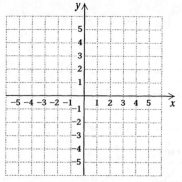

10. $f(x) = (x+3)^2$

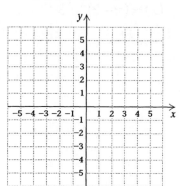

11. $f(x) = -\dfrac{1}{2}x^2$

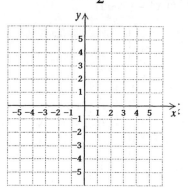

12. $f(x) = (x-1)^2 - 2$

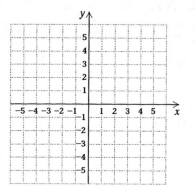

13. $f(x) = 2x^2 - 16x + 29$

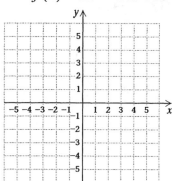

14. $f(x) = 2(x-1)^2$

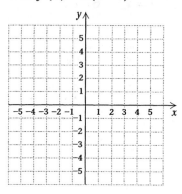

15. $f(x) = (x+3)^2$

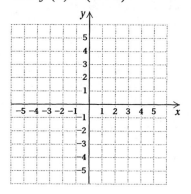

EXTRA PRACTICE 26
Graphing Quadratic Functions
Use after Section 8.6

Name

16. $f(x) = 1.5x^2$

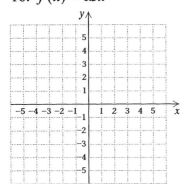

17. $f(x) = -2(x+3)^2 + 4$

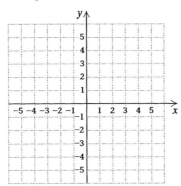

18. $f(x) = x^2 - 2x + 3$

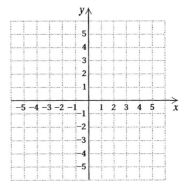

19. $f(x) = (x+1)^2$

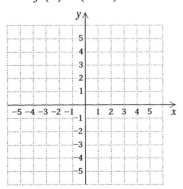

20. $f(x) = (x+2)^2 - 3$

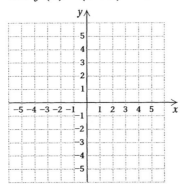

21. $f(x) = -4x^2 + 24x - 35$

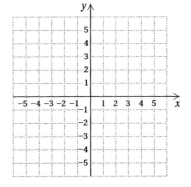

22. $f(x) = -\dfrac{1}{2}(x+1)^2 + 4$

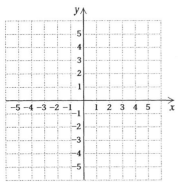

23. $f(x) = -2(x-1)^2$

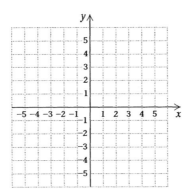

24. $f(x) = -4x^2$

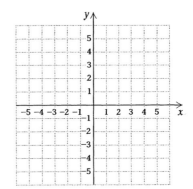

441

EXTRA PRACTICE 26
Graphing Quadratic Functions
Use after Section 8.6

25. $f(x) = x^2 + 6x + 7$

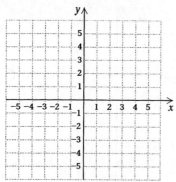

26. $f(x) = -(x+4)^2$

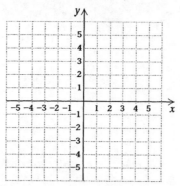

27. $f(x) = 4x^2 - 4x + 1$

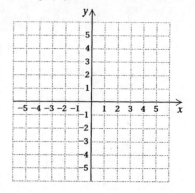

28. $f(x) = \frac{1}{3}(x+3)^2$

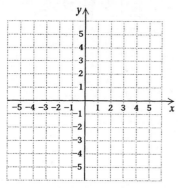

29. $f(x) = -x^2 - 4x - 9$

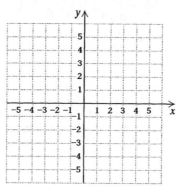

30. $f(x) = 2x^2 - 4x - 3$

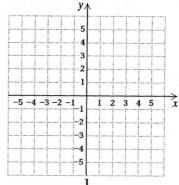

31. $f(x) = (x+2)^2 - 1$

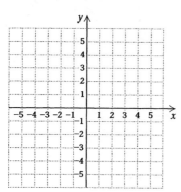

32. $f(x) = -4(x-1)^2 + 1$

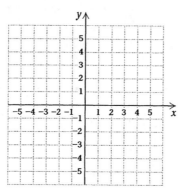

33. $f(x) = \frac{1}{4}(x-4)^2$

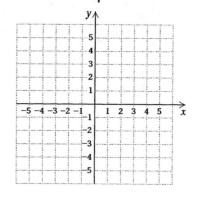

EXTRA PRACTICE 27
Polynomial and Rational Inequalities
Use After Section 8.9

Name _____

See Section 8.10 for Examples.

Solve.

1. $(x-6)(x+10) < 0$ _____

2. $(x-8)(x+3) > 0$ _____

3. $(x+5)(x-2) \geq 0$ _____

4. $(x-3)(x+7) \leq 0$ _____

5. $x^2 + x - 12 > 0$ _____

6. $x^2 - x - 20 > 0$ _____

7. $x^2 - x - 6 \leq 0$ _____

8. $x^2 + 3x - 54 < 0$ _____

9. $(x+1)(x+6)(x-9) > 0$ _____

10. $(x-5)(x+7)(x-10) < 0$ _____

11. $(x+5)(x-2)(x+3) \leq 0$ _____

12. $(x+2)(x-6)(x+11) > 0$ _____

13. $(x-6)(x+3)(x-8) > 0$ _____

14. $(x-7)(x+12)(x+1) \geq 0$ _____

15. $(x-10)(x+4)(x-4) \leq 0$ _____

16. $(x+8)(x+9)(x-8) > 0$ _____

EXTRA PRACTICE 27
Polynomial and Rational Inequalities
Use After Section 8.9

Solve.

17. $\dfrac{4}{x-8} > 0$ _____

18. $\dfrac{2}{x+3} < 0$ _____

19. $\dfrac{x}{x+6} \geq 0$ _____

20. $\dfrac{x+1}{x-9} \leq 0$ _____

21. $\dfrac{x-4}{x+5} < 0$ _____

22. $\dfrac{x-6}{x+1} > 0$ _____

23. $\dfrac{x+3}{x-8} < 0$ _____

24. $\dfrac{x+7}{x-4} < 0$ _____

25. $\dfrac{2x+1}{x+4} \geq 0$ _____

26. $\dfrac{x-4}{x-3} > 0$ _____

27. $\dfrac{3x+2}{x-4} < 0$ _____

28. $\dfrac{x+2}{x+1} > 0$ _____

29. $\dfrac{(x+1)(x+2)}{x-3} > 0$ _____

30. $\dfrac{(x+3)(x-4)}{x+5} \geq 0$ _____

EXTRA PRACTICE 28
Solving Exponential and Logarithmic Equations
Use after Section 9.6   Name_____

Examples. Solve.

a) $7^{x-1} = 343$
$7^{x-1} = 7^3$
$x - 1 = 3$
$x = 4$

b) $6^x = 15$
$\log 6^x = \log 15$
$x \log 6 = \log 15$
$x = \dfrac{\log 15}{\log 6}$
$x \approx \dfrac{1.1761}{0.7782}$
$x \approx 1.5113$

c) $e^{-3t} = 0.04$
$\ln e^{-3t} = \ln 0.04$
$-3t \ln e = \ln 0.04$
$-3t = \ln 0.04$
$t = \dfrac{\ln 0.04}{-3}$
$t \approx \dfrac{-3.2189}{-3}$
$t \approx 1.073$

Solve.

1. $3^{5x} = 81$ _____

2. $e^{4t} = 120$ _____

3. $4^x = 6$ _____

4. $6^x = 2$ _____

5. $e^{-2t} = 0.6$ _____

6. $5^{3x+2} = 625$ _____

7. $8^{x+1} = 16$ _____

8. $10^x = 7$ _____

9. $7^x = 1520$ _____

10. $e^{0.04t} = 10$ _____

11. $e^{5t} = 5$ _____

12. $6^x = 7.1$ _____

13. $6^{x+3} = 36$ _____

14. $4^{x-1} = 3$ _____

15. $12^{2x-3} = 16$ _____

16. $10^{5-x} = 1000$ _____

EXTRA PRACTICE 28
Solving Exponential and Logarithmic Equations
Use after Section 9.6

Example. Solve: $\log_2(x+1) - \log_2(x-1) = 4$

$\log_2(x+1) - \log_2(x-1) = 4$

$\log_2 \dfrac{x+1}{x-1} = 4$

$\dfrac{x+1}{x-1} = 16$

$x + 1 = 16x - 16$

$17 = 15x$

$\dfrac{17}{15} = x$

The solution is $\dfrac{17}{15}$.

Check:
$\log_2(x+1) - \log_2(x-1) = 4$

$\log_2\left(\dfrac{17}{15}+1\right) - \log_2\left(\dfrac{17}{15}-1\right) \mid 4$

$\log_2 \dfrac{32}{15} - \log_2 \dfrac{2}{15}$

$\log_2\left(\dfrac{32}{15} \div \dfrac{2}{15}\right)$

$\log_2 16$

4

Solve.

17. $\log x + \log(x+15) = 2$

18. $\log(x+2) - \log x = 3$

19. $\log_3(2x-7) = 4$

20. $\log_5(x-11) = 2$

21. $\log x + \log(x-21) = 2$

22. $\log_2(x-2) + \log_2(x+2) = 5$

23. $\log(3x+4) = 1$

24. $\log(x+33) - \log x = 2$

25. $\log x - \log(x+5) = -1$

26. $\log_4(x+3) - \log_4 x = 3$

27. $\log_4(x-6) + \log_4(x+6) = 3$

28. $\log_6 x + \log_6(x-9) = 2$

29. $\log x + \log(x - 0.21) = -2$

30. $\log(x-48) + \log x = 2$

31. $\log_7 x + \log_7(4x+21) = 3$

32. $\log_2(5-x) = 4$

EXTRA PRACTICE 29
Solving Nonlinear Systems of Equations
Use after Section 10.4   Name_____

Example. Solve: $4y^2 = 16 - 3x^2$,
$\qquad\qquad\qquad 2x^2 = y^2 + 7$.

$3x^2 + 4y^2 = 16 \rightarrow \quad 3x^2 + 4y^2 = 16$
$2x^2 - y^2 = 7 \rightarrow \quad \underline{8x^2 - 4y^2 = 28}$ (Multiplying by 4)
$\qquad\qquad\qquad\qquad 11x^2 = 44$ (Adding)
$\qquad\qquad\qquad\qquad\quad x^2 = 4$
$\qquad\qquad\qquad\qquad\quad\ x = \pm 2$

If $x = 2$, $x^2 = 4$, and if $x = -2$, $x^2 = 4$, so substituting 2 or -2 in $2x^2 = y^2 + 7$, we have
$2 \cdot 4 = y^2 + 7$
$\quad 1 = y^2$
$\pm 1 = y$

The solutions are $(2,1)$, $(-2,1)$, $(2,-1)$, and $(-2,-1)$.

Solve.

1. $x^2 + y^2 = 20$,
   $xy = 8$

2. $x^2 + y^2 = 49$,
   $x^2 - y^2 = 49$

3. $x^2 - 25 = -y^2$,
   $y - x = 1$

4. $x^2 + y^2 = 82$,
   $xy = -9$

5. $x^2 + 4y^2 = 16$,
   $2y = 4 - x$

6. $x^2 + y^2 = 41$,
   $5x - 4y = 0$

7. $25x^2 + y^2 = 100$,
   $10x + 2y = 20$

8. $x^2 = 36 - y^2$,
   $x^2 = 36 + y^2$

9. $y^2 - 3x^2 = 25$,
   $3x^2 + y^2 = 25$

10. $x^2 + y^2 = 34$,
    $y - x = 2$

11. $x^2 - y = 8$,
    $x^2 + y^2 = 20$

12. $x^2 + y^2 = 64$,
    $y^2 = x + 8$

EXTRA PRACTICE 29
Solving Nonlinear Systems of Equations
Use after Section 10.4

13. A rectangle has perimeter 170 cm, and the length of a diagonal is 65 cm. What are its dimensions? _____

14. The area of a rectangle is $12\sqrt{2}$ m². The length of a diagonal is $\sqrt{34}$ m. Find the dimensions. _____

15. The product of two numbers is -44. The sum of their squares is 137. Find the numbers. _____

16. The sum of the squares of two positive numbers is 89. Their difference is 3. What are the numbers? _____

17. The sum of the squares of two positive integers is 58. Their difference is 4. What are the integers? _____

18. The perimeter of a rectangle is 44 m and the area is 105 m². What are the dimensions of the rectangle? _____

19. The product of two numbers is $\dfrac{1}{6}$. The sum of their squares is $\dfrac{13}{36}$. Find the numbers. _____

20. The area of a rectangle is 0.48 cm². The length of a diagonal is 1.0 cm. Find the dimensions of the rectangle. _____

EXTRA PRACTICE 30
The Binomial Theorem
Use after Section 11.4   Name_____

Examples:
a) $5! = 5 \cdot 4 \cdot 3 \cdot 2 \cdot 1 = 120$

b) $\binom{7}{3} = \dfrac{7!}{(7-3)!\,3!} = \dfrac{7 \cdot 6 \cdot 5 \cdot 4 \cdot 3 \cdot 2 \cdot 1}{4 \cdot 3 \cdot 2 \cdot 1 \cdot 3 \cdot 2 \cdot 1} = 35$

c) Expand: $(2x - 5y)^4$.

Using the binomial theorem $(a+b)^n = \binom{n}{0}a^n + \binom{n}{1}a^{n-1}b + \binom{n}{2}a^{n-2}b^2 + \cdots + \binom{n}{n}b^n$.

$a = 2x,\ b = -5y,\ n = 4$

$= \binom{4}{0}(2x)^4 + \binom{4}{1}(2x)^3(-5y) + \binom{4}{2}(2x)^2(-5y)^2 + \binom{4}{3}(2x)(-5y)^3 + \binom{4}{4}(-5y)^4$

$= 1(16x^4) + 4(8x^3)(-5y) + 6(4x^2)(25y^2) + 4(2x)(-125y^3) + 1(625y^4)$

$= 16x^4 - 160x^3y + 600x^2y^2 - 1000xy^3 + 625y^4$

Simplify.

1. $7!$ _____

2. $9!$ _____

3. $8!$ _____

4. $\dfrac{5!}{3!}$ _____

5. $\binom{4}{3}$ _____

6. $\binom{9}{6}$ _____

7. $\binom{12}{8}$ _____

8. $\binom{25}{25}$ _____

9. $\binom{12}{2}$ _____

10. $\binom{18}{1}$ _____

EXTRA PRACTICE 30
The Binomial Theorem
Use after Section 11.4

Expand each of the following.

11. $(x+y)^5$ _____

12. $(x-y)^6$ _____

13. $(2x+y)^3$ _____

14. $(3x-2y)^3$ _____

15. $(x-3y)^5$ _____

16. $(2x+y)^6$ _____

17. $(x+y)^8$ _____

18. $(3x-2y)^4$ _____

19. $(-x+y)^3$ _____

20. $(2x-y)^5$ _____

21. $\left(x+\dfrac{1}{y}\right)^5$ _____

22. $\left(2x-\dfrac{5}{y}\right)^3$ _____

Find the indicated term of the binomial expansion.

23. 5th, $(x+y)^7$ _____

24. 6th, $(2x-3y)^9$ _____

25. 8th, $(2x+3y^2)^{10}$ _____

26. 7th, $(3x+2y)^8$ _____

ANSWER KEYS FOR EXTRA PRACTICE SHEETS

Extra Practice 1
1. -32 2. -6.2 3. $-\dfrac{1}{4}$ 4. -24.6 5. $-\dfrac{23}{20}$ 6. -32 7. $-\dfrac{4}{15}$ 8. -34 9. 13.22
10. -12 11. 0 12. 18 13. -1.2 14. $-\dfrac{1}{5}$ 15. 0 16. $-\dfrac{1}{6}$ 17. -86 18. 2.09 19. 71
20. $\dfrac{3}{5}$ 21. 0 22. -20.4 23. -57 24. $-\dfrac{7}{8}$ 25. $\dfrac{1}{18}$ 26. 9.7 27. $\dfrac{23}{35}$ 28. 0 29. $-\dfrac{1}{4}$
30. 84 31. $-\dfrac{39}{40}$ 32. -26 33. -20.87 34. -5.115 35. -1 36. $-\dfrac{2}{45}$ 37. 7
38. 0.17 39. -29 40. $\dfrac{1}{12}$ 41. -72 42. 0.097 43. -3 44. 19 45. -19 46. 3
47. 0 48. 38 49. -38 50. 0 51. 18 52. -107 53. 58 54. -40 55. $-\dfrac{13}{10}$ 56. 5.2
57. 89 58. -13.7 59. 0 60. -50 61. 44 62. -21.1 63. $\dfrac{23}{24}$ 64. -55
65. $9\dfrac{1}{3}$, or $\dfrac{28}{3}$ 66. 4.6 67. -24.03 68. -33 69. $-\dfrac{2}{3}$ 70. $-\dfrac{4}{15}$ 71. -22.5
72. -24 73. -134 74. $-\dfrac{1}{5}$ 75. 4.82 76. -7.67 77. 31 78. $-\dfrac{1}{42}$ 79. 85
80. $-\dfrac{25}{12}$ 81. -53 82. $\dfrac{33}{38}$ 83. -2.4 84. $\dfrac{5}{4}$

Extra Practice 2
1. 6 2. $15°$ 3. 11 in.   4. $5, 7$ 5. -2 6. -10 7. 2 8. $18, 19, 20, 21$ 9. 8 10. 36 in., 12 in.   11. 16 ft by 26 ft   12. $\$595$ 13. 7 14. $35°, 45°, 100°$ 15. $18, 20, 22$ 16. 3.5 hr

Extra Practice 3
1. 3^6 2. 4^7 3. 9^5 4. f^9 5. r^8 6. $-e^8$ 7. $-8r^3z^5$ 8. $10x^3y$ 9. $12n^5$ 10. $18x^6$
11. $-12a^9b^3$ 12. $-40x^4y^9$ 13. $12n^2xy^3$ 14. $-5x^5y^5$ 15. b^2 16. $\dfrac{4x}{3}$ 17. $5x^2$
18. $\dfrac{13n^3}{7}$ 19. $\dfrac{3n}{2}$ 20. $2y$ 21. $-\dfrac{5a^6}{6}$ 22. $-3n^4$ 23. $\dfrac{11x^2}{9}$ 24. $-7x^2z$ 25. $6a^2b^5c^5$
26. $\dfrac{16}{5}a^2b^2c^5$ 27. 1 28. -1024 29. -125 30. $\dfrac{1}{216}$ 31. $\dfrac{1}{16}$ 32. -512 33. $\dfrac{-1}{216}$
34. 6561 35. 729 36. $\dfrac{1}{1024}$ 37. $\dfrac{1}{2187}$ 38. 1 39. $\dfrac{1}{a^3}$ 40. $\dfrac{1}{5x^4y^2}$ 41. $\dfrac{1}{x^4y^2}$
42. $\dfrac{1}{64x^3y^4}$ 43. $\dfrac{1}{16c^{10}x^4}$ 44. $\dfrac{1}{2x^2y}$ 45. $\dfrac{1}{d^4f^{12}}$ 46. $\dfrac{y^2z}{x^2}$ 47. $\dfrac{3f^3}{n^7}$ 48. $\dfrac{1}{m^{12}y^{18}}$

Extra Practice 3 (continued)

49. $\dfrac{5m^4c^2y^2}{x^3}$ 50. 2^{-4} 51. $3\cdot 2^{-2}$ 52. $(-3)^{-3}$ 53. $\dfrac{5}{x^2}$ 54. $\dfrac{1}{n^{-5}}$ 55. y^{-2} 56. $-3(2y)^{-3}$

57. $(-2x)^{-2}$ 58. $\dfrac{1}{5}$ 59. 6^6 60. $\dfrac{1}{x}$ 61. $12x^4y^3$ 62. $8b^4c$ 63. $5xy^5z$ 64. $\dfrac{3y^2}{2x}$

65. $\dfrac{4bc^3}{3a^2m^3}$ 66. $\dfrac{3}{64x^2y^2}$ 67. $\dfrac{z^3}{2x^2y}$ 68. $4a^8b^{12}$ 69. $\dfrac{1}{36x^6y^4}$ 70. $8x^9y^6$ 71. 4^9

72. $\dfrac{-z}{4x^2y^2}$ 73. $\dfrac{d^5}{-2a^6b^3}$ 74. $\dfrac{64y^{21}}{27x^3}$ 75. $\dfrac{4x^4z^4}{y^2}$ 76. $\dfrac{64}{27a^3b^3c^9}$ 77. 4^6x^{12} 78. 4^x

79. $\dfrac{1}{5^6x^{18}y^{12}}$ 80. $1000a^6c^3x^3$

Extra Practice 4

1.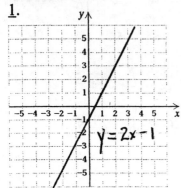
2.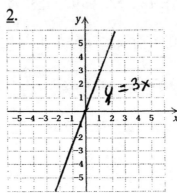
3.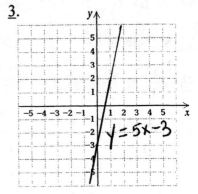
4.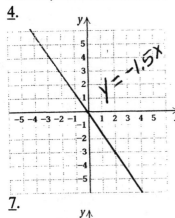
5.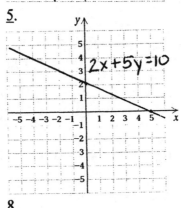
6.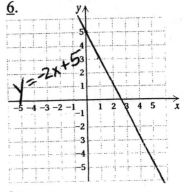
7.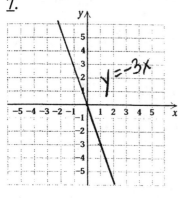
8.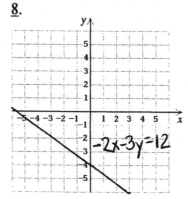
9.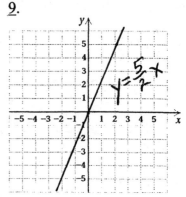

Extra Practice 4 (continued)

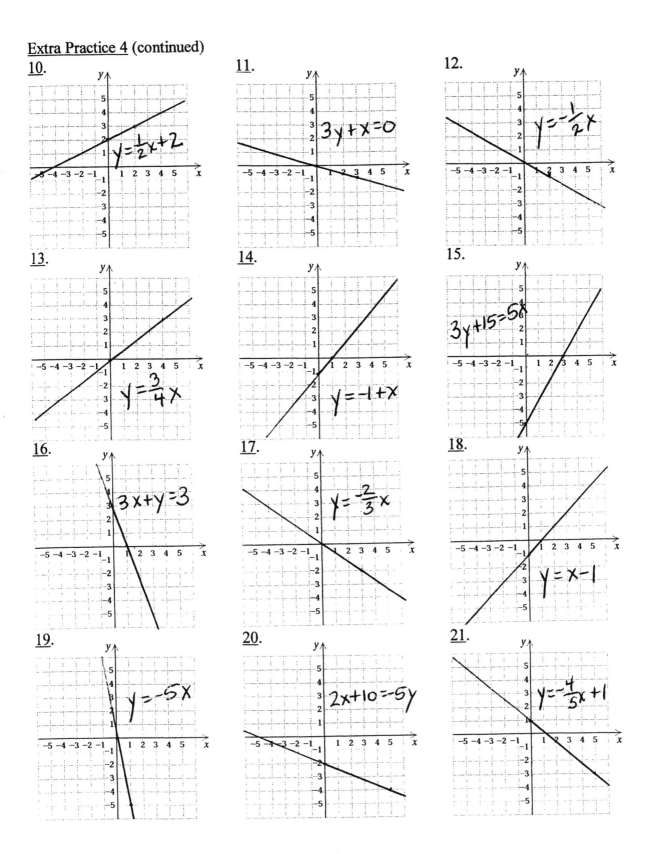

Extra Practice 4 (continued)

22.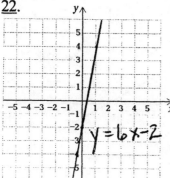
23.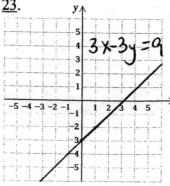
24.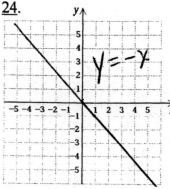
25.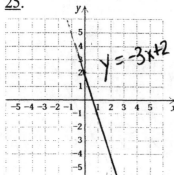
26.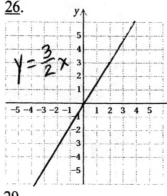
27.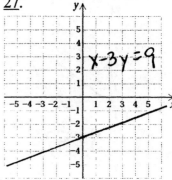
28.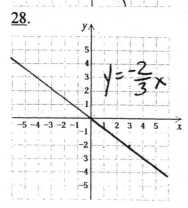
29.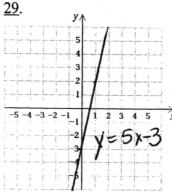
30.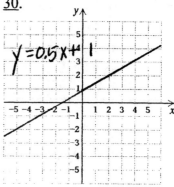

Extra Practice 5
1. $(-2,3)$ 2. $(5,4)$ 3. $(1,3)$ 4. $(5,-1)$ 5. $(-3,4)$ 6. $(2,2)$ 7. $(-1,-3)$ 8. $(4,-5)$
9. $(3,3)$ 10. $(6,-2)$ 11. $(-1,-1)$ 12. $(2,-6)$ 13. $(8,1)$ 14. $(-4,-4)$ 15. $(2,1)$
16. $(-5,3)$ 17. $\left(\frac{11}{3},-\frac{5}{6}\right)$ 18. $(2,4)$ 19. $\left(\frac{1}{4},\frac{1}{3}\right)$ 20. $(18,24)$

Extra Practice 6
1. 7, −18   2. $8500   3. 18 nickels, 24 dimes   4. 7   5. 21, 35   6. Length is 29 ft; width is 13 ft   7. 321 adult's tickets; 227 child's tickets   8. 540 mi   9. 39, −18   10. Length is 39 ft; width is 16 ft   11. 40 gallons of 20%, 80 gallons of 65%   12. $8500 at 11%; $14,500 at 8%   13. 25°, 65°   14. 20 mph

Extra Practice 7
1. (1,3,–2) 2. (1,1,6) 3. (3,1,9) 4. (9,3,9) 5. (8,9,1) 6. (7,8,3) 7. (14,7,–1)
8. (6,7,1) 9. (–3,–4,2) 10. (2,5,7) 11. (1,9,1) 12. No solution 13. Dependent
14. (9,9,0) 15. Dependent 16. No solution 17. (1,4,10) 18. (–3,2,8) 19. (8,4,10)
20. (6,3,–1) 21. (3,4,5) 22. (8,11,–5) 23. (4,7,–3) 24. (20°,120°,40°) 25. (35°,68°,77°)
26. (38°,91°,51°) 27. Albert: 150; Beth: 213; Cathy: 97 28. Apple: 8; blueberry: 7; chocolate: 15

Extra Practice 8
1. $\{x|x \leq 4\}$ 2. $\{y|y \leq \frac{9}{5}\}$ 3. $\{y|y < -\frac{4}{3}\}$ 4. $\{x|x < 1.4\}$ 5. $\{x|x \leq 5\}$ 6. $\{y|y > 2\}$
7. $\{x|x < \frac{11}{7}\}$ 8. $\{y|y \geq -1\}$ 9. $\{y|y < 16\}$ 10. $\{y|y < 0\}$ 11. $\{x|x \leq 8\}$ 12. $\{y|y \leq 7\}$
13. $\{x|x < 9\}$ 14. $\{y|y \leq -14\}$ 15. $\{y|y < 22\}$ 16. $\{x|x < \frac{1}{4}\}$ 17. $\{x|x \leq -\frac{1}{8}\}$
18. $\{x|x \geq 12.5\}$ 19. $\{y|x > \frac{10}{3}\}$ 20. $\{x|x \geq \frac{2}{3}\}$ 21. $\{y|x \geq 15\}$ 22. $\{x|x < -4\}$
23. $\{x|x \geq 2\}$ 24. $\{x|x < -1\}$ 25. $\{x|x > \frac{25}{9}\}$ 26. $\{y|y \geq 1\}$ 27. $\{x|x < 3\}$ 28. $\{y|y \geq 7\}$
29. $\{y|y > -20\}$ 30. $\{x|x \geq 60\}$

Extra Practice 9
1. $\{x|x < -\frac{9}{4} \text{ or } x > 3\}$ 2. $\{y|-5 \leq y \leq 9\}$ 3. $\{x|-\frac{31}{5} < x < 3\}$ 4. $\{2,7\}$ 5. $\{-4,4\}$
6. $\{y|y \leq 0 \text{ or } y \geq 12\}$ 7. $\{y|-11 \leq y \leq -7\}$ 8. $\{y|y < -\frac{5}{3} \text{ or } y > 1\}$
9. $\{x|x < -\frac{5}{2} \text{ or } x > 4\}$ 10. $\{x|-16 < x < 16\}$ 11. $\{-0.34, 0.6\}$ 12. $\{x|x \leq -\frac{3}{2} \text{ or } x \geq 6\}$
13. $\{x|x < -26 \text{ or } x > 8\}$ 14. $\{-1, -\frac{1}{2}\}$ 15. $\{y|y < -2 \text{ or } y > 20\}$ 16. $\{y|-\frac{1}{5} \leq y \leq \frac{1}{5}\}$
17. $\{y|y < -1 \text{ or } y > 1\}$ 18. $\{1,5\}$ 19. $\{x|x \leq -\frac{13}{5} \text{ or } x \geq \frac{17}{5}\}$ 20. $\{x|-\frac{3}{2} < x < 10\}$
21. $\{-48, 54\}$ 22. $\{x|x < 0 \text{ or } x > 38\}$ 23. $\{x|-\frac{1}{4} \leq x \leq 1\}$ 24. $\{y|-3 < y < 12\}$
25. $\{y|y < -9 \text{ or } y > 9\}$ 26. $\{y|\frac{1}{27} \leq y \leq \frac{1}{3}\}$ 27. $\{y|-9 < y < \frac{43}{3}\}$ 28. $\{x|x \leq -7 \text{ or } x \geq 2\}$

Extra Practice 9 (continued)
29. $\left\{x \mid x \leq -\dfrac{2}{9} \text{ or } x \geq \dfrac{2}{3}\right\}$ 30. $\{y \mid 5 \leq y \leq 12\}$

Extra Practice 10
1. Yes   2. No   3. Yes   4. Yes   5. No   6. Yes

7. 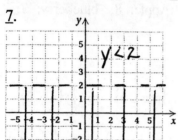 $y < 2$

8. 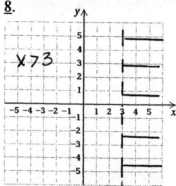 $x > 3$

9. 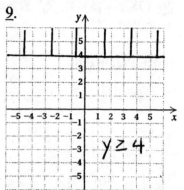 $y \geq 4$

10. 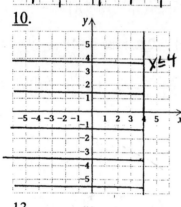 $x \leq 4$

11. 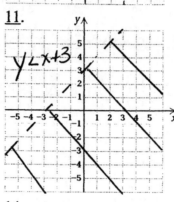 $y < x+3$

12. 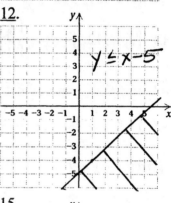 $y \leq x-5$

13. 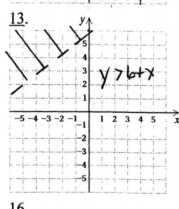 $y > 6+x$

14. 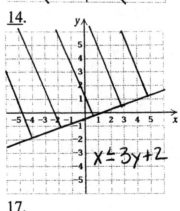 $x \leq 3y+2$

15. 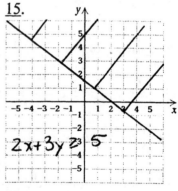 $2x+3y \geq 5$

16. 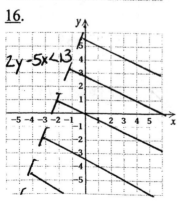 $2y-5x < 3$

17. 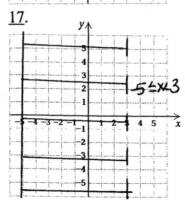 $-5 \leq x \leq 3$

18. 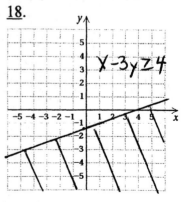 $x-3y \geq 4$

Extra Practice 10 (continued)

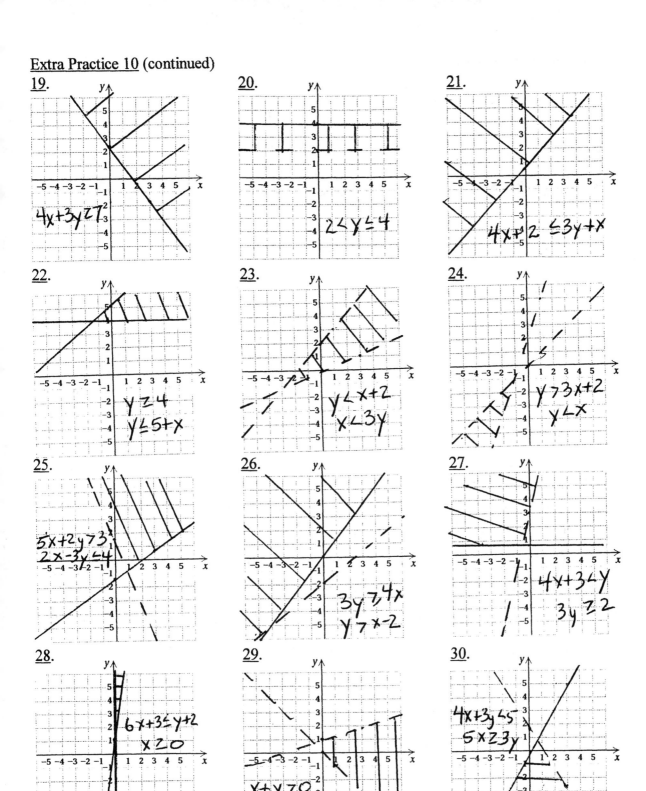

Extra Practice 10 (continued)

31.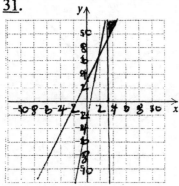
32.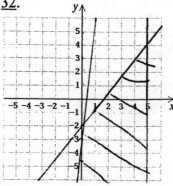
33.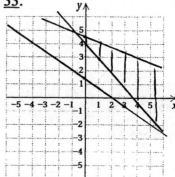
34.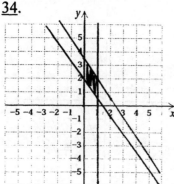
35.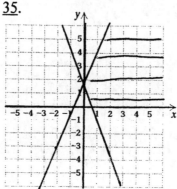
36.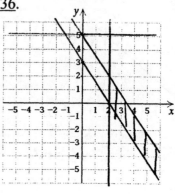

Extra Practice 11

1. $(5x-1)(25x^2+5x+1)$ 2. $(w+8)(w-8)$ 3. $(y-6)^2$ 4. $(x-12)(x+4)$
5. $a(a-4)(a-3)$ 6. Not factorable   7. $(x-3)(2x+3)$ 8. $6(x+1)^2$ 9. $(y-9)(y-2)$
10. $(8-b)(5+b)$ 11. $3x^2(x^3-4)$ 12. $2(5x+1)(25x^2-5x+1)$
13. $7x(y^2+z^2)(y+z)(y-z)$ 14. $y^2(2y-3)(y+4)$ 15. $(8x-5)(3x+1)$
16. $(y-2)(y+16)$ 17. $(0.2w+0.7)^2$ 18. $4x(x+2)(x+8)$ 19. $(4y+3)(16y^2-12y+9)$
20. $\left(\frac{1}{9}+x\right)\left(\frac{1}{9}-x\right)$ 21. Not factorable   22. $(x-7)(x^2+7x+49)$ 23. $4(2y+3)(5y-4)$
24. $b(3a-5c+d)$ 25. $(2c^2-5d^2)(4c^4+10c^2d^2+25d^4)$ 26. $(9-z)^2$ 27. $x^2(x+5)^2$
28. $(x-y)(z-w)$ 29. $(y-4)(y+9)$ 30. $(x-14)(x+3)$ 31. $7(a+b)(a-b)$
32. $(6-a)(36+6a+a^2)$ 33. $(9+y)^2$ 34. $(b-7)(b+2)$ 35. $q^2(q-3)(q-7)$

36. $y^2(3x+5y)(3x-5y)$ 37. $(7+x)(15-x)$ 38. Not factorable   39. $6(y+2)(y^2-2y+4)$
40. $a(a-7)^2$ 41. $(y-12)(3y+2)$ 42. $(a+4)^2$ 43. $(y+11)(y-11)$ 44. $(7-a)(6+a)$
45. $x(3x-4)^2$ 46. $\left(x-\frac{1}{2}\right)\left(x^2+\frac{1}{2}x+\frac{1}{4}\right)$ 47. $(5w-3)(2w+7)$ 48. $(2x+7)(8x-1)$
49. $(9x+2)(3x-4)$ 50. $(x+1)(x-1)(x^2-x+1)(x^2+x+1)$ 51. $(x-0.3)^2$
52. $(x-5)(4x+7)$ 53. Not factorable   54. $x(7x-1)^2$ 55. $(8y+3)(5y-1)$

Extra Practice 11 (continued)
56. $15(w+2)(w-3)$ 57. $(0.2a+0.7b)(0.2a-0.7b)$ 58. $xy^2(x-2)(x+9)$
59. $2(x^2-3y^2)(x^4+3x^2y^2+9y^4)$ 60. $\left(\frac{1}{2}x-5\right)^2$

Extra Practice 12
1. $\{-1,1\}$ 2. $\{-3,4\}$ 3. $\{-3,8\}$ 4. $\{-5,2\}$ 5. $\{-9,-3\}$ 6. $\{1\}$ 7. $\{-4,2\}$
8. $\{-6,-3\}$ 9. $\{2,7\}$ 10. $\{-5,-2\}$ 11. $\{-1,10\}$ 12. $\{-2,0\}$ 13. $\{3,10\}$ 14. $\{-7,-4\}$
15. $\left\{-\frac{5}{6},8\right\}$ 16. $\left\{-8,\frac{7}{3}\right\}$ 17. $\{-1,8\}$ 18. $\{-3,-2\}$ 19. $\left\{-\frac{9}{2},-2\right\}$ 20. $\{5,6\}$
21. $\left\{-4,-\frac{3}{4}\right\}$ 22. $\left\{-\frac{8}{3},4\right\}$ 23. $\{-3,8\}$ 24. $\{9\}$ 25. $\{-4,3\}$ 26. $\{-2,5\}$ 27. $\{3\}$
28. $\{-9,3\}$ 29. $\{-4,0\}$ 30. $\{-2,7\}$ 31. $\{x|x \in R \text{ and } x \neq -2 \text{ and } x \neq 3\}$
32. $\{x|x \in R \text{ and } x \neq -3 \text{ and } x \neq 4\}$ 33. $\{x|x \in R \text{ and } x \neq -8 \text{ and } x \neq -3\}$
34. $\{x|x \in R \text{ and } x \neq -5 \text{ and } x \neq 4\}$ 35. $\{x|x \in R \text{ and } x \neq -8 \text{ and } x \neq 3\}$
36. $\{x|x \in R \text{ and } x \neq -6 \text{ and } x \neq 2\}$ 37. 14   38. -9 or 8 39. 12 and 13   40. 14 and 16
41. 9 in. by 11 in.   42. 2ft, 3 ft   43. 36 ft^2 44. 5 ft x 12 ft   45. 9, 11, 13   46. (22,23,24)
47. 196 in.^{2} 48. 25 ft by 35 ft

Extra Practice 13
1. $8x$ 2. $2x$ 3. $2x^2+8$ 4. $\dfrac{2x+1}{3x^2}$ 5. $\dfrac{13+12x}{3}$ 6. $\dfrac{x}{3x^2+5}$ 7. $\dfrac{x+3}{x-4}$ 8. $\dfrac{x+3}{x-5}$
9. $\dfrac{x+3}{x+9}$ 10. $\dfrac{x+2}{x-5}$ 11. $\dfrac{x+7}{x+3}$ 12. $\dfrac{4x+3}{2x+3}$ 13. $\dfrac{6y^2}{5x}$ 14. $6ab^2$ 15. $\dfrac{x-1}{2}$ 16. x
17. $x^2(3x+4)$ 18. $\dfrac{(2x+5)(6x+5)}{(2x+3)(3x+5)}$ 19. $\dfrac{(x-9)(3x+2)}{(3x+4)(2x-9)}$ 20. $(x-1)(x+2)$
21. $\dfrac{(x-2)(x-1)}{(2x-1)(x+2)}$ 22. $\dfrac{2x+5}{(x+6)(3x+5)}$ 23. $\dfrac{2x+3}{x-4}$ 24. $\dfrac{x+2}{(x-2)(x-1)}$ 25. $\dfrac{x-5}{2x+5}$
26. $\dfrac{x(x+10)}{(x+5)(x-5)}$ 27. $\dfrac{(x-1)(x-5)}{(x-3)(x-3)}$

Extra Practice 14

1. $\dfrac{4x+6}{(x-2)(x+2)}$ 2. $\dfrac{x}{5y}$ 3. $\dfrac{2x-3}{x-y}$ 4. $\dfrac{2ab-b^2}{3(a+b)}$ 5. $\dfrac{-x+3}{(x-1)(2x+5)(3x+2)}$ 6. $5a-5b$

7. $\dfrac{xy+y^2+5x-15}{(x-3)y}$ 8. $\dfrac{y^2+5x+5y}{y(x+y)}$ 9. $\dfrac{ab-3b+a}{ab}$ 10. $\dfrac{2x}{4x-3}$ 11. $\dfrac{-4}{(x+8)(2x+1)}$

12. $\dfrac{11x^2+x-1}{(x-2)(x+2)(5x-1)}$ 13. $\dfrac{x^2-x-14}{(x-3)(x+3)}$ 14. $\dfrac{6x^2+11x}{(x+4)(x+1)(3x-1)}$ 15. 0

Extra Practice 14 (continued)

16. $\dfrac{50}{(a-5)(a+5)}$ 17. $\dfrac{x-5}{5x}$ 18. $\dfrac{2}{b+7}$ 19. $\dfrac{x^3-17x-3}{(x-4)(x+4)}$ 20. $\dfrac{x-10}{(x-4)(x-3)(x+3)}$

Extra Practice 15

1. $\dfrac{1}{x-1}$ 2. $x+y$ 3. $\dfrac{2b-a}{5a-3b}$ 4. $\dfrac{4xy}{x+y}$ 5. $\dfrac{1-x}{1-2x}$ 6. $\dfrac{a^2b-3a^2}{b^2-ab^2}$ 7. $-xy(x+y)$

8. $\dfrac{-7x^2}{4x^2+4}$ 9. $\dfrac{9}{a-c}$ 10. $\dfrac{x}{x-1}$ 11. $ab(a+b)$ 12. $\dfrac{x}{x-y}$ 13. $\dfrac{x^2+5x-6}{3x^2+4x-4}$

14. $\dfrac{1-3y^2}{1+9y^2}$ 15. $\dfrac{8x-2}{8x-1}$ 16. $\dfrac{x-5}{x^2-2x+4}$

Extra Practice 16

1. $-2, \dfrac{1}{5}$ 2. $\dfrac{62}{5}$ 3. $\dfrac{4}{3}$ 4. $-4, -\dfrac{2}{3}$ 5. $\dfrac{13}{4}$ 6. 4 7. No solution 8. -30 9. 5

10. 16 11. $-4, -1$ 12. $8\dfrac{4}{7}$ min 13. 125 mph 14. $1\dfrac{3}{7}$ hr 15. 480 mi 16. $1\dfrac{31}{32}$ hr

17. 250 mph 18. 3 hr 19. 4 mph 20. 32 min

Extra Practice 17

1. $5x^3+4x^2-7x$ 2. $5x+1+\dfrac{-10}{3x-5}$ 3. $x+9$ 4. $7x^2+2x-5+\dfrac{12}{x+1}$

5. $2y^2-4y+2+\dfrac{-8}{5y+2}$ 6. $6+2x^3-3x$ 7. y^3-2y^2+4y-8

8. $a^3+3a^2+11a+33+\dfrac{84}{a-3}$ 9. $x^4-3x^3+9x^2-18x+36+\dfrac{-74}{x+2}$ 10. $x^2-6x+36$

11. $7x^3+8x^5-5x^2$ 12. $x^6-x^4+x^2-1$ 13. $3b+2$ 14. $5x+7$ 15. $4a+1$

16. $5x+7+\dfrac{16}{x-2}$ 17. $4y^2-y+2+\dfrac{-11}{8y+1}$ 18. $4x^2+3x$ 19. $b^2-4b+9+\dfrac{-27}{b+2}$

20. $2x^2+5x-3$ 21. $8x^2-24x+83+\dfrac{-257}{x+3}$ 22. $x^6+x^4+6x^2+12$

23. $x^5-x^4+x^3+2x^2-7x+7+\dfrac{-7}{x+1}$ 24. $x^6+3+\dfrac{16}{x^6-4}$

Extra Practice 18

1. $x^2 + 6x + 16 + \dfrac{53}{x-3}$ 2. $x^2 + x + 6 + \dfrac{9}{x-2}$ 3. $x^2 - 5x - 24$ 4. $2x^2 - 13x + 20$

5. $2x^2 - 7x + 3 + \dfrac{3}{x+8}$ 6. $x^2 + x - 6 + \dfrac{5}{x-4}$ 7. $3x^2 - 13x + 80 + \dfrac{-483}{x+6}$

8. $2x^2 + 6x + 14 + \dfrac{45}{x-3}$ 9. $x^3 + 2x^2 + 4x + 8$ 10. $x^2 - 6x + 24 + \dfrac{-93}{x+4}$

Extra Practice 18 (continued)

11. $x^2 + 6x + 15 + \dfrac{34}{x-2}$ 12. $x^2 + 2x + \dfrac{-4}{x+1}$ 13. $2x^2 + 5x + 44 + \dfrac{306}{x-7}$ 14. $x^2 - 3 + \dfrac{19}{x+6}$

15. $x^2 + 7x + 7 + \dfrac{10}{x-2}$ 16. $x^2 + 2x + 18 + \dfrac{134}{x-7}$ 17. $x^2 - 6x + 16 + \dfrac{-52}{x+3}$

18. $x^2 - 8x + 70 + \dfrac{-555}{x+8}$ 19. $x^2 + 6x + 30 + \dfrac{123}{x+3}$ 20. $x^2 - 12x + 63 + \dfrac{-321}{x+5}$ 21. 1579

22. 319 23. 101 24. 4746 25. 412 26. 494 27. 1721 28. 4666 29. 1899 30. 26,566

Extra Practice 19

1. $\sqrt{11}$, $\sqrt{23}$, $\sqrt{32} = 4\sqrt{2}$, does not exist 2. $\sqrt{14}$, does not exist, $\sqrt{20} = 2\sqrt{5}$, $\sqrt{8} = 2\sqrt{2}$
3. 4, 0, $\sqrt{20} = 2\sqrt{5}$, does not exist 4. $\sqrt[3]{16} = 2\sqrt[3]{2}$, $\sqrt[3]{-24} = -2\sqrt[3]{3}$, $\sqrt[3]{21}$, $\sqrt[3]{48} = 2\sqrt[3]{6}$
5. $\sqrt[3]{-10}$, $\sqrt[3]{14}$, $\sqrt[3]{20}$, $\sqrt[3]{-40} = -2\sqrt[3]{5}$ 6. $\sqrt[4]{3}$, does not exist, $\sqrt[4]{5}$, $\sqrt[4]{7}$ 7. 5 8. $4|x|$
9. $9|x|\sqrt{x}$ 10. $7x$ 11. $4t$ 12. -3 13. $-z$ 14. $|2x+3|$ 15. $|2x+3|$ 16. $-6x$
17. $\dfrac{6}{7}$ 18. $-\dfrac{5}{9}$ 19. $\dfrac{2}{3}$ 20. $4ab$ 21. $\{x|x \geq 3\}$, or $[3,\infty)$ 22. $\{x|x \in \text{Reals}\}$, or $(-\infty,\infty)$
23. $\{x|x \geq 9\}$, or $[9,\infty)$ 24. $\{x|x \in \text{Reals}\}$, or $(-\infty,\infty)$ 25. $\{x|x \geq 5\}$, or $[5,\infty)$
26. $\{x|x \in \text{Reals}\}$, or $(-\infty,\infty)$ 27. $\sqrt[6]{y}$ 28. $\sqrt{4}$, or 2 29. $\sqrt[5]{32}$, or 2 30. $\sqrt{16^3}$, or 64
31. $\sqrt[6]{xy}$ 32. $\sqrt{169x^3}$, or $13x\sqrt{x}$ 33. $\sqrt[4]{64^5}$, or $128\sqrt[4]{4}$ 34. $\sqrt[3]{16xy^2z}$ or $2\sqrt[3]{2xy^2z}$
35. $\sqrt[4]{625^3}$, or 125 36. $16^{\frac{1}{3}}$ 37. $x^{\frac{2}{5}}$ 38. $(x^2y)^{\frac{1}{3}}$ 39. $(4xy)^{\frac{3}{2}}$ 40. $(10ab^3)^{\frac{1}{4}}$
41. $(abc^2)^{\frac{3}{5}}$ 42. $(x^2y^3z)^{\frac{1}{4}}$ 43. $(3x^2y^3)^{\frac{1}{2}}$ 44. $(x^3y^2z^5)^{\frac{1}{5}}$ 45. $\dfrac{1}{\sqrt[4]{y^3}}$ 46. $\dfrac{1}{\sqrt{(10x^2y)^3}}$
47. $\sqrt{64}$, or 8 48. $\left(\dfrac{6bc}{5a}\right)^{\frac{2}{3}}$ 49. $\dfrac{1}{\sqrt[10]{x^{11}}}$ 50. $x^{19/10}$ 51. $\sqrt[18]{\dfrac{y^4}{x^{15}}}$ 52. $x^{5/18}$
53. $\sqrt[4]{a^{3/5}b^{1/2}}$, or $\sqrt[40]{a^6b^5}$ 54. $6^{4/5}$, or $\sqrt[5]{6^4}$ 55. $\dfrac{1}{\sqrt[10]{4^{11}}}$

Extra Practice 20

1. $2xz\sqrt{5xy}$ 2. $4x\sqrt[3]{2xy^2}$ 3. a^4b^3 4. $\dfrac{7a\sqrt{a}}{b^2}$ 5. $3ac\sqrt{5ab}$ 6. 64 7. $\dfrac{2x\sqrt[3]{2x^2}}{y^2}$

8. $2ab^3\sqrt[4]{4a^3}$ 9. $5ab^2\sqrt{2b}$ 10. $8x^6$ 11. $\dfrac{4x\sqrt{x}}{9}$ 12. $10xz^5\sqrt{5yz}$ 13. 36

14. $\dfrac{4a^2\sqrt[3]{a}}{3}$ 15. $2xy\sqrt[3]{30xy^2}$ 16. $xy^2z^3\sqrt[4]{x^3y}$ 17. $\dfrac{2x\sqrt{6x}}{5}$ 18. 64 19. $8a^3b^6$

20. $9a^2\sqrt[3]{4}$ 21. $5(x+2)\sqrt[3]{x+2}$ 22. $4a^2\sqrt{b}$ 23. $6x$ 24. $4x^2\sqrt[3]{x^2y^2}$ 25. $2x^2y\sqrt{6y}$

26. $3ab^2\sqrt[3]{a}$ 27. $5xy\sqrt[3]{x^2}$ 28. $3(x+3)^2\sqrt{2}$ 29. $36a\sqrt[3]{6b^2}$ 30. $3y^2$ 31. $6xy\sqrt[5]{x^2y^4}$

Extra Practice 20 (continued)
32. $2(y-3)^2\sqrt[3]{y-3}$

Extra Practice 21
1. 1, 2   2. 36   3. 7   4. 7   5. 0, 1   6. −3, 0   7. 9   8. 30   9. 28   10. 6   11. No solution
12. −4, 0   13. 29   14. 16   15. 6   16. 0   17. No solution   18. 2   19. 8   20. −3

Extra Practice 22
1. $c = 5$ 2. $c = 13$ 3. $b = 8$ 4. $a = \sqrt{180} = 6\sqrt{5}$ 5. $b = 5$ 6. $b = \sqrt{208} = 4\sqrt{13}$
7. $a = \sqrt{7}$ 8. $c = \sqrt{200} = 10\sqrt{2}$ 9. $b = \sqrt{132} = 2\sqrt{33}$ 10. $a = \sqrt{108} = 6\sqrt{3}$ 11. $\sqrt{449}$ ft
12. $\sqrt{369}$ ft   13. $2\sqrt{61} + 4\sqrt{5}$ ft   14. $\sqrt{280} = 2\sqrt{70}$ in.   15. $c = 5$ 16. $b = \sqrt{288} = 12\sqrt{2}$
17. $a = 4; b = \sqrt{32} = 4\sqrt{2}$ 18. $b = 12$ 19. 8   20. $a = \frac{5\sqrt{3}}{3}, b = \frac{10\sqrt{3}}{3}$
21. $a = 8\sqrt{3}, b = 16$ 22. 8   23. $\sqrt{128} = 8\sqrt{2}$ 24. $\sqrt{272} = 4\sqrt{17}$ 25. $8\sqrt{2}$ ft   26. 50 mi
27. $4\sqrt{2}$ m   28. 28 ft

Extra Practice 23
1. $\frac{-2 \pm 6i}{5}$ 2. $2 \pm i\sqrt{2}$ 3. $\frac{-1 \pm \sqrt{6}i}{6}$ 4. $\frac{7 \pm \sqrt{29}}{2}$ 5. $\frac{3 \pm i\sqrt{7}}{2}$ 6. $\frac{1 \pm \sqrt{41}}{10}$ 7. $\frac{7 \pm \sqrt{13}}{6}$
8. $\frac{-3 \pm 3i\sqrt{3}}{2}$ 9. $\frac{1 \pm i\sqrt{31}}{8}$ 10. $\frac{3 \pm i\sqrt{23}}{4}$ 11. $1 \pm \sqrt{3}$ 12. $3 \pm \sqrt{14}$ 13. $\frac{1 \pm i\sqrt{59}}{10}$
14. $3 \pm i\sqrt{3}$ 15. $\frac{3 \pm \sqrt{33}}{2}$ 16. $\frac{-1 \pm \sqrt{11}}{4}$ 17. $\frac{-1 \pm i\sqrt{14}}{3}$ 18. $5 \pm \sqrt{13}$ 19. $3 \pm \sqrt{5}$
20. $-\frac{1}{3}, \frac{1}{2}$ 21. $\frac{2 \pm \sqrt{14}}{10}$ 22. $\frac{1 \pm \sqrt{15}}{7}$ 23. $\frac{5 \pm \sqrt{17}}{3}$ 24. $\frac{-4 \pm \sqrt{46}}{3}$ 25. $\frac{3 \pm \sqrt{41}}{8}$
26. $\frac{-9 \pm \sqrt{61}}{2}$ 27. $4 \pm \sqrt{13}$ 28. $2 \pm \sqrt{5}$ 29. $\frac{1 \pm 2i\sqrt{5}}{7}$ 30. $\frac{1 \pm \sqrt{151}}{15}$

Extra Practice 24
1. 8%   2. 7.8 sec   3. 6%   4. 9.1 sec   5. 5.8%   6. 15 mph   7. 30 hours   8. 30 mph
9. Gary: 10 hr; Marsha: 15 hr   10. 54 mph   11. 3.4 hours   12. 8.5 mph

Extra Practice 25
1. 81   2. $\pm\sqrt{2}, \pm\sqrt{6}$ 3. $-\frac{1}{3}, \frac{1}{4}$ 4. $-\frac{4}{3}, \frac{4}{3}$ 5. 1, 81   6. 4, 169   7. $\pm\sqrt{2}, \pm 2$
8. 16, 81   9. $-2, -1, 3, 4$ 10. $\pm\sqrt{3}, \pm i\sqrt{7}$ 11. $-1, 1, 4, 6$ 12. 4, 100   13. $\frac{3}{4}$
14. 144, 225   15. $\pm\sqrt{3}, \pm 2$ 16. $-\frac{1}{5}, \frac{2}{3}$

Extra Practice 26

1.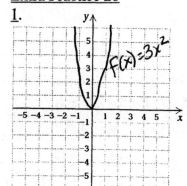
$f(x)=3x^2$

2.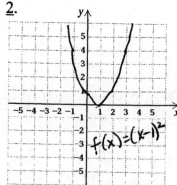
$f(x)=(x-1)^2$

3.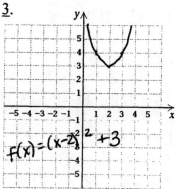
$f(x)=(x-2)^2+3$

4.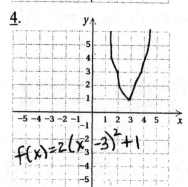
$f(x)=2(x-3)^2+1$

5.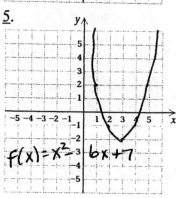
$f(x)=x^2-6x+7$

6.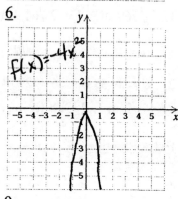
$f(x)=-4x^2$

7.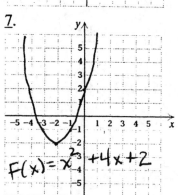
$f(x)=x^2+4x+2$

8.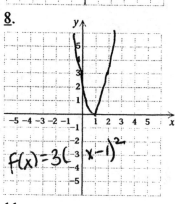
$f(x)=3(x-1)^2$

9.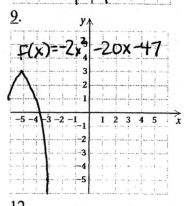
$f(x)=-2x^2-20x-47$

10.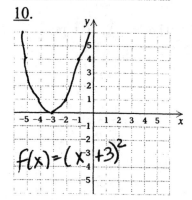
$f(x)=(x+3)^2$

11.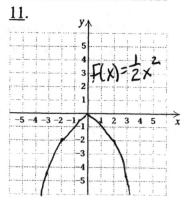
$f(x)=\frac{1}{2}x^2$

12.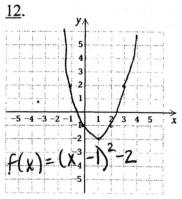
$f(x)=(x-1)^2-2$

Extra Practice 26 (continued)

13.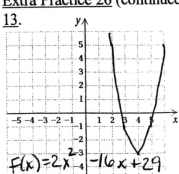
$f(x) = 2x^2 - 16x + 29$

14.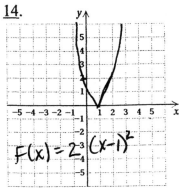
$f(x) = 2(x-1)^2$

15.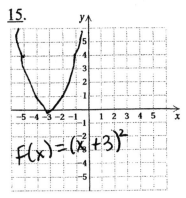
$f(x) = (x+3)^2$

16.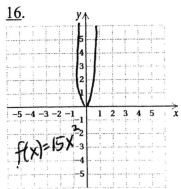
$f(x) = 15x$

17.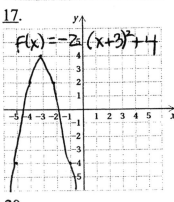
$f(x) = -2(x+3)^2 + 4$

18.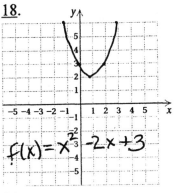
$f(x) = x^2 - 2x + 3$

19.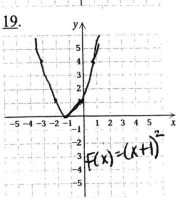
$f(x) = (x+1)^2$

20.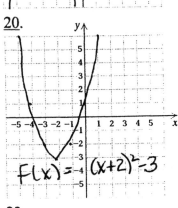
$f(x) = (x+2)^2 - 3$

21.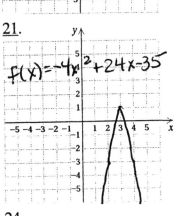
$f(x) = -4x^2 + 24x - 35$

22.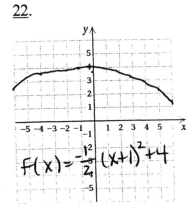
$f(x) = -\frac{1}{2}(x+1)^2 + 4$

23.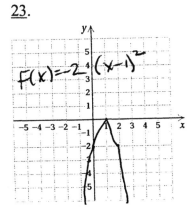
$f(x) = -2(x-1)^2$

24.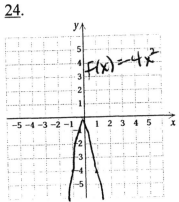
$f(x) = -4x^2$

Extra Practice 26 (continued)

25.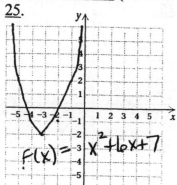
26.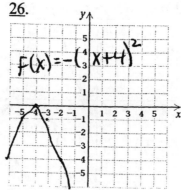
27.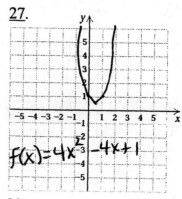
28.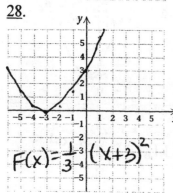
29.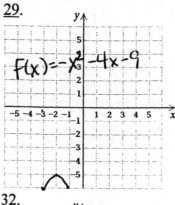
30.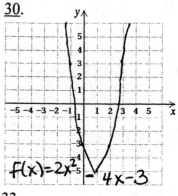
31.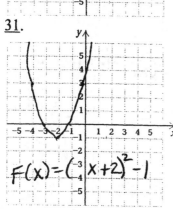
32.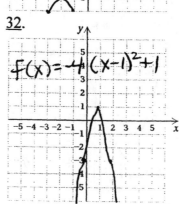
33.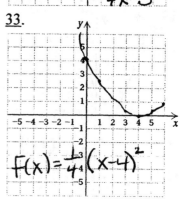

Extra Practice 27

1. $\{x|-10 < x < 6\}$, or $(-10, 6)$ 2. $\{x|x < -3 \text{ or } x > 8\}$, or $(-\infty, -3) \cup (8, \infty)$
3. $\{x|x \leq 5 \text{ or } x \geq 2\}$, or $(-\infty, -5) \cup [2, \infty)$ 4. $\{x|-7 \leq x \leq 3\}$, or $[-7, 3]$
5. $\{x|x < -4 \text{ or } x > 3\}$, or $(-\infty, -4) \cup (3, \infty)$ 6. $\{x|x < -4 \text{ or } x > 5\}$, or $(-\infty, -4) \cup (5, \infty)$
7. $\{x|-2 \leq x \leq 3\}$, or $[-2, 3]$ 8. $\{x|-9 < x < 6\}$, or $(-9, 6)$
9. $\{x|-6 < x < -1 \text{ or } x > 9\}$, or $(-6, -1) \cup (9, \infty)$
10. $\{x|x < -7 \text{ or } 5 < x < 10\}$, or $(-\infty, -7) \cup (5, 10)$
11. $\{x|x \leq -5 \text{ or } -3 \leq x \leq 2\}$, or $(-\infty, 5] \cup [-3, 2]$
12. $\{x|-11 < x < -2 \text{ or } x > 6\}$, or $(-11, -2) \cup (6, \infty)$
13. $\{x|-3 < x < 6 \text{ or } x > 8\}$, or $(-3, 6) \cup (8, \infty)$

Extra Practice 27 (continued)

14. $\{x|-12 \leq x \leq -1 \text{ or } x \geq 7\}$, or $[-12,-1] \cup [7,\infty)$
15. $\{x \leq -4 \text{ or } 4 \leq x \leq 10\}$, or $(-\infty,-4] \cup [4,10]$
16. $\{x|-9 < x < -8 \text{ or } x > 8\}$, or $(-9,-8) \cup (8,\infty)$ 17. $\{x|x > 8\}$, or $(8,\infty)$
18. $\{x|x < -3\}$, or $(-\infty,-3)$ 19. $\{x|x < -6 \text{ or } x \geq 0\}$, or $(-\infty,-6) \cup [0,\infty)$
20. $\{x|-1 \leq x < 9\}$, or $[-1,9)$ 21. $\{x|-5 < x < 4\}$, or $(-5,4)$
22. $\{x|x < -1 \text{ or } x > 6\}$, or $(-\infty,-1) \cup (6,\infty)$ 23. $\{x|-3 < x < 8\}$, or $(-3,8)$
24. $\{x|-7 < x < -4\}$, or $-7,-4$ 25. $\{x|x < -4 \text{ or } x \geq -\frac{1}{2}\}$, or $(-\infty,-4) \cup [-\frac{1}{2},\infty)$
26. $\{x|x < 3 \text{ or } x > 4\}$, or $(-\infty,3) \cup (4,\infty)$ 27. $\{x|-\frac{2}{3} < x < 4\}$, or $(-\frac{2}{3},4)$
28. $\{x|x < -2 \text{ or } x > -1\}$, or $(-\infty,-2) \cup (-1,\infty)$
29. $\{x|x-2 < x < -1 \text{ or } x > 3\}$, or $(-2,-1) \cup (3,\infty)$
30. $\{x|-5 < x \leq -3 \text{ or } x \geq 4\}$, or $(-5,-3] \cup [4,\infty)$

Extra Practice 28

1. $\frac{4}{5}$ 2. 1.1969   3. 1.2925   4. 0.3869   5. 0.2554   6. $\frac{2}{3}$ 7. $\frac{1}{3}$ 8. 0.8451   9. 3.7651
10. 57.5646   11. 0.3219   12. 1.0939   13. -1 14. 1.7925   15. 2.0579   16. 2   17. 5
18. $\frac{2}{999}$ 19. 44   20. 36   21. 25   22. 6   23. 2   24. $\frac{1}{3}$ 25. $\frac{5}{9}$ 26. $\frac{1}{21}$ 27. 10   28. 12
29. $\frac{1}{4}$ 30. 50   31. 7   32. -11

Extra Practice 29

1. $(-4,-2), (-2,-4), (2,4), (4,2)$ 2. $(-7,0), (7,0)$ 3. $(-4,3), (3,4)$
4. $(-9,1), (-1,9), (9,-1), (1,-9)$ 5. $(0,2), (4,0)$ 6. $(-4,-5), (4,5)$ 7. $(0,10), (2,0)$
8. $(-6,0), (6,0)$ 9. $(0,-5)(0,5)$ 10. $(-5,-3), (3,5)$
11. $(-2,-4), (2,-4), (-\sqrt{11},3), (\sqrt{11},3)$ 12. $(-8,0), (7,-\sqrt{15}), (7,\sqrt{15})$
13. 25 cm by 60 cm   14. 4 m by $3\sqrt{2}$ m   15. -4 and 11 or 4 and -11 16. 5 and 8
17. 3 and 7   18. 7 m by 15 m   19. $\frac{1}{3}$ and $\frac{1}{2}$ or $-\frac{1}{3}$ and $-\frac{1}{2}$ 20. 0.6 cm by 0.8 cm

Extra Practice 30

1. 5040   2. 362,880   3. 40,320   4. 20   5. 4   6. 84   7. 495   8. 1   9. 66   10. 18
11. $x^5 + 5x^4y + 10x^3y^2 + 10x^2y^3 + 5xy^4 + y^5$
12. $x^6 - 6x^5y + 15x^4y^2 - 20x^3y^3 + 15x^2y^4 - 6xy^5 + y^6$ 13. $8x^3 + 12x^2y + 6xy^2 + y^3$
14. $27x^3 - 54x^2y + 36xy^2 - 8y^3$ 15. $x^5 - 15x^4y + 90x^3y^2 - 270x^2y^3 + 405xy^4 - 243y^5$

Extra Practice 30 (continued)

16. $64x^6 + 192x^5y + 240x^4y^2 + 160x^3y^3 + 60x^2y^4 + 12xy^5 + y^6$

17. $x^8 + 8x^7y + 28x^6y^2 + 56x^5y^3 + 70x^4y^4 + 56x^3y^5 + 28x^2y^6 + 8xy^7 + y^8$

18. $81x^4 - 216x^3y + 216x^2y^2 - 96xy^3 + 16y^4$ 19. $-x^3 + 3x^2y - 3xy^2 + y^3$

20. $32x^5 - 80x^4y + 80x^3y^2 - 40x^2y^3 + 10xy^4 - y^5$ 21. $x^5 + \dfrac{5x^4}{y} + \dfrac{10x^3}{y^2} + \dfrac{10x^2}{y^3} + \dfrac{5x}{y^4} + \dfrac{1}{y^5}$

22. $8x^3 - \dfrac{60x^2}{y} + \dfrac{150x}{y^2} - \dfrac{125}{y^3}$ 23. $35x^3y^4$ 24. $-126(2x)^4(3y)^5$, or $-489,888x^4y^5$

25. $120(2x)^3(3y^2)^7$, or $2,099,520x^3y^{14}$ 26. $28(3x)^2(2y)^6$, or $16,128x^2y^6$

Intermediate Algebra: Concepts and Applications, Sixth Edition
Bittinger/Ellenbogen

Fifth Edition to Sixth Edition Correlation Chart

Fifth Edition to Sixth Edition TOC changes are indicated in gray

CHAPTER	IN FIFTH EDITION	IN SIXTH EDITION
1	**Algebra and Problem Solving**	**Algebra and Problem Solving**
SECTION		
1.1	Some Basics of Algebra	Some Basics of Algebra
1.2	Operations and Properties of Real Numbers	Operations and Properties of Real Numbers
1.3	Solving Equations	Solving Equations
1.4	Introduction to Problem Solving	Introduction to Problem Solving
1.5	Formulas, Models, and Geometry	Formulas, Models, and Geometry
1.6	Properties of Exponents	Properties of Exponents
1.7	Scientific Notation	Scientific Notation
	SUMMARY AND REVIEW	SUMMARY AND REVIEW
	TEST	TEST

CHAPTER	IN FIFTH EDITION	IN SIXTH EDITION
2	**Graphs, Functions, and Linear Equations**	**Graphs, Functions, and Linear Equations**
SECTION		
2.1	Graphs	Graphs
2.2	Functions	Functions
2.3	Linear Functions: Graphs and Models	Linear Functions: Slope, Graphs, and Models
2.4	Another Look at Linear Graphs	Another Look at Linear Graphs
2.5	Other Equations of Lines	Other Equations of Lines
2.6	The Algebra of Functions	The Algebra of Functions
	SUMMARY AND REVIEW	SUMMARY AND REVIEW
	TEST	TEST

CHAPTER	IN FIFTH EDITION	IN SIXTH EDITION
3	**Systems of Equations and Problem Solving**	**Systems of Linear Equations and Problem Solving**
SECTION		
3.1	Systems of Equations in Two Variables	Systems of Equations in Two Variables

3.2	Solving by Substitution or Elimination	Solving by Substitution or Elimination	
3.3	Solving Applications: Systems of Two Equations	Solving Applications: Systems of Two Equations	
3.4	Systems of Equations in Three Variables	Systems of Equations in Three Variables	
3.5	Solving Applications: Systems of Three Equations	Solving Applications: Systems of Three Equations	
3.6	Elimination Using Matrices	Elimination Using Matrices	
3.7	Determinants and Cramer's Rule	Determinants and Cramer's Rule	
3.8	Business and Economic Applications	Business and Economic Applications	
	SUMMARY AND REVIEW	SUMMARY AND REVIEW	
	TEST	TEST	
	CUMULATIVE REVIEW: CHAPTERS 1–3	CUMULATIVE REVIEW: CHAPTERS 1–3	

CHAPTER	IN FIFTH EDITION	IN SIXTH EDITION
4	**Inequalities and Problem Solving**	**Inequalities and Problem Solving**
SECTION		
4.1	Inequalities and Applications	Inequalities and Applications
4.2	Intersections, Unions, and Compound Inequalities	Intersections, Unions, and Compound Inequalities
4.3	Absolute-Value Equations and Inequalities	Absolute-Value Equations and Inequalities
4.4	Inequalities in Two Variables	Inequalities in Two Variables
4.5	Applications Using Linear Programming	Applications Using Linear Programming
	SUMMARY AND REVIEW	SUMMARY AND REVIEW
	TEST	TEST

CHAPTER	IN FIFTH EDITION	IN SIXTH EDITION
5	**Polynomials and Polynomial Functions**	**Polynomials and Polynomial Functions**
SECTION		
5.1	Introduction to Polynomials and Polynomial Functions	Introduction to Polynomials and Polynomial Functions
5.2	Multiplication of Polynomials	Multiplication of Polynomials
5.3	Common Factors and Factoring by Grouping	Common Factors and Factoring by Grouping

5.4	Factoring Trinomials	Factoring Trinomials
5.5	Factoring Perfect-Square Trinomials and Differences of Squares	Factoring Perfect-Square Trinomials and Differences of Squares
5.6	Factoring Sums or Differences of Cubes	Factoring Sums or Differences of Cubes
5.7	Factoring: A General Strategy	Factoring: A General Strategy
5.8	Applications of Polynomial Equations	Applications of Polynomial Equations
	SUMMARY AND REVIEW	SUMMARY AND REVIEW
	TEST	TEST

CHAPTER	IN FIFTH EDITION	IN SIXTH EDITION
6	**Rational Expressions, Equations, and Functions**	**Rational Expressions, Equations, and Functions**
SECTION		
6.1	Rational Expressions and Functions: Multiplying and Dividing	Rational Expressions and Functions: Multiplying and Dividing
6.2	Rational Expressions and Functions: Adding and Subtracting	Rational Expressions and Functions: Adding and Subtracting
6.3	Complex Rational Expressions	Complex Rational Expressions
6.4	Rational Equations	Rational Equations
6.5	Solving Applications Using Rational Equations	Solving Applications Using Rational Equations
6.6	Division of Polynomials	Division of Polynomials
6.7	Synthetic Division	Synthetic Division
6.8	Formulas and Applications	Formulas, Applications, and Variation
	SUMMARY AND REVIEW	SUMMARY AND REVIEW
	TEST	TEST
	CUMULATIVE REVIEW: CHAPTERS 1–6	CUMULATIVE REVIEW: CHAPTERS 1–6

CHAPTER	IN FIFTH EDITION	IN SIXTH EDITION
7	**Exponents and Radicals**	**Exponents and Radicals**
SECTION		
7.1	Radical Expressions and Functions	Radical Expressions and Functions
7.2	Rational Numbers as Exponents	Rational Numbers as Exponents

7.3	Multiplying, Adding, and Subtracting Radical Expressions	Multiplying Radical Expressions
7.4	Multiplying, Dividing, and Simplifying Radical Expressions	Dividing Radical Expressions
7.5	More with Multiplication and Division	Expressions Containing Several Radical Terms
7.6	Solving Radical Equations	Solving Radical Equations
7.7	Geometric Applications	Geometric Applications
7.8	The Complex Numbers	The Complex Numbers
	SUMMARY AND REVIEW	SUMMARY AND REVIEW
	TEST	TEST

CHAPTER	IN FIFTH EDITION	IN SIXTH EDITION
8	**Quadratic Functions and Equations**	**Quadratic Functions and Equations**
SECTION		
8.1	Quadratic Equations	Quadratic Equations
8.2	The Quadratic Formula	The Quadratic Formula
8.3	Applications Involving Quadratic Equations	Applications Involving Quadratic Equations
8.4	Studying Solutions of Quadratic Equations	Studying Solutions of Quadratic Equations
8.5	Equations Reducible to Quadratic	Equations Reducible to Quadratic
8.6	Variation and Problem Solving	Quadratic Functions and Their Graphs
8.7	Quadratic Functions and Their Graphs	More About Graphing Quadratic Functions
8.8	More About Graphing Quadratic Functions	Problem Solving and Quadratic Functions
8.9	Problem Solving and Quadratic Functions	Polynomial and Rational Inequalities
8.10	Polynomial and Rational Inequalities	N/A
	SUMMARY AND REVIEW	SUMMARY AND REVIEW
	TEST	TEST

CHAPTER	IN FIFTH EDITION	IN SIXTH EDITION
9	**Exponential and Logarithmic Functions**	**Exponential and Logarithmic Functions**
SECTION		
9.1	Exponential Functions	Composite and Inverse Functions
9.2	Composite and Inverse Functions	Exponential Functions
9.3	Logarithmic Functions	Logarithmic Functions

9.4	Properties of Logarithmic Functions	Properties of Logarithmic Functions
9.5	Common and Natural Logarithms	Common and Natural Logarithms
9.6	Solving Exponential and Logarithmic Equations	Solving Exponential and Logarithmic Equations
9.7	Applications of Exponential and Logarithmic Functions	Applications of Exponential and Logarithmic Functions
	SUMMARY AND REVIEW	SUMMARY AND REVIEW
	TEST	TEST
		CUMULATIVE REVIEW: CHAPTERS 1–9

CHAPTER	IN FIFTH EDITION	IN SIXTH EDITION
10	**Conic Sections**	**Conic Sections**
SECTION		
10.1	Conic Sections: Parabolas and Circles	Conic Sections: Parabolas and Circles
10.2	Conic Sections: Ellipses	Conic Sections: Ellipses
10.3	Conic Sections: Hyperbolas	Conic Sections: Hyperbolas
10.4	Nonlinear Systems of Equations	Nonlinear Systems of Equations
	SUMMARY AND REVIEW	SUMMARY AND REVIEW
	TEST	TEST

CHAPTER	IN FIFTH EDITION	IN SIXTH EDITION
11	**Sequences, Series, and the Binomial Theorem**	**Sequences, Series, and the Binomial Theorem**
SECTION		
11.1	Sequences and Series	Sequences and Series
11.2	Arithmetic Sequences and Series	Arithmetic Sequences and Series
11.3	Geometric Sequences and Series	Geometric Sequences and Series
11.4	The Binomial Theorem	The Binomial Theorem
	SUMMARY AND REVIEW	SUMMARY AND REVIEW
	TEST	TEST
	CUMULATIVE REVIEW: CHAPTERS 1–11	CUMULATIVE REVIEW: CHAPTERS 1–11

ENDMATTER	IN FIFTH EDITION	IN SIXTH EDITION
Appendix	The Graphing Calculator	The Graphing Calculator
Answers	[yes]	[yes]
Index	[yes]	[yes]
Videotape and CD Index	[no]	[yes]

Intermediate Algebra: Concepts and Applications, Sixth Edition
Bittinger/Ellenbogen
Video/Exercise Index

Tape Number	Section Number	Section Title	Exercises Used
	Ch. 1	**Algebra and Problem Solving**	
1	1.1	Some Basics of Algebra	17, 35, 41, 47
1	1.2	Operations and Properties of Real Numbers	21, 31, 49, 69, 85, 125, 135
1	1.3	Solving Equations	7, 17, 61, 71
2	1.4	Introduction to Problem Solving	5
2	1.5	Formulas, Models, and Geometry	10
2	1.6	Properties of Exponents	9, 77
2	1.7	Scientific Notation	5, 55
	Ch. 2	**Graphs, Functions, and Linear Equations**	
3	2.1	Graphs	43
3	2.2	Functions	7, 9, 41b, 41d, 41f
3	2.3	Linear Functions: Slope, Graphs, and Models	19, 45
4	2.4	Another Look at Linear Graphs	5, 15, 21, 37, 51, 63, 67, 69
4	2.5	Other Equations of Lines	15, 23, 53, 71
4	2.6	The Algebra of Functions	51, 57
	Ch. 3	**Systems of Equations and Problem Solving**	
5	3.1	Systems of Equations in Two Variables	3, 9
5	3.2	Solving by Substitution or Elimination	13
5	3.3	Solving Applications: Systems of Two Equations	37
5	3.4	Systems of Equations in Three Variables	1, 10, 29, 31
6	3.5	Solving Applications: Systems of Three Equations	5
6	3.6	Elimination Using Matrices	9
6	3.7	Determinants and Cramer's Rule	3, 9, 11, 19
6	3.8	Business and Economic Applications	9
	Ch. 4	**Inequalities and Problem Solving**	
7	4.1	Inequalities and Applications	none
7	4.2	Intersections, Unions, and Compound Inequalities	11, 23
8	4.3	Absolute-Value Equations and Inequalities	none
8	4.4	Inequalities in Two Variables	3, 13, 45
8	4.5	Applications Using Linear Programming	12
	Ch. 5	**Polynomials and Polynomial Functions**	
9	5.1	Introduction to Polynomials and Polynomial Functions	3, 37, 73
9	5.2	Multiplication of Polynomials	23, 45, 59
9	5.3	Common Factors and Factoring by Grouping	15, 27
9	5.4	Factoring Trinomials	55
10	5.5	Factoring Perfect-Square Trinomials and Differences of Squares	9, 29, 41
10	5.6	Factoring Sums or Differences of Cubes	17, 31
10	5.7	Factoring: A General Strategy	15, 45
10	5.8	Applications of Polynomial Equations	1, 37
	Ch. 6	**Rational Expressions, Equations, and Functions**	
11	6.1	Rational Expressions and Functions: Multiplying and Dividing	45

Tape Number	Section Number	Section Title	Exercises Used
11	6.2	Rational Expressions and Functions: Adding and Subtracting	37
11	6.3	Complex Rational Expressions	7, 21, 33
11	6.4	Rational Equations	12, 33
12	6.5	Solving Applications Using Rational Equations	10, 21, 24
12	6.6	Division of Polynomials	36
12	6.7	Synthetic Division	7
12	6.8	Formulas, Applications, and Variation	1, 9, 31, 43, 61, 63
	Ch. 7	**Exponents and Radicals**	
13	7.1	Radical Expressions and Functions	1, 33, 37, 69, 75, 89
13	7.2	Rational Numbers as Exponents	11, 23, 27, 39, 49, 55, 65, 85
13	7.3	Multiplying Radical Expressions	1, 9, 21, 39, 43, 51, 67
13	7.4	Dividing Radical Expressions	25, 31, 61
14	7.5	Expressions Containing Several Radical Terms	25, 37, 71, 81
14	7.6	Solving Radical Equations	34
14	7.7	Geometric Applications	15, 17, 27
14	7.8	The Complex Numbers	17, 73, 79
	Ch. 8	**Quadratic Functions and Equations**	
15	8.1	Quadratic Equations	63
15	8.2	The Quadratic Formula	9
15	8.3	Applications Involving Quadratic Equations	11, 25
15	8.4	Studying Solutions of Quadratic Equations	7, 43
15	8.5	Equations Reducible to Quadratic	2
16	8.6	Quadratic Functions and Their Graphs	14, 39
16	8.7	More About Graphing Quadratic Functions	7
16	8.8	Problem Solving and Quadratic Functions	9, 29
16	8.9	Polynomial and Rational Inequalities	17, 33
	Ch. 9	**Exponential and Logarithmic Functions**	
17	9.1	Composite and Inverse Functions	17, 19
17	9.2	Exponential Functions	18, 21, 37
17	9.3	Logarithmic Functions	3, 7, 15, 27, 39, 43, 55, 59, 73, 77
17	9.4	Properties of Logarithmic Functions	7
18	9.5	Common and Natural Logarithms	11, 27
18	9.6	Solving Exponential and Logarithmic Equations	25, 41
18	9.7	Applications of Exponential and Logarithmic Functions	7
	Ch. 10	**Conic Sections**	
19	10.1	Conic Sections: Parabolas and Circles	39, 53
19	10.2	Conic Sections: Ellipses	13, 21, 41
19	10.3	Conic Sections: Hyperbolas	5, 9
19	10.4	Nonlinear Systems of Equations	31, 41
	Ch. 11	**Sequences, Series, and the Binomial Theorem**	
20	11.1	Sequences and Series	7, 23, 39
20	11.2	Arithmetic Sequences and Series	17, 43
20	11.3	Geometric Sequences and Series	5, 37, 53, 59
20	11.4	The Binomial Theorem	39

TEST AID: NUMBER LINES

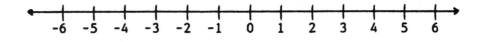

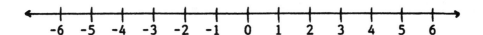

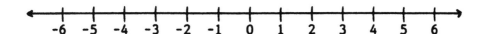

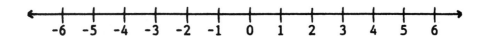

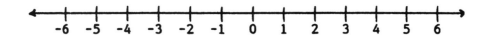

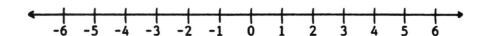

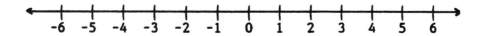

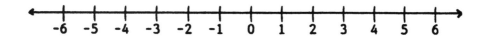

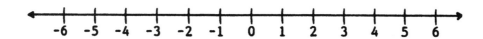

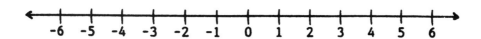

TEST AID: RECTANGULAR COORDINATE GRIDS

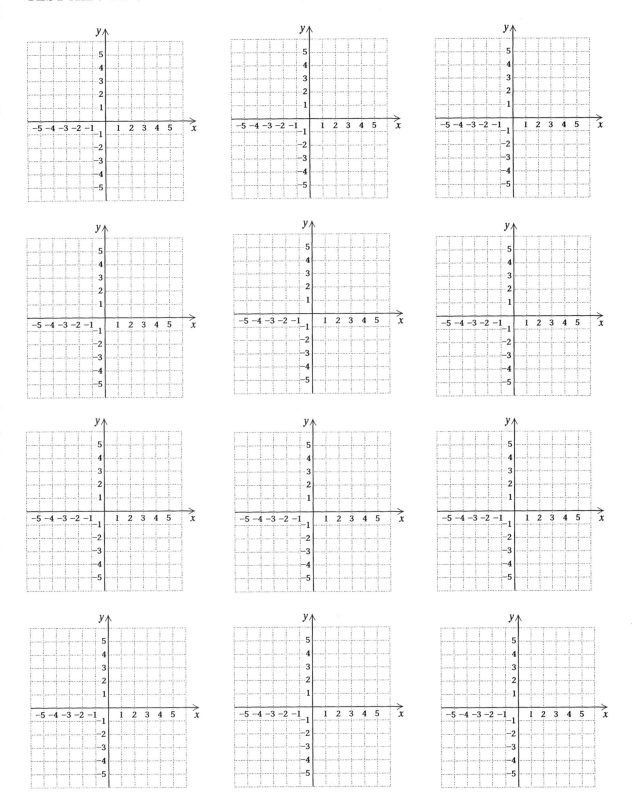

TEST AID: RECTANGULAR COORDINATE GRIDS

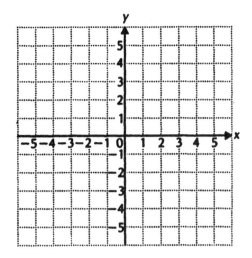
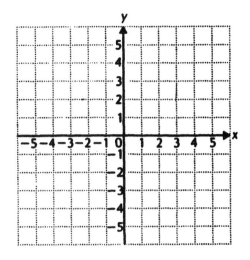
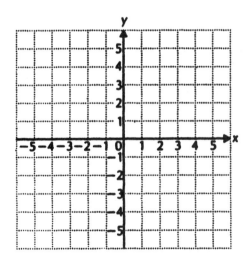
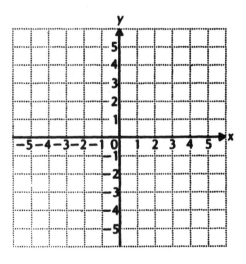

TRANSPARENCY MASTER: NUMBER LINES

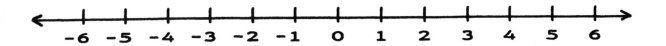

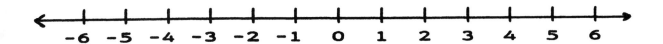

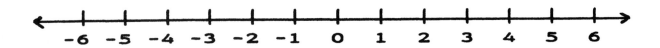

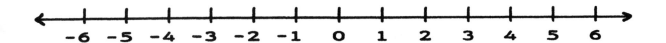

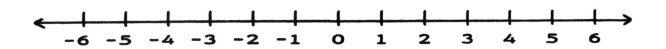

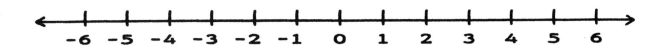

TRANSPARENCY MASTER: RECTANGULAR COORDINATE GRIDS

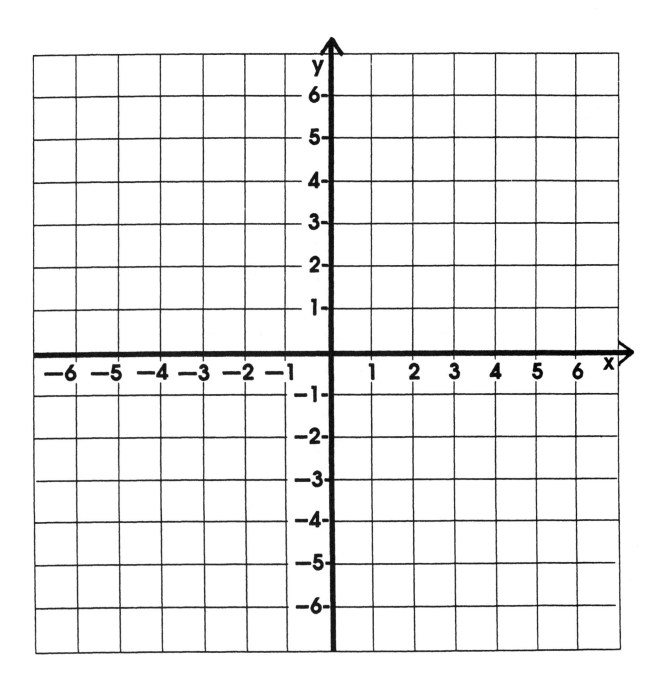